SATELLITE COMMUNICATIONS
Mobile and Fixed Services

THE KLUWER INTERNATIONAL SERIES
IN ENGINEERING AND COMPUTER SCIENCE

COMMUNICATIONS AND INFORMATION THEORY

Consulting Editor:
Robert Gallager

Other books in the series:

Digital Communication, Edward A. Lee, David G. Messerschmitt
    ISBN: 0-89838-274-2
An Introduction to Cryptology, Henk C.A. van Tilborg
    ISBN: 0-89838-271-8
Finite Fields for Computer Scientists and Engineers, Robert J. McEliece
    ISBN: 0-89838-191-6
An Introduction to Error Correcting Codes With Applications, Scott A. Vanstone and
Paul C. van Oorschot
    ISBN: 0-7923-9017-2
Source Coding Theory, Robert M. Gray
    ISBN: 0-7923-9048-2
Adaptive Data Compression, Ross N. Williams
    ISBN: 0-7923-9085
Switching and Traffic Theory for Integrated Broadband Networks, Joseph Y. Hui
    ISBN: 0-7923-9061-X
Advances in Speech Coding, Bishnu Atal, Vladimir Cuperman and Allen Gersho
    ISBN: 0-7923-9091-1
Source and Channel Coding: An Algorithmic Approach, John B. Anderson and
Seshadri Mohan
    ISBN: 0-7923-9210-8
Third Generation Wireless Information Networks, Sanjiv Nanda and David J.
Goodman
    ISBN: 0-7923-9128-3
Vector Quantization and Signal Compression, Allen Gersho and Robert M. Gray
    ISBN: 0-7923-9181-0
Image and Text Compression, James A. Storer
    ISBN: 0-7923-9243-4
*Digital Satellite Communications Systems and Technologies: Military and Civil
Applications,* A. Nejat Ince
    ISBN: 0-7923-9254-X
Sequence Detection for High-Density Storage Channel, Jaekyun Moon and L.
Richard Carley
    ISBN: 0-7923-9264-7
Wireless Personal Communications, Martin J. Feuerstein and Theodore S. Rappaport
    ISBN: 0-7923-9280-9
Applications of Finite Fields, Alfred J. Menezes, Ian F. Blake, XuHong Gao, Ronald
C. Mullin, Scott A. Vanstone, Tomik Yaghoobian
    ISBN: 0-7923-9282-5
Discrete-Time Models for Communication Systems Including ATM, Herwig Bruneel
and Byung G. Kim
    ISBN: 0-7923-9292-2
Wireless Communications: Future Directions, Jack M. Holtzman and David J.
Goodman
    ISBN: 0-7923-9316-3

SATELLITE COMMUNICATIONS
Mobile and Fixed Services

edited
by

Michael J. Miller
University of South Australia

Branka Vucetic
University of Sydney

Les Berry
Royal Melbourne Institute of Technology

KLUWER ACADEMIC PUBLISHERS
Boston / Dordrecht / London

Distributors for North America:
Kluwer Academic Publishers
101 Philip Drive
Assinippi Park
Norwell, Massachusetts 02061 USA

Distributors for all other countries:
Kluwer Academic Publishers Group
Distribution Centre
Post Office Box 322
3300 AH Dordrecht, THE NETHERLANDS

Library of Congress Cataloging-in-Publication Data

Satellite communications : mobile and fixed services / edited by
    Michael J. Miller, Branka Vucetic, Les Berry.
           p.   cm. -- (The Kluwer international series in engineering and
    computer science.   Communications and Information Theory)
      Includes bibliographical references and index.
      ISBN 0-7923-9333-3
      1. Artificial satellites in telecommunication.   2. Mobile
    communication systems. I. Miller, Michael J. II. Vucetic, Branka
    III. Berry, Les. IV. Series.
    TK5104.S3644   1993
    621.382'5--dc20   93-10183
                                           CIP

Printed on acid-free paper.

Printed in the United States of America

TABLE OF CONTENTS

viii

AUTHORS

Les Berry
Sanjay Bose
William G. Cowley
Nick Hart
Zoran M. Markovic
Michael J. Miller
Andrew Perkis
Jean-Luc Thibault
Branka Vucetic

PREFACE

This is a satellite communications textbook for the 1990's. It is based on the premise that designers of future satellite systems must take account of the strong competition that satellites face from optical fibres. In future years, satellites will continue to be commerically viable media for telecommunications only if systems designers take account of the unique features that satellites have to offer.

Accordingly this book places more emphasis on satellite mobile services and broadcasting, and less emphasis on fixed point-to-point high capacity services than traditional textbooks in the field. The particular attention given to methods of design of satellite mobile communications systems should make it an indispensable resource for workers in that field.

This book is intended as a textbook for students in senior undergraduate and graduate courses. It is designed to be suitable for courses in Satellite Communications, albeit with a different emphasis to traditional approaches. It also contains some recent results of propagation modelling and system design studies which should be of particular value to researchers and designers of satellite systems for mobile communications services.

An emphasis is given in the book to design issues. Numerous illustrative system design examples and numerical problems are provided.

The authors represent a consortium of experts in the field. The authorship team is a combination of university academics together with researchers and system designers from industry. They represent contributions from four continents.

A great deal of effort has been put into ensuring consistency and continuity between chapters. Although it is a multi-author text, it should be transparent to the reader. Enjoy it!

Printing Specifications

The document has been typeset entirely in $\LaTeX$™. The typesetting was done on a SUN™ Workstation, the layout on a NeXT™ computer, using $\TeX$™view and Adobe Illustrator™. The camera-ready version of the document was printed on a NeXT™ 400dpi Laserprinter.

Special Thanks

We would like to thank everyone who has been involved in the process of writing, proof reading and publishing this book. In particular, we would like to thank Frank Butler for drawing the figures, Feng Rice for translating the manuscripts into $\LaTeX$, Peter Asenstorfer for his availability for technical advice and Karin Schlegel for doing the layout.

SATELLITE COMMUNICATIONS
Mobile and Fixed Services

Chapter 1

INTRODUCTION TO SATELLITE COMMUNICATIONS

by Michael Miller
Australian Space Centre for Signal Processing
University of South Australia

We begin our study of satellite communications by examining some of the special features that satellite systems have to offer in comparison with optical fibres and other communication media. The opportunity to provide efficient mobile communications via satellite is seen as likely to have a major influence on the design of future satellite systems.

This Chapter then proceeds to review the basic elements in communication satellites with particular emphasis on link design procedures. The Chapter examines design equations needed for satellite broadcasting and point-to-point communications applications.

1.1 A TECHNOLOGY UNDER SIEGE

1.1.1 The Evolution of "Extra Terrestrial Relays"

As the twentieth century enters its final years, the field of telecommunications presents a dazzling array of opportunities and challenges. New services, new network configurations, and new terminal devices present ever expanding opportunities for humans to communicate with each other and with machines. Whether they are in one place or on the move, people have come to expect to be able to have access to information and to be able to communicate with anyone anywhere in the rest of the world.

The challenge for the designers of telecommunication systems and networks is to have the vision and creativity to build on all the advances of the past and to reach out to create new ones. To meet this challenge, telecommunications engineers have in their repertoire a wide variety of transmission media and network structures.

In the midst of this variety, one particular class of transmission medium stands out as unique. It is a testament to the ability of engineers to translate dreams into reality. That is the field of *satellite communications systems.* They will be the subject of our journey through this text.

Such dreams first become public when Arthur C. Clarke's famous paper "Extra Terrestrial Relays" was published in 1945 (See Reference [1] for details). It was generally regarded as an interesting piece of science fiction. Yet within 25 years, Clarke's extraordinary ideas had materialized into a relatively mature technology which had a vital part to play in the world's information networks. Satellite communications systems enabled for the first time, any nation to communicate with any other nation anywhere on the globe without discrimination as to the distance between them.

The era of satellite communications opened in the form of many experiments in the late fifties and early sixties with satellites in low altitude orbits. Operating in orbits a few hundred kilometres above the earth, these satellites are now referred to as *low earth orbit* (LEO) satellites. Some difficulties are experienced in sending and receiving communications signals from LEO satellites. These resulted from the fact that such satellites do not remain fixed over any given point on the earth. As a result, it is necessary to utilize relatively complex antenna tracking systems at the earth stations. Worst still, a LEO satellite after passing in the vicinity of any given earth station, eventually disappears for a time over the horizon. As a result, communications are interrupted unless some other LEO satellite can be arranged to appear over the horizon in time to take over the communications task.

By the mid-1960's, the *geostationary equatorial orbit* (GEO) with its unique characteristics had become increasingly attractive and presented a better solution for communications systems.

Despite the disadvantage that satellites have to be placed as far out as 36,000km from the earth to be geostationary, their advantage is that they remain stationary with respect to a particular point on the surface of the earth. That is, having an orbital period of 24 hours, the geographic area covered by the satellite does not vary. Hence constant use of expensive tracking equipment is not required. Furthermore, if the antenna on board the geostationary satellite is designed for so-called "global beam coverage", then its beamwidth of approximately 17° will enable interconnection between any earth stations located within an area corresponding to one-third of the earth's surface. That is, three such satellites can provide complete global coverage with the flexibility to connect together any pair of earth stations anywhere on the globe.

The era of *geostationary satellites* was inaugurated by the launch by the US National Aeronautics and Space Administration (NASA)NASA) in 1964 of the SYNCOM 3 satellite [1]. One of a series of experimental satellites to be launched by NASA, this satellite demonstrated for the first time, the feasibility of utilizing the geostationary orbit.

Almost all commercial communications satellites now lie in the geostationary orbit.

During the 1970's and 1980's GEO satellites became by far the most popular medium for providing commercial satellite-based communications services. Most of them operated (and many continue to operate) in the 4/6 GHz frequency band, that is, 4 GHz downlink and 6 GHz uplink . As such, they share the frequency bands also occupied by line-of-sight terrestrial microwave systems.

A number of *global satellite systems* for communications were developed by international consortia. The largest global satellite system is operated by INTELSAT (International Telecommunications Satellite Organisation), an organisation jointly owned by a consortium of over 100 countries. INTELSAT was established in 1964 and immediately appointed one of its signatories, the US Communications Satellite Corporation (COMSAT) as its project manager. Its first satellite, INTELSAT I was launched in 1965 by NASA. It carried a total of 240 telephone circuits or one television transmission between Europe and North America. Despite being designed to operate for only eighteen months, it provided full-time service for over three and a half years. In 1967 it was joined in orbit by three new satellites, known as INTELSAT II satellites. These also provided service to the Pacific Ocean region.

In 1969, the next generation of three satellites, INTELSAT III, were established in orbit and provided the first world-wide satellite service via their global beams. The satellites were located over the Atlantic, Pacific and Indian Oceans respectively, and were each capable of carrying 1200 telephone circuits or 700 telephone circuits plus a television channel. The telephone circuits were carried by analogue FDM-FM-FDMA transmission techniques. That is, they used a frequency division multiple access (FDMA) technique to multiplex carriers which in turn, had been frequency modulated (FM) by frequency division multiplex (FDM) assemblies of 24, 60 or 120 telephone channels [1].

Later generations of larger, higher capacity INTELSAT satellites were launched in 1971 (INTELSAT IV), in 1980 (INTELSAT V), and in 1988-89 (INTELSAT VI). Each INTELSAT VI satellite is capable of carrying over 30,000 telephone circuits and four television transmissions.

In the period since 1964, many nations have established their own domestic satellite systems consisting of one or more satellites in geostationary orbit. Each satellite usually has one or more antenna systems that radiate *spot beams* covering only the whole or part of the owner-nation and its adjoining regions. Some of the countries that developed domestic or regional satellite systems were: USSR - 1965 - MOLNIYA satellite system, Canada - 1972 - ANIK, USA - 1974 - WESTAR, Indonesia - 1976 - PALAPA, India - 1983 - INSAT, Japan - 1983 - CS, Australia - 1985 - AUSSAT, Brazil - 1985 - BRASILSAT, France - 1984 - TELECOM, UK - 1989 - BSB, and Germany - 1989 - DFS.

Many different types of earth stations are required for satellite communications. The size of the earth station antenna is the primary feature distinguishing one earth station from another.

- *Large earth stations*, that is, stations with large antennas 10-30 metres in diameter, are required to provide for high capacity telephone, data or television transmission. In general, the larger the antenna, the greater the traffic capacity of the earth station.

- *Small earth stations*, with antenna diameters of 1-2 m, are now a common sight on the roofs or in the gardens of commercial and domestic buildings to provide capabilities for reception of broadcast television. Other small earth stations located in remote regions are designed to provide connection to thin-route *telephony systems*, that is, those capable of carrying a small number of telephone or data services.

- *VSAT networks* are networks of satellite earth terminals which each have antenna diameters in the region of 1 metre. That is, each such antenna is said to have a very small aperture (VSAT). Such earth stations make inefficient use of satellite power and bandwidth but are attractive because they are very cheap. VSAT networks are usually arranged in a star configuration in which small aperture terminals each communicate via the satellite to a large central earth station known as a *hub station*.

The applications of communication satellites have increased remarkably in the period since 1965. Some of the advantages of satellite communication techniques over terrestrial radio and cables are the following:

- *Earth coverage* – A single satellite can provide coverage over a large geographical area. Three satellites 120 degrees apart in geostationary orbit can provide almost world wide coverage (if the polar regions are excluded).

- *Flexibility* – Satellite systems have the inherent flexibility to connect a large number of users separated over a wide geographical area and to readily facilitate reconfiguration to meet changes in user location or traffic requirements. The cost of communication via satellite is relatively insensitive to distance.

- *Broadcast capability* – Satellites readily provide broadcasting sevices from one transmitter to many receivers. Likewise they can provide a report-back capability, namely connections from many transmitters to one receiver. They also can be used to facilitate conferencing among geographically separated users.

1.1.2 The Threat of Extinction

In the 1980's, it seemed that almost everyone had something to gain from the benefits that satellites could offer. Indeed, despite the fact that most adults had been born before the first satellite networks were commissioned, yet it had become difficult to imagine life without them.

However, by the 1990's, a new view of the future of satellite communications was evolving. Increasingly, satellite communications engineers were being confronted by developments that suggested that, despite the euphoria of the 1970's and 1980's, the prospects for communications satellites were no longer as rosy as they once appeared.

Competing terrestrial technologies such as cable, microwave radio and particularly, fibre optic systems, were staging a battle for the telecommunications marketplace and threatening the position of satellites. This was becoming true even in the very market areas where satellites had made their first breakthrough, namely that of trans-oceanic links.

By the 1990's, satellite communications engineers had to admit that "the game appeared to be up". Satellites could no longer compete with terrestrial systems in the conventional tasks of carrying high volume telephone, data or television traffic over point-to-point trunk routes.

The main reason for this demise lay in the emergence of cheaper optical fibre systems. The availability of potentially enormous bandwidths in fibres greatly increases the number of telephone and other circuits that can be carried in a single cable installation [2].

Geostationary satellite systems are also under threat because of their inherently long transmission paths. Consider the geometrical situation illustrated in Figure 1.1.

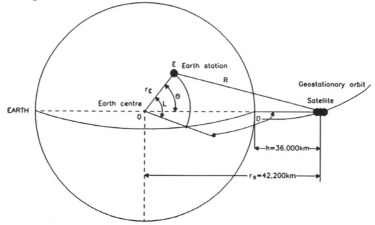

Figure 1.1: Distance relationships for a geostationary satellite.

The distance of a geostationary satellite (S) from the centre of the earth (O) is 42,176km. It is shown in Figure 1.1 as $r_S = 42,200$km approximately. The value of r_S ensures that the satellite orbit period is equal to that of a sidereal day, that is the period of a complete rotation of the earth relative to the fixed stars. The radius of the earth is 6,378km at the equator. Let r_E denote the radius of the earth at the earth station E. Let the earth station E have latitude L (degrees) and difference D (degrees) in longitude between E and S. Then it follows from Figure 1.1 that the distance R between E and S is related by

$$R^2 = r_S^2 + r_E^2 - 2r_S r_E \cos \theta \tag{1.1}$$

where

$$\cos \theta = \cos L . \cos D \tag{1.2}$$

The distance R is a minimum when the satellite is directly overhead the earth station ($\theta = 0°$). It is a maximum for earth stations on the horizon as viewed from the satellite.

Exercise 1.1
Determine the minimum and maximum distances respectively between an earth station and a geostationary satellite. Evaluate for $r_S = 42,200km$ and $r_E = 6,200km$.
Solution
Let

$$h = r_S - r_E \cos \theta \qquad (1.3)$$

as illustrated in Figure 1.1. Then when we obtain

$$R_{\min} = r_S - r_E = 36,000 km. \qquad (1.4)$$

Likewise, θ is a maximum (on the horizon) when

$$\cos \theta = \frac{r_E}{r_E + h} \qquad (1.5)$$

and we obtain

$$R_{\max} = h\sqrt{1 + \frac{2r_E}{h}} = 42,000 km \qquad (1.6)$$

The long transmission distances inherent in geostationary satellite systems give rise to at least two major problems that count against geosatellite systems in their competition for survival against optical fibre systems.

1. *The propagation paths have high losses* - losses can be as high as 200 dB at frequencies in the region of 14GHz (the commonly used uplink frequency in 12/14 GHz Ku-band satellite systems). If large high gain antennas are used to overcome these losses then not only are the antennas expensive and bulky but antenna pointing problems are significant.

2. *The transmission delay is a problem* - the transmission delay for a signal travelling at the speed of light ($c = 3 \times 10^8$ m/s) to the satellite on the uplink and then from the satellite to a receiving earth station on the downlink, is approximately 0.25 seconds total. This represents a significant annoyance to telephone customers.

Satellite communications engineers have now come to realize that "their game will be up" unless they can establish an appropriate niche for satellites in the total telecommunications systems spectrum [3].

1.1.3   The Fight Back

Not surprisingly, satellite communications engineers are not taking these threats lying down. They are currently addressing two issues. Firstly, they are busily searching out applications where satellite systems have something unique to offer. That is, they are looking for commercially attractive "niches". Secondly, they are giving high priority to those features which can reduce the costs and maximize the potential advantages of satellite services and hence make them more competitive with terrestrial links.

Mobile Satellite Systems

In regard to the first of the above issues being addressed by satellite communications engineers, that is to find unique niches for the use of satellites, it has become realized that one area in which satellites can play a unique role is in long distance *mobile communications.* Satellites can offer a unique service to users in rural and remote areas which are outside the range of existing telecommunications systems.

Of all other known communications media, only high frequency (HF) radio systems have been tried for this type of service. Whilst useful as an emergency service, HF radio has proven unreliable as a medium for good quality voice or data transmission because of its dependence on the reflective properties of the ionosphere. These can vary somewhat unpredictably from one hour of the day to another and from one location to another.

Mobile communications services provided via satellite represent a relatively new field of investigation for satellite communications engineers in many countries. Such services have or will soon become available for aeronautical, maritime and land mobile applications, providing communications between aircraft, shipping or land vehicles and the existing terrestrial networks. Through the use of digital mobile satellite systems, users worldwide will be able to access global telephone, data and facsimile networks, from their automobiles or whilst travelling as passengers in aircraft, trains or on board ships.

One motivation for this new emphasis on mobile communications via satellite is a recognition of our apparently unquenchable desire for mobile communications of all types. A landmark in the development of *terrestrial* mobile communications was the evolution of cellular radio systems. Within a few years of the beginning of cellular telephone services in the 1980's, the user demand far outstripped market projections. By 1990, the number of cellular telephones exceeded 7 million - the number anticipated in 1983 that would only be achieved by the year 2000. More recent projections are that there may be 10 million cellular telephones in use worldwide by 2000.

Mobile satellite communications will not replace cellular telephone networks, but rather will extend mobile communication coverage area to the entire globe. Land mobile satellite services (LMSS's) for telephony will commence in 1993 with the commencement of the Australian Mobilesat service. Later that year or early in 1994, the international Inmarsat Standard M system will be placed in service, These will be the first digital LMSS telephone systems. LMSS systems are also expected to be operational in North America in 1993 or 1994. It is estimated that the growth in demand for LMSS terminals for the Inmarsat service will reach 300,000 p.a. by the year 2000 and 500,000 p.a. by the year 2005 [3]. The number of LMSS mobile terminals in Canada and the US is expected to reach 1,000,000 by the year 2000.

In view of this dramatic development, considerable emphasis will be given to the theme of mobile communications in this book. That is, particular attention will be given to the emerging role of mobile applications in shaping the design of satellite communications systems.

In the design of future communications satellites, provision for mobile service requirements will however, only share part of the total spacecraft payload, together with conventional provision for high capacity trunk relays between fixed earth stations. Many communications satellites may also incorporate facilities for sound and/or television broadcasting. We will address each of these applications in following chapters.

LEO Satellite Networks

In regard to the second of the big issues facing satellite communications engineers, one way of reducing launch costs and transmission delays is to use *low earth orbit* (LEO) satellites. Since their inception in the 1960's, LEO satellites have continued to be used for remote sensing applications, for predicting weather patterns and for surveillance. However they were tried but discarded as a vehicle for commercial communications. The complexity of the tracking and control systems required for LEO satellite networks were major factors prohibiting their use for communications.

However, by the 1990's, the situation had changed. The ready availability of powerful computing and signal processing devices had significantly reduced the concern about the complexity issues. This has evolved to such an extent that once again, LEO satellites have become attractive as a means for providing communications services. As a result, several communications networks utilizing LEO satellites have been proposed for launch in the latter half of the 1990's.

One system of LEO satellites suggested is a network of 77 satellites conceived by Motorola Satellite Communications. The system has been named IRIDIUM after the chemical element of that name which has 77 electrons encircling the nucleus. Figure 1.2 illustrates a possible constellation of 77 satellites which will be arranged in seven polar orbital planes, each plane containing 11 satellites.

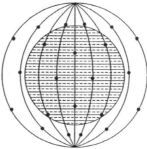

Figure 1.2:   Motorola Iridium 77 satellite constellation.

Recently, Motorola have indicated that a 66 satellite constellation of 11 satellites in in each of 6 planes may be preferred [4]. Each satellite performs the functions of a base station in a cellular mobile network. Because of the low earth orbit's, LEO satellites can provide a means of reducing earth terminal size down to that of a hand-held cellular-telephone-like unit. Network control and satellite hand-off procedures associated with LEO satellites may involve more complex signal processing in the terminals. However the saving on antenna costs will more than offset this disadvantage.

1.2   SATELLITE COMMUNICATIONS SYSTEMS

In this section, we will review the various components required for satellite communications.

A communications satellite can be considered to comprise two main modules:

- The *spacecraft bus* (or "space platform" or "service module")

- The *communications subsystem* (or "communications payload", also known as "module").

The satellite may also include an apogee motor as an integral part if it is to be put into orbit by a multi-stage launcher.

1.2.1 Spacecraft Bus

The spacecraft bus provides all of the support services, such as structural support, power supply, thermal control, and satellite positioning. These support services are required to enable the communications module to function effectively. The subsystems of the spacecraft bus include the following:

1. *Structural subsystem* - This comprises a mechanical skeleton on which the equipment modules are mounted. It also includes a skin or shield which protects the satellite from the effects of micro-meteorites and from the extremes of heat and cold. Also incorporated is a satellite stabilization system. Most communication satellites are either box-shaped (for example INTELSAT V) or cylindrical (for example INTELSAT VI). Box-shaped satellites are stabilized by means of inertia wheels spinning within the body. They are said to be body-stabilized. Cylindrical satellites are spin-stabilized. That is, they are stabilized by spinning the whole body. In these cases, the antenna systems are usually maintained stationary. They are said to be de-spun.

2. *Telemetry, tracking and command (TT&C) subsystem* - This is a system for monitoring the state of the on-board equipment, for transmitting telemetry signals to a control earth station, and for receiving and executing commands from the ground control centre. The telemetry system multiplexes data from many sensors on the spacecraft and transmits them via a digital communications link to the controlling earth station. This station incorporates a satellite tracking system to monitor changes in the satellite orbit. If required, the control station is able to transmit remote control functions to the satellite via an uplink telemetry system. Control functions on the satellite might include firing the apogee motor or using thruster jets to correct satellite attitude and position. Other functions include control of certain on-board communications subsystems, such as operating switching matrices to direct communications signals to specified antennas.

3. *Power subsystem* - Solar arrays provide the primary source of power for a communications satellite. Most of the power is used by the high power amplifiers in the communications module. Back-up batteries are

sometimes also provided to cope with those periods when the satellite passes through the earth's shadow. Fortunately, the earth's axis is tilted relative to its orbital plane so that a geostationary satellite is always in sunlight at the summer and winter solstices. However it suffers a period of eclipse each day for a total of 84 days each year around the equinoxes.

4. *Thermal control subsystem* - This is a combination of bimetallic louvres, electric heaters and surface finishes designed to protect electronic equipment from operating at extreme temperatures. Note that parts of the satellite will be in direct sunlight while other parts are facing cold space at a temperature close to $-270°C$.

5. *Attitude and orbit control subsystem* - This includes a system of small rocket motors (thrusters) required to keep the satellite in the correct orbital position and to ensure that its antennas are pointing in the right directions. Note that once located in orbit, the satellite position and attitude will be disturbed by gravitational attractions of the sun and moon and by the non-uniformity of the earth's gravitational field. Because the earth is not a perfect sphere, the gravitational field has maxima and minima so that geostationary satellites would, if left uncontrolled, drift to one of two points of maximum gravitational attraction at longitudes of approximately $75°E$ and $255°E$.

1.2.2   Communications Subsystem

The *communications subsystem* on a communications satellite consists of a number of repeaters which amplify the signals received from the uplink and condition them in preparation for transmission back to earth on the downlink. Each repeater may be designed to be one of the following types:

1. *transparent repeater* (also referred to as a "non-regenerative" or "bent-pipe" repeater)

2. *on-board processing repeater* (also known as "regenerative" or "switching regenerative" repeater).

Transparent Repeaters

The communications subsystems in civilian communications satellites are currently of the transparent type. In this type of communications subsystem, the repeaters are referred to as *transponders.*

Uplink communication signals from an earth station received at a satellite usually consist of multiple frequency division multiplexed (FDM) signals known as *carriers*. For example, for satellites designed for 4/6 GHz band operation, the uplink carriers for the high capacity fixed services occupy a total transmission bandwidth of 500 MHz in the vicinity of 6 GHz. The 500 MHz bandwidth is divided up into a number of *channels*, each typically 36 or 40 or 72 MHz wide. The uplink signals are handled by a separate transponder for each channel. (Details on frequency plans are given in the following Section.)

The signals are received at the satellite by a receiver antenna the output of which is connected to the transponders. Each transponder performs the processes of signal amplification, selection of one or more received signals (using a bandpass filter), translation of the signals to a new frequency band and amplification of them to a high power level for retransmission. Two types of transponders are illustrated in Figure 1.3.

Figure 1.3(a) illustrates the basic elements of a typical *transponder* used for 4/6 GHz band operation. The transponder consists of the following:

- a *receiver* - which includes a low noise amplifier and a downconvertor,

- an *input multiplexer* - in which a bandpass filter selects the channel frequency components assigned to that transponder,

- a *high power amplifier* (HPA) - which consists of a solid state power amplifier (SSPA) or a travelling wave tube amplifier (TWTA), and

- an *output multiplexer* - in which a number of transponder output signals are combined before being fed to the satellite's transmitting antennas.

It is usually possible to electronically switch the output of a transponder to one of a number of downlink antennas. This enables the satellite system operator to adapt the satellite resources to meet changing traffic demands.

Figure 1.3(b) illustrates a typical transponder used for 11/14 GHz operation. Note that a double conversion system is used. This is because it is easier to design filters, amplifiers, and equalizers at an intermediate frequency (IF) such as 1100 MHz rather than at 11 or 14 GHz. Hence the incoming 14 GHz carrier is translated to the IF frequency where it is amplified and filtered before being translated back to 11 GHz in preparation for amplification by the HPA. An electronically controlled attenuator inserted prior to the HPA, enables the overall gain of the transponder to be adjusted by remote control from a controlling earth station.

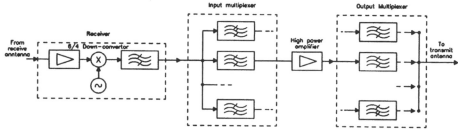

(a) Typical 4/6 GHz transponder.

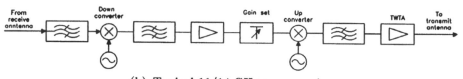

(b) Typical 11/14 GHz transponder.

Figure 1.3: Transponder subsystems.

We will examine the transmission performance characteristics of transponders in greater detail in a later Section.

Figure 1.4 is a simplified block diagram of the whole communication subsystem in a typical satellite such as would be used to provide coverage over a single nation.

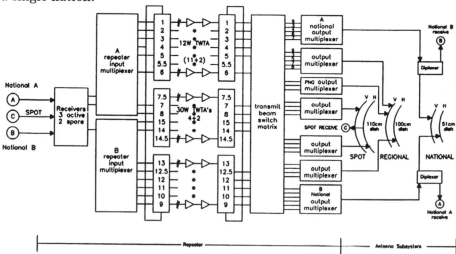

Figure 1.4: Block diagram of a typical communications subsystem.

As shown at the right of the figure, the satellite has three antennas, two of them (National A and National B) providing beams over the whole nation, and a third providing a spot beam over a selected area. Each antenna operates in a transmit and receive mode, using vertical (V) and horizontal (H) polarization respectively. (See Section 1.2.4 below for details on the use of dual polarization).

The signals received on the satellite uplinks are shown connected from the antennas to the input of a number of receiver systems, shown at the left of Figure 1.4. It is left to the reader to identify the thirteen 12W transponders and the six 30W transponders, two of each type being assigned as spares. The transponder outputs are switched by the output multiplexers to selected transmit antennas. The reader is challenged to work out from Figure 1.4, a table of possible connections between the transponders and the transmit antennas.

Satellites of the "transparent repeater" type have the advantage that their transponders impose minimal constraints on the characteristics of the communications transmission signals. For example, they are "transparent" to different modulation types, whether they be analogue or digital. Currently, it is common to transmit television image signals using analogue modulation techniques, whereas telephony, data and other signals are transmitted using digital modulation techniques (to be described later in this Chapter). Satellites with transparent repeaters therefore offer maximum flexibility for designers of the larger communications network of which the satellite is a part.

On the other hand, as it becomes more and more common that satellite systems are used to transmit all-digital signals, then it is likely that many future satellites will be of the "on-board processing repeater" type.

On-board processing repeater types of communication subsystems perform signal processing functions which include:

1. *Signal regeneration* - coherent reception and regeneration of the digital signals from the received uplink signal. On-board regenerators can lead to improved overall bit error rate in a complete earth station to earth station link.

2. *"Switchboard in the sky"* functions - a term used to describe a system for circuit switching between electronically hopping spot beam antennas and optical or radio links between satellites.

3. *Concentrator functions* - a number of digital signals at low bit rates received from several VSAT terminals may, after regeneration, be multiplexed in a single time division multiplex (TDM) signal for transmission at high bit rate to a larger central earth station.

A significant study into on-board processing satellites being undertaken by NASA involves the development of the Advanced Communications Technology Satellite (ACTS). Launched in 1991, ACTS may pave the way for the design of future communication satellites. For more details see Reference [4].

1.2.3   Frequency bands

When satellite communications first became commercially and technically realizable, all of the radio frequency spectrum up to 40GHz had already been allocated to terrestrial point-to-point radio services. In 1963 the International Telecommunications Union (ITU) decided that the best solution to this problem was not to reallocate to satellite services, frequencies which had until then been assigned to terrestrial radio systems. Rather it was decided to require that the satellite services share use of frequency bands with the terrestrial microwave radio services, both services being required to coexist on the same frequency bands.

Of the frequency bands allocated for shared use, two pairs of bands are of great importance for commercial *fixed satellite service* (FSS) operations (that is, services to fixed earth stations). They are as follows:

1. *4/6 GHz Bands:* The bands 3700 - 4200MHz and 5925 - 6425MHz are referred to as the 4 and 6GHz bands. The 4GHz band is commonly used for downlink satellite services, the 6GHz band being used for the paired uplinks. In the jargon stemming from early radar developments, these bands are collectively referred to as C-band frequencies.

2. *11/14 GHz Bands:* These represent the bands 10,950 - 11,200MHz, 11,450 - 11,700MHz and 14,000 - 14,500MHz, which are referred to as Ku-band frequencies. The frequencies in the nominal 11GHz region are commonly used as satellite downlink frequencies. They are paired with 14GHz frequencies which are used on the associated uplinks.

Frequencies for land *mobile satellite services* (MSS) were proposed by the ITU at a World Administrative Radio Conference (WARC) in 1987. These were as follows:

1.5GHz band: For satellite to mobile (downlink) transmission, the frequency band 1545 - 1559MHz was allocated. For mobile to satellite (uplink) transmission, the band 1646.5 - 1660.5MHz was allocated. This is illustrated in Figure 1.5, this pair of bands being referred to as L-band frequencies.

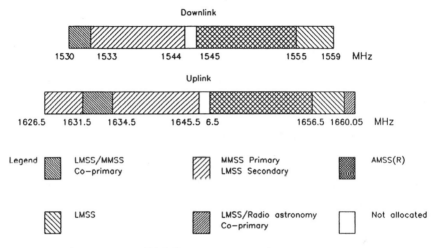

Figure 1.5: WARC-87 frequency plan for L-band.

Note that a bandwidth of only 14MHz is allocated to land mobile services for each of the uplink and downlink bands. Adjacent L-band frequencies were also allocated in 1987 to aeronautical services (20MHz), and maritime services (14MHz).

Subsequently, in the U.S., the domestic regulating authority, the Federal Communications Commission (FCC) decided to reject the ITU recommendations choosing instead to allocate 28MHz of bandwidth at L-band for the combined operation of all forms of mobile satellite services in the belief that a generic system serving all users would be more spectrum efficient.

For *direct broadcasting services* (DBS) from satellites to homes, the ITU allocated frequencies in the 12 GHz band as follows:

12GHz band: The broadcast frequency band of 11.7 - 12.5GHz was divided into channels each with a bandwidth of 27MHz. Each nation was allocated four or five TV channels together with one of the geostationary satellite orbit positions. These are positions designated at spacings of 6° intervals in the geostationary orbit.

A summary of frequency allocations for satellite communications use can be found in reference [5].

1.2.4   Transponder characteristics

The electrical characteristics of the satellite transponders have a major bearing on the performance of satellite communications systems using transparent repeater techniques. Since all the carriers from one antenna must pass through a low-noise amplifier(LNA), an amplifier failure could be catastrophic. Because of this, some *redundancy* is usually provided. That is, the received signals are taken to two parallel LNA's and recombined at their output. If either LNA fails, the other one can still carry all the traffic.

In early satellites such as INTELSAT I and II, only one or two wideband transponders (250 MHz bandwidth) were used. This proved unsatisfactory because the nonlinearity of the transponder HPA's gave rise to significant intermodulation between all the carriers involved. Later satellites used larger numbers of transponders (typically between 12 and 24), each with lower bandwidth to avoid excessive intermodulation products. Transponder bandwidths in common use are 36, 40, 45 and 72 MHz. That is, the 500 Mhz bandwidth is divided up into *channels*, each a few tens of megahertz wide, and each handled by a separate transponder.

A transponder channel frequency plan for a typical domestic satellite service using Ku-band frequencies is shown in Figure 1.6.

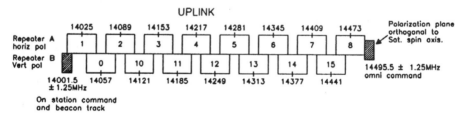

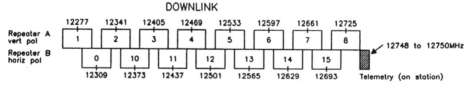

Figure 1.6: Typical satellite frequency/polarization plan.

In this example, there are 15 pairs of uplink and downlink channels occupying the following frequency bands:

- Uplink 14,000-14,500 MHz

- Downlink 12,250-12,750 MHz.

The channel allocation assumes that the antenna systems are able to support two *orthogonal polarization* modes, namely horizontal and vertical polarization respectively. At the receiving antenna, a polarization separation filter can isolate signals of one polarization from those of another.

As shown in Figure 1.6, channels 1-8 are transmitted using horizontal polarization and channels 9-15 are assigned vertical polarization. The centre frequencies of transponders using the same polarization are separated by 64 MHz. The centre frequencies of cross-polarized transponders are offset by only 32 MHz. Note that this is less than the bandwidth of each transponder which is 45 MHz. That is, the use of orthogonal polarization allows closer spacing between adjacent channels than if a single mode of polarization were used.

Many satellites operating in the 6/4 GHz band carry 24 active transponders, each of bandwidth 36 MHz. The centre frequencies are spaced 40 MHz apart to allow guard bands for the 36 MHz wide filter skirts. With a total of 500 MHz available, only 12 transponders could be accomodated on a single polarization satellite. However orthogonal polarizations allow *frequency reuse* and hence 24 transponders can be accomodated in the 500 MHz bandwidth. Stringent requirements are placed on the following properties of transponders:

1. *Gain-frequency response* - The transponder must provide good rejection of unwanted frequency signals and intermodulation products. Figure 1.7 illustrates a gain versus frequency response for a representative 45 MHz transponder. The transponder must not only provide sufficient rejection in the stopband, but also have very low gain variations across its passband. The shaded *mask* in Figure 1.7 shows how the maximum variation in passband ripple may be specified.

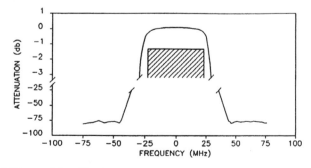

Figure 1.7: Typical transponder frequency response.

2. *Group delay distortion* - Transponders are also required to have very small phase variations across the passband. The slope $d\phi/d\omega$ of the phase shift versus frequency characteristic is known as the *group delay*. Ideally, the group delay should be constant over the frequency band occupied by the signal. Otherwise when an angle modulated carrier such as a phase shift keyed (PSK) digital signal or a frequency modulated (FM) television signal is passed through the transponder, significant signal distortion can occur. Figure 1.8 illustrates a typical group delay versus frequency response for a 45 MHz transponder.

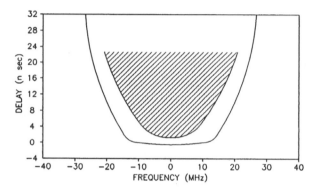

Figure 1.8: Typical transponder group delay response.

In the design of a transponder filter, it is frequently necessary to include in the filter an equalizer which smooths out group delay distortion in the region of the passband.

3. *Transponder power output* - The saturated power output of transponder output amplifiers used for fixed service digital transmission ranges from about 2 W to about 30 W. For mobile satellite services, and for television broadcast services, the transponder output power at saturation is typically 200 W.

4. *Amplitude and Phase Transfer Characteristics* - The high power amplifier (HPA) at the output of the transponder may exhibit a nonlinear *power transfer characteristic*, especially when operated near saturation. In practice, the operating point of the transponder may be set so that the output power is less than the saturation value to limit the effects of nonlinearity. This reduction in output power below saturation is called the *output backoff* and is denoted $B_o(dB)$. Likewise

the equivalent reduction of input power is termed the *input backoff*, denoted $B_i(dB)$. Typical output backoff values vary from as little as 0.2 dB for an amplifier carrying a single time division multiple access (TDMA) signal to 10 dB for a transponder carrying many carriers in frequency division multiple access (FDMA) form. Figure 1.9 shows a typical power transfer characteristic for an HPA.

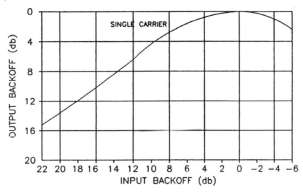

Figure 1.9: Transponder power transfer (AM/AM) characteristic.

Figure 1.10 illustrates a typical transponder *phase transfer characteristic*. This shows how the relative phase shift through the transponder is affected by backoff values.

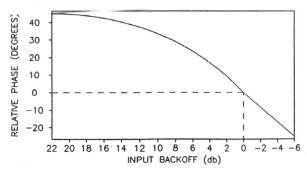

Figure 1.10: Transponder phase transfer characteristic.

If the HPA in a transponder were distortionless, the power transfer characteristic would be linear. That is, the output power P_o would be related to the input received power P_r by $P_o = a\, P_r$ where a is a constant. For the HPA illustrated in Figure 1.9, the power transfer characteristic becomes more nonlinear as the output power approaches *saturation* (the maximum power of which the HPA is capable).

The power transfer characteristic can be represented by a polynomial of the form

$$P_o = aP_r + bP_r^2 + cP_r^3 + \ldots \tag{1.7}$$

where $a, b, c, \ldots$ are constants. If the input is a single signal at frequency f_1, the terms $bP_r^2, cP_r^3, \ldots$, generate output components at harmonic frequencies $2f_1, 3f_1, \ldots$. On the other hand, when the input comprises two or more signals, then the terms $bP_r^2, cP_r^3, \ldots$, generate intermodulation products at frequencies which are sums and differences of the input frequencies, and their harmonics. This is illustrated in Figures 1.11(a) and 1.11(b) which represent typical power spectra as would be displayed on a spectrum analyser monitoring transponder input and output for the case where two sinusoidal tones are transmitted to the satellite. Figure 1.11(a) shows the input spectrum. The output spectrum shown in Figure 1.11(b) clearly exhibits many unwanted intermodulation components.

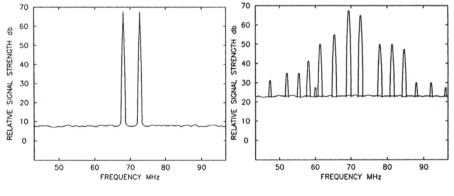

(a) Two-tone Input Spectrum   (b) Output Spectrum

Figure 1.11:   Effect of transponder nonlinearity.

Nonlinearities in the transponder power transfer characteristic have a major impact on the system performance. They also influence the choice of the types of modulation to be used for satellite communications. To see why nonlinearities are so important for *digital transmission* systems, consider a nonlinear HPA with a narrowband input signal, and bandpass filtering of the HPA output, as as illustrated in Figure 1.12.

Figure 1.12   Nonlinear amplifier model.

For any general modulation type (which might involve amplitude and/or phase modulation), the input signal x(t) can be written

$$x(t) = R(t)\cos[\omega_c t + \phi(t)] \tag{1.8}$$

where $R(t)$ and $\phi(t)$ represent amplitude and phase modulation respectively. Let the nonlinear system output (before filtering) be written as

$$y(t) = F\{x(t)\} \tag{1.9}$$

where $F\{\cdot\}$ is a complex memoryless nonlinear function. There will usually be a bandpass filter at the nonlinear system output so that we obtain to a good approximation the output after filtering as

$$z(t) = f(R(t))\cos[\omega_c t + g(R(t)) + \phi(t)] \tag{1.10}$$

where $f(R(t))$ and $g(R(t))$ are nonlinear functions of R(t). Note that

- $f(R(t))$ is called the *AM/AM conversion function*, and

- $g(R(t))$ is called the *AM/PM conversion function*.

That is, the nonlinear HPA will distort the digital signal by means of amplitude-to-amplitude (AM/AM) and amplitude-to-phase (AM/PM) conversions. The following special cases should be noted:

1. Linear system: For this case

$$f(R(t)) = KR(t), K \text{ a constant, and } g(R(t)) = 0 \tag{1.11}$$

2. Hard limiter: For this case

$$f(R(t)) = K, \text{ and } g(R(t)) = 0 \tag{1.12}$$

An important conclusion can be drawn from Equation (1.11). Consider the case where the input is a *constant envelope* signal with say, $R(t)) = R_o$, where R_o is a constant. Then the system output is

$$z(t) = f(R_o)\cos[\omega_c t + g(R_o) + \phi(t)] \tag{1.13}$$

where $f(R_o)$ and $g(R_o)$ are constants. That is, the output has a constant envelope. The constant phase term $g(R_o)$ can be easily removed in the earth station receiver. This means that if the modulated signal at the HPA input has constant envelope, then the resultant output signal remains unaffected by the nonlinearity.

This will not be so for digital amplitude and phase modulation types which have varying envelopes. In such cases, $R(t)$ is not constant and so the resultant term $g(R(t))$ in the output phase will represent intersymbol interference in the demodulated digital signal. In turn, this will lead to degraded bit error rate performance. In that situation, it is often necessary to operate the earth station and satellite transmitter power amplifiers at power levels as much as 20dB below their maximum value to avoid the intersymbol interference. That is, it is necessary to use a significant output power backoff.

Because of the above, designers of digital satellite communication systems often prefer to use constant envelope modulation schemes. These include *M-ary phase shift keyed* (M-PSK) modulation and *continuous phase frequency shift keyed* (CPFSK) modulation. Digital modulation techniques will be discussed in more detail later in this chapter.

1.3 SATELLITE LINK DESIGN

1.3.1 Communications Performance Requirements

A single-hop satellite communications link between fixed earth stations consists of an uplink from a transmitting earth station to the satellite, and a downlink from the satellite to the receiving earth station. The choice by the link designer of the uplink and downlink parameters (transmit power, antenna gain, path loss, etc.), determines the overall performance of the link.

Any communication system must be designed to meet certain minimum performance standards within limitations of transmitter power and bandwidth. For *digital transmission*, the most important performance criterion is the *bit error rate* (BER). This is the ratio between the number of bits received in error to the number of bits transmitted. Examples of typical minimum BER performance requirements for satellite communications systems carrying data or digitally encoded telephone speech, are as follows:

For data: The BER shall not be greater than 10^{-6}, (sometimes expressed as 1 in 10^{-6}), for at least 99% of any month

For speech: The BER shall not be greater than 10^{-4} for at least 99% of any month.

For *analogue transmission*, the basic measure of system performance is the signal-to-noise ratio (SNR) in the information channel at the link output. In this text, it will be assumed that digital transmission is used unless explicitly stated otherwise.

The BER for a satellite link depends on a number of factors, particularly:

- the carrier-to-noise ratio (C/N) of the RF signal in the uplink receiver (on the satellite), and in the downlink receiver (in the receiving earth station), and

- the type of modulation and error correction coding schemes used for transmission.

We will consider some of the most popular modulation schemes next.

1.3.2 Modulation Schemes

M-PSK modulation

The most popular forms of modulation used in satellite communications are the M-PSK schemes, where $M = 2^b$, and where it is usual to choose $b = 1, 2$, or 3. In M-PSK modulation, the input binary information stream is first divided into b-bit blocks. Then each block is transmitted as one of M possible *symbols*, each symbol being a carrier frequency sinusoid having one of M possible phase values.

Let T_s represent the duration of a symbol. Then for the kth M-tuple say, the M-PSK signal can be written

$$s(t) = A\cos(\omega_c t + \phi_k) \qquad (k-1)T_s \leq t \leq kT_s \qquad (1.14)$$

where $\phi_k \in \{0, 2\pi/M, 4\pi/M, \ldots, 2(M-1)\pi/M\}$.

The most important M-PSK modulation schemes for satellite communications are the following.

- *binary PSK* - usually referred to as BPSK, or 2-PSK, in which $M = 2$. In this case, a binary input data stream modulates a constant amplitude, constant frequency carrier in such a way that two phase values differing by 180° represent the binary symbols 0 and 1, respectively. A BPSK symbol may be written

$$s(t) = v(t)[A\cos\omega_c t] \qquad (1.15)$$

where $v(t)$ is a baseband non-return-to-zero (NRZ) pulse waveform consisting of random binary rectangular pulses taking values 1 or -1, each having period T_s.

- *four-phase PSK* - referred to as QPSK (quadrature PSK), or 4-PSK, in which the phase of the carrier can take on one of four values 45°, 135°, 225°, or 315°. Each transmitted symbol represents two input bits as follows:

Input bits	Transmitted symbol	Input bits	Transmitted symbol
00	$A\cos(\omega_c t + 45°)$	11	$A\cos(\omega_c t + 235°)$
01	$A\cos(\omega_c t + 135°)$	10	$A\cos(\omega_c t + 315°)$

- *eight-phase PSK* - usually denoted 8-PSK, in which a transmitted carrier symbol takes on one of eight possible values, each representing three input bits.

Figure 1.13 shows signal space *phasor diagrams* for BPSK and QPSK signals.

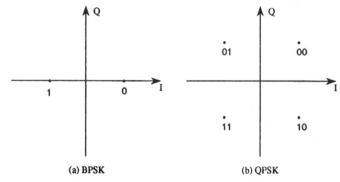

(a) BPSK   (b) QPSK

Figure 1.13: BPSK and QPSK signal vector diagrams.

In M-PSK modulation schemes, the conversion from binary symbols to phase angles is usually done using Gray coding.

The essential idea of *Gray coding* is to permit only one binary number change in the assignment of binary symbols to adjacent phase angles. This minimizes the number of bit errors that result from a demodulation error. This is because when noise gives rise to an error, the most probable type of error is that in which the digital receiver incorrectly selects a symbol adjacent to the correct one.

Figure 1.14 shows a block diagram of a *quadrature modulator* which can be used for any form of M-PSK modulation. Consider the functions required for QPSK modulation. The multiplexer converts the binary input stream into two parallel, half rate signals, $v_I(t)$ and $v_Q(t)$, referred to as the I (in-phase) and Q (quadrature) signals.

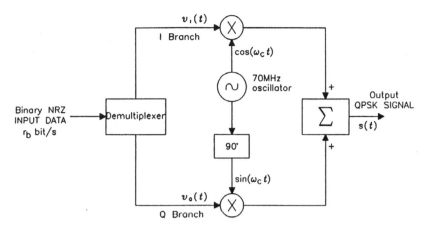

Figure 1.14: QPSK modulation.

These signals take values $+A/\sqrt{2}$ or $-A/\sqrt{2}$ in any symbol interval. They are fed to two balanced modulators with input carriers of relative phase 0° and 90°, respectively.

Then the QPSK signal can be written

$$s(t) = v_I(t) \cos \omega_c t + v_Q(t) \sin \omega_c t \qquad (1.16)$$

The waveforms shown in Figure 1.15 illustrate the QPSK modulation process. The top waveform represents conventional unfiltered QPSK. The bottom diagram shows the effect of passing the I and Q signals through lowpass filters (see later discussion of Nyquist filtering).

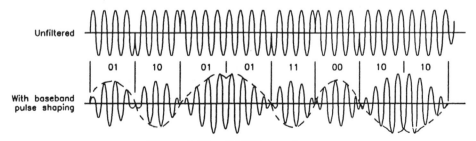

Figure 1.15: QPSK waveforms.

Exercise 1.2

For the QPSK modulator in Figure 1.14, if in a given T_s time interval, $v_I = +A/\sqrt{2}$ and $v_Q = -A/\sqrt{2}$, find the output $s(t)$.

Solution

It is a simple trigonometry exercise to show that

$$s(t) = A \cos \left(\omega_c t - \frac{\pi}{4} \right) \tag{1.17}$$

Frequency Spectra of M-PSK Signals

For the M-PSK signal represented by equation (1.14), it can be shown that for any value of $M = 2^b$, the power spectral density is given by

$$S(f) = A^2 T_s \left\{ \frac{\sin[\pi T_s (f - f_c)]}{[\pi T_s (f - f_c)]} \right\}^2 \quad V^2/Hz \tag{1.18}$$

In this equation, f_c is the unmodulated carrier frequency, A is the carrier amplitude, and T_s is the symbol interval. Note that if T_b is the input binary bit interval, then

$$T_s = T_b \log_2 M \tag{1.19}$$

The power spectral density is shown in Figure 1.16.

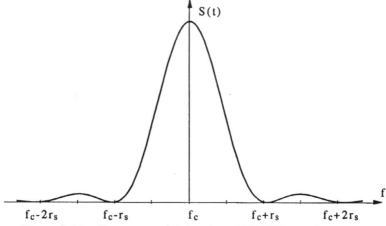

Figure 1.16:   Power spectral density of M-PSK signals.

As illustrated in Figure 1.16, the power spectral density of an unfiltered M-PSK signal occupies a bandwidth which is a function of symbol rate $r_s = (1/T_s)$. For a given transmitter symbol, the power spectrum for any M-PSK signal remains the same, independent of the number M of symbol levels being used. That is BPSK, QPSK, and 8-PSK signals each have the same spectral shape if T_s is the same in each case.

Nyquist Filtering and Spectral Efficiency

In an M-ary PSK scheme, each transmitted symbol represents $\log_2 M$ bits. Therefore, for a fixed input bit rate, as the value of M increases, so the transmitter symbol rate decreases. As a result, there is an equivalent reduction in the width of the resulting signal spectrum. That is, there is an increase in *spectral efficiency* for larger M. The spectral efficiency of any digital modulation scheme is defined as the ratio of the input data rate (r_b) and the allocated channel bandwidth (B). That is, the spectral efficiency is given by

$$\eta = \frac{r_b}{B} \quad (bit/s/Hz) \tag{1.20}$$

The spectral efficiency of 8-PSK, for example, will be three times as great as that for BPSK. This can only be achieved however, at the expense of increased error probability, as will be discussed below.

It remains to be explained how the allocated RF channel bandwidth (B) of an M-PSK signal is to be defined. Examination of Figure 1.16 indicates that the M-PSK spectrum rolls off relatively slowly. Theoretically, an infinite bandwidth would be necessary to accomodate the M-PSK spectrum. In practice, it is necessary to filter the M-PSK signal so that its spectrum is limited to a finite bandpass channel region in order to avoid *adjacent channel interference*. However, filtering causes distortion of the transmitted pulses. This distortion usually occurs in the form of a spreading of a given pulse in time, so that it interferes with signals in adjacent time intervals. That is, it gives rise to intersymbol interference (ISI). This in turn, may degrade the BER.

A class of filtering which can limit the modulated signal spectrum without causing ISI is known as *Nyquist filtering* or *raised-cosine filtering*. This is the method most commonly adopted to limit the spectral width of digitally modulated signals.

Nyquist filtering requires that the voltage spectrum of the signal received at the demodulator input must have one of the shapes illustrated in Figure 1.17. Note that in each case, the spectrum rolls off with a raised-cosine shape either side of the carrier frequency. Alternative raised-cosine spectra are characterized by a factor α, known as the *excess bandwidth factor* or the *roll-off factor*. The excess bandwidth factor α, lies in the range 0-1, and specifies the excess bandwidth of the spectrum compared to that of an ideal bandpass spectrum ($\alpha = 0$) for which the bandwidth would be

$$B = r_s \tag{1.21}$$

For example, if a filter with $\alpha = 0.5$ is used, then the output spectrum would have bandwidth of $1.5r_s$ (Hz). In practice, α values of 0.3-0.5 are commonly used.

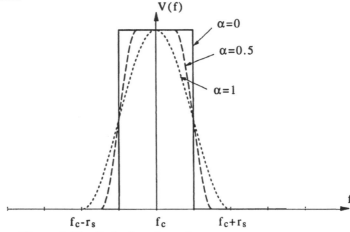

Figure 1.17:   Raised-cosine voltage spectrum.

In general for M-PSK transmission using Nyquist filtering with roll-off factor α, the bandwidth required will be

$$B = r_s(1 + \alpha) \tag{1.22}$$

It follows that the maximum bit rate in terms of the transmission bandwidth B, and roll-off factor α, is given by

$$r_b = \frac{B \log_2 M}{1 + \alpha} \tag{1.23}$$

Hence, for an M-PSK system with ideal Nyquist filtering ($\alpha = 0$), the signal spectrum is centred on f_c, is constant over the bandwidth $B = 1/T_s$ and is zero outside that band. For this case, the transmitted bandwidth for an M-PSK signal can be written

$$B = \frac{1}{T_b \log_2 M} \tag{1.24}$$

It follows that for M-PSK, with ideal Nyquist filtering, ($\alpha = 0$) the spectral efficiency is

$$r_b/B = \log_2 M \text{ (bit/s/Hz)}. \tag{1.25}$$

Exercise 1.3

Given that binary data is to be transmitted over a satellite link at 34 Mbit/s in a channel bandwidth of 26 MHz, determine which M-PSK scheme is appropriate.

Solution

From Equation 1.24 for ideal Nyquist filtering, the transmitted signal bandwidth is $B = 1/T_s$. For BPSK, $T_s = T_b$ so that the minimum bandwidth required is $B = 1/T_b = 34MHz$. For QPSK, $T_s = 2T_b$ so that $B = 1/(2T_b) = 17MHz$, and for 8-PSK, $B = 1/(3T_b) = 11.3MHz$. Hence, if we choose QPSK, we allow a margin for attenuation of the sideband components at frequencies beyond $f_c \pm r_s/2$. For example, if Nyquist filtering with excess bandwidth factor $\alpha = 0.5$ is used, then the required QPSK bandwidth is $B = 25.5MHz$.

Nyquist spectral shaping can be achieved in practice by means of either low-pass filtering of the baseband signal prior to modulation, or by bandpass filtering after modulation. For M-PSK modulation with $M = 4, 8, \ldots$, it is convenient to perform Nyquist filtering at baseband.

Referring to Figure 1.14, the baseband I and Q signals from the demultiplexer are first passed through Nyquist filters before being modulated on to orthogonal carriers. This is referred to as *pre-modulation filtering*.

In general, the Nyquist filtering on a satellite link is equally divided between the transmitter and the receiver at each end of the link. In that case, the spectrum at the output of the transmitter should be of the form of the square root of the raised cosine response. This is somes referred to as *root-Nyquist filtering*. A further root-Nyquist filter at the receiver then is required to provide the overall raised-cosine spectrum.

Envelope Variations and Spectral Regrowth

It is important to bear in mind that when a constant envelope M-PSK signal is bandlimited by filtering, then the envelope becomes non-constant. As an example, Figure 1.18(a) shows how the envelope of a filtered QPSK signal may vary with different phase transitions between adjacent symbol intervals.

It will be seen that a transition between adjacent sectors of the signal phasor diagram in Figure 1.18(b) (such as from symbol '00' to symbol '10') causes only a relatively minor variation in the envelope. On the other hand, a transition between diametrically opposite sectors of the phasor diagram (such as from '00' to '11'), causes the envelope to go to zero.

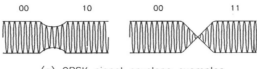

(a) QPSK signal envelope examples

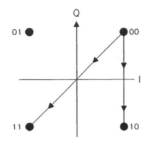

(b) QPSK phase transitions

Figure 1.18   Envelope fluctuations of filtered QPSK signals.

If an M-PSK signal with significant envelope fluctuations passes through a nonlinear amplifier, then two effects can occur.

- ISI may result, as discussed in Section 1.2.4 above.

- *Spectral regrowth* may occur. That is, the AM-AM and AM-PM characteristics of the amplifier may produce spreading of the spectrum beyond the filtered bandwidth.

Pre-modulation filtering may be easier to implement than RF filtering after the modulator and HPA. However, the former has the disadvantage that, if a nonlinear power amplifier follows the modulator, then the spectrum is broadened and at least partially restored to its original shape.

There are a number of modulation methods which are especially designed to produce RF signals with little or no envelope fluctuations, even when filtered by a Nyquist filter. This includes a class of schemes known as continuous phase frequency shift keyed (CPFSK) modulation, which includes minimum shift keyed (MSK) modulation. See Reference [6] for details.

QPSK Variants

A number of modified QPSK schemes have also been developed with the aim of minimizing the signal envelope fluctuations. We noted that, as shown in Equation (1.15), QPSK may be visualized as the sum of two independent BPSK signals whose carriers are in phase quadrature. In conventional

QPSK, the signals v_I and v_Q that modulate these carriers each make step changes at the same time.

Offset QPSK modulation is a technique in which the quadrature modulator I and Q signal phase transitions are staggered so that v_I makes step changes at the beginning of each symbol period, and v_Q makes changes at the midpoint of each symbol period. It is easy to show that this staggering of the carrier phase transitions results in a sequence of signal vector carrier phase states which eliminate all 180° transitions (through the origin of the signal phasor diagram). As a result, the signal envelope never goes to zero.

Another modulation scheme known as $\pi/4$ QPSK avoids phase state transitions through the origin of the signal phasor diagram. In this scheme, alternate QPSK symbols in a transmitted sequence are chosen from two alternate signal constellations. As illustrated in Figure 1.19, the two constellations are rotated 45° with respect to each other. The resultant set of possible phase state transitions are also illustrated in Figure 1.19.

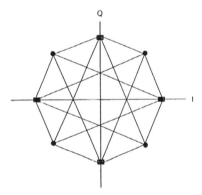

Figure 1.19: Signal vectors for $\pi/4$ QPSK.

The relative merits of QPSK, offset QPSK and $\pi/4$ QPSK will be discussed further in Chapter 3.

Scramblers

It should be noted that the form of the M-PSK spectrum represented by Equation (1.18) is strictly only valid if the input data sequence is a random one. If this is not so, such as during the transmission of simple idling patterns of all 1's or alternating 1's and 0's, then the spectrum can consist of large discrete spectral lines. In practice, this could give rise to undesirable levels of interference to adjacent channels.

To prevent this, *scramblers* are usually employed to randomize the data prior to modulation. These add a pseudorandom binary sequence to the data input before modulation. A matching de-scrambler is required at the receiving end so that, after demodulation, the pseudorandom sequence is stripped off. This is done by the addition of an identical pseudorandom sequence to the received scrambled sequence.

1.3.3 BER Performance and Error Correction Coding

In receiving systems for digital satellite communications, the demodulator generally employed is a *coherent demodulator*. A coherent demodulator for M-PSK signals is illustrated in Figure 1.20.

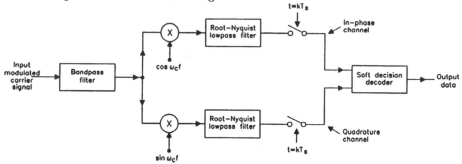

Figure 1.20: M-PSK demodulator and decoder.

A coherent demodulator is the optimum form of demodulator, (that is, it results in minimum BER), provided that the link can be modelled as an additive white Gaussian noise (AWGN) channel [6]. This model is usually appropriate for a satellite system.

The demodulator includes a *quadrature detector* consisting of two balanced multipliers with carrier inputs in phase quadrature. This is followed by root-Nyquist filters in the output I and Q arms. The resultant I and Q signals are then sampled at the centre of each symbol to produce the demodulator output I and Q signals which are delivered to the decoder.

In a coherent demodulator, it is required that the demodulator carrier be locked in phase with the incoming carrier signal. Furthermore, the I and Q signals must be sampled at times exactly coinciding with the peaks of the received data signal values.

The most challenging aspect of the design of a coherent demodulator is the realization of the carrier synchronization and symbol timing synchronization circuits. This will be discussed further in Chapter 5.

Forward error correction (FEC) coding is usually used in digital satellite communications to combat the effect of errors caused by noise at the receiver systems. That is, at the transmitter, the input data is first encoded with an FEC code before being modulated on a carrier for transmission. Likewise, at the receiving earth station, the demodulator is usually followed by an error correction decoder. FEC techniques are discussed in detail in Chapter 7.

The demodulator in Figure 1.20 is known as a *soft-decision demodulator* because it does not attempt to convert the recovered I and Q signals back to a binary data form (known as *hard-decisions*). Instead it simply passes the values of I and Q signal estimates (*soft decisions*) to the decoder for error correction. Soft decision systems are generally used because they perform approximately 2-3dB better than hard-decision systems. That is, they require 2-3dB less signal power for the same BER performance. Soft decision decoders will be discussed in Chapter 7.

Probability of Error

It is possible to compute the bit error probability at the output of a coherent receiver system for the case where an AWGN channel model is assumed. First, let us consider the case where a *hard decision demodulator* is used. Furthermore, let us assume that no FEC coding is used. We will consider only M-PSK schemes here since they are by far the most popular for satellite communications.

Consider the case where the transmitter sends an M-PSK symbol, which for convenience, we represent by its I and Q vector values. At the receiver, the demodulator must process the received waveform and estimate which transmitted I and Q symbol is represented by that waveform. A demodulator error occurs when a given symbol is transmitted and the channel noise causes the demodulated signal value to lie closer to some I, Q symbol other than that which was sent.

An M-PSK modulator produces symbols with one of M phase values spaced $2\pi/M$ apart. Each signal is demodulated correctly at the receiver if the phase is within π/M radians of the correct phase at the demodulator sampling instant. In the presence of noise, evaluation of the probability of error involves a calculation of the probability that the received phase lies outside the angular segment within π/M radians of the transmitted symbol at the sample time.

The probability that a demodulator error occurs is referred to as the *symbol error probability* P_s. In an M-ary modulation scheme with $M = 2^b$ bits, each symbol represents b bits. The most probable symbol errors are those that choose an incorrect symbol adjacent to the correct one. If Gray coding is used, then only one bit error results from such a symbol error. Then the bit error probability P_b is related to the symbol error probability by

$$P_b = P_s/m. \tag{1.26}$$

Consider the evaluation of symbol error probability for the QPSK case. Symbol errors occur when noise pushes the received phasor into the wrong quadrant. This is illustrated in Figure 1.21. Where it is assumed that the transmitted symbol had a phase of $\pi/4$ rad, corresponding to demodulator I and Q values (in the noise-free case) of $v_I = V$ and $v_Q = V$ volts.

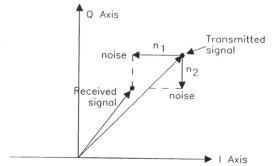

Figure 1.21: Transmitted and received signal vectors.

The additive narrowband Gaussian noise may be resolved into independent orthogonal components that are respectively in phase and in quadrature with any arbitrary phasor as shown in Figure 1.21. We will consider the noise phasors n_1 and n_2 to be pointing in directions that are most likely to cause symbol errors. Then a symbol error will occur if either n_1 or n_2 exceeds V.

For simplicity we assume that a QPSK signal is transmitted without Nyquist filtering and demodulated by a demodulator in which hard-decisions are made at the demodulator output. In the presence of additive Gaussian noise, the probability that the demodulated symbol value is correct is equal to the product of the probabilities that each demodulator low-pass filter output lies in the correct quadrant. That is the probability that the demodulated symbol value is correct is

$$P_c = (1 - P_{e1})(1 - P_{e2}) \tag{1.27}$$

where P_{e1} and P_{e2} are the probabilities that the two filter output sample values lie in the wrong quadrant.

It is not difficult to show that if the low-pass filters are equivalent to integrators (the optimum choice if Nyquist filtering is not employed), then

$$P_{e1} = P_{e2} = Q\left(\sqrt{\frac{E_s}{N_o}}\right) \tag{1.28}$$

where $E_s = \frac{A^2 T_s}{2}$ is the energy per symbol, and where $N_o/2$ is the two-sided noise power spectral density (in V^2/Hz) at the demodulator input.

The function $Q(x)$ is defined

$$Q(x) = \frac{1}{\sqrt{2\pi}} \int_x^\infty e^{-t^2/2} dt. \tag{1.29}$$

An upper bound on $Q(x)$ which is useful for high x values is

$$Q(x) \le \frac{1}{x\sqrt{2\pi}} e^{-x^2/2} \quad x \gg 1 \tag{1.30}$$

Because $P_{e1} = P_{e2}$, we can write the symbol error probability as

$$P_s = 1 - P_c = 2P_{e1} - P_{e1}^2 \tag{1.31}$$

which becomes

$$P_s \approx 2P_{e1} \quad \text{for} \quad P_{e1} \ll 1 \tag{1.32}$$

and hence for the QPSK system, the symbol error probability is

$$P_s \approx 2Q\left(\sqrt{\frac{E_s}{N_o}}\right) \tag{1.33}$$

For QPSK, $E_s = 2E_b$ where E_b is the energy per bit. Then using Equation (1.26) we obtain the bit error rate for QPSK to a good approximation as

$$P_b = Q\left(\sqrt{\frac{2E_b}{N_o}}\right) \tag{1.34}$$

The above analysis assumed that no Nyquist filtering is used. It can be shown that Equation (1.34) also holds for the case where root-Nyquist filters are used at the transmitter and receiver respectively, and for the condition that the demodulator input energy per bit (E_b) and noisy density (N_o) are the same in both cases.

By a similar analysis, it can be shown that, for BPSK, the bit error probability is the same as that given for QPSK by Equation (1.34). Bit error probalities for other modulation schemes can be found in Reference [6].

Figure 1.22 illustrates some of the above results by plotting bit error rate versus E_b/N_o for selected M-PSK schemes, with and without FEC coding. Note that the performance of uncoded BPSK and QPSK are identical. For other uncoded M-MSK systems, as the number of symbol vectors M increases, so the regenerator decision regions become smaller and the probability of error increases.

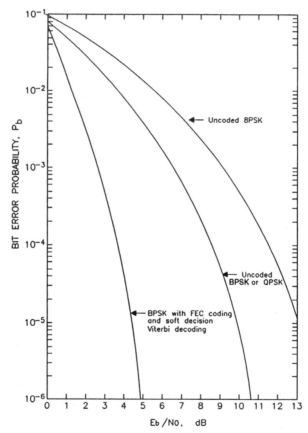

Figure 1.22: Bit error probability performance of various M-PSK schemes.

The effect of FEC coding is shown in Figure 1.22 by considering the most popular FEC coding scheme used in satellite communications. The particular code is referred to as a half-rate, constraint length 7, convolutional code and it is assumed to be decoded by a soft decision Viterbi decoder. Details will be given in Chapter 7.

Exercise 1.4

Compare the transmission parameters of uncoded BPSK, QPSK and 8-PSK respectively, if it is required that transmission is at a data rate of 2 Mbit/s with a bit error rate equal to 10^{-4}. In each case, determine the minimum transmission bandwidth and E_b/N_o ratio required.

Solution

The symbol rate for BPSK is $f_s = 2 \times 10^6$ symbol/s. For QPSK for 8-PSK, it is 1×10^6 and $= 0.67 \times 10^6$ symbol/s respectively Hence, the minimum bandwidths for Nyquist filtering ($\alpha = 0$) are 2 MHz for BPSK, 1 MHz for QPSK and 0.67 MHz for 8-PSK.

The E_b/N_o ratio required in each case for BER= 10^{-4} is obtained from Figure 1.22 as :

BPSK and QPSK: $E_b/N_o = 8.4$ dB
8-PSK: $E_b/N_o = 11.8$ dB.

1.3.4 Link Signal to Noise Ratios

Now we are in a position to study techniques for the design of a satellite communications link between two fixed earth stations. The overall link is made up of an uplink to the satellite and a downlink to the earth receiving station. Unless otherwise stated in this text, we assume the satellite is of the "transparent" transponder type.

Satellite link design involves an analytical approach to the selection of link subsystem variables in such a way that the overall performance requirements are met. One important decision is the choice of modulation type (as discussed in the previous Section).

The main variable parameters associated with the uplink are the transmitting earth station power output and antenna charateristics, and the satellite receiver and receiving antenna characteristics. For the downlink, the main variables are the satellite transponder power output and transmit antenna characteristics, as well as the characteristics of the receiving antenna and the receiver at the receiving earth station.

The main performance criterion for a digital satellite system is the overall BER. The design of the uplink and downlink must be such that in cascade, the overall BER is less than the specified performance limit.

The BER for a satellite link depends primarily on the carrier energy per bit to noise power density ratio E_b/N_o, at the receiver in the satellite (for the uplink) and at the receiver in the receiving earth station (for the downlink).

Note that there are a number of alternative ways of describing receiver signal to noise ratios. In practice, the following alternative forms are used:

- *bit energy to noise density ratio* (E_b/N_o): the ratio of the received signal energy per bit (E_b) to the one-sided noise power spectral density (N_o)

- *carrier-to-noise ratio* (C/N): defined as the ratio of the received carrier power (C) at the receiver input, to the total equivalent receiving system noise power at the receiver input (N).

- *carrier to noise density ratio* (C/N_o):the ratio of the received signal power to the noise power spectral density ratio.

- *carrier to noise temperature ratio* (C/T): the ratio of the received carrier power to the receiving system noise temperature.

For digital satellite transmission, the BER performance is best expressed in terms of the E_b/N_o ratio. For example, by examination of Figure 1.22, if QPSK modulation is used, then the theoretical value of E_b/N_o required to give BER $= 10^{-6}$ is 10.5 dB.

Note that the four ratios listed above are interrelated. Apart from C/N, the other three, namely $E_b/N_o, C/N_o$, and C/T, are independent of receiver bandwidth. Interrelationships can be deduced by noting that the equivalent receiver input noise density is

$$N_o = kT \qquad (1.35)$$

where $k = 1.38 \times 10^{-23}$ is Boltzmann's constant. Also for an M-PSK modulation scheme, each transmitted symbol represents $\log_2 M$ bits. If T_s is the symbol interval, then the carrier power is related to the energy per bit by

$$C = (E_b \log_2 M)/T_s \qquad (1.36)$$

Furthermore the equivalent receiver input noise power (N) is related to the noise density (N_o) by

$$N = N_o B \qquad (1.37)$$

where B is the noise bandwidth of the receiver. Then it is usually assumed for analytical convenience, that the receiver bandwidth (B) is considered equal to the ideal minimum Nyquist bandwidth. In that case,

$$B = 1/T_s. \tag{1.38}$$

Then the noise power N, is the noise power in the bandwidth $B = r_s$. Hence it follows that

$$C/N = (E_b/N_o)log_2 M \tag{1.39}$$

1.3.5 Link Budgets

In order to meet a specified BER requirement, a satellite link must be designed to ensure a sufficiently large value of E_b/N_o, (or equivalently, C/N) at the receiver. The usual design approach is to prepare a *link budget* which calculates link equations for power and noise, respectively. These are often referred to as the *link power budget* and *link noise budget*, respectively. In order to develop the *link power budget* equations, consider the satellite to ground downlink illustrated in Figure 1.23.

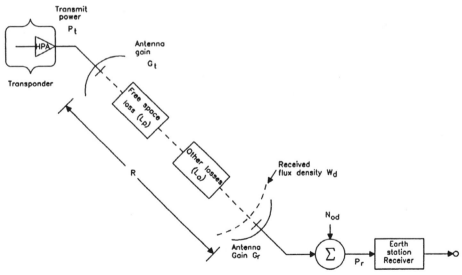

Figure 1.23: Satellite downlink model.

Let the satellite transponder output power be P_t (W). At this stage let us ignore any losses between the output of the transponder HPA and the transmit antenna. That is, the power delivered to the antenna is P_t.

Let the transmit antenna have gain $G(\theta)$ in a direction θ where θ is the angle from the antenna boresight. The antenna gain $G(\theta)$ is defined as the ratio of the power per solid angle radiated in a given direction to the average power radiated in the same solid angle by an isotropic source. A typical plot of the antenna polar radiation function $G(\theta)$ is shown in Figure 1.24. The antenna gain at $\theta = 0°$ is a maximum and is denoted G_t.

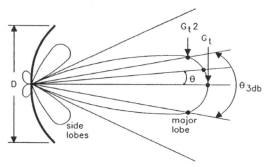

Figure 1.24:   Antenna radiation pattern (polar Diagram).

If the antenna were isotropic, that is $G(\theta) = 1$ for all θ, then the received flux density at any distance R from the transmitter would be given by

$$W_d = \frac{P_t}{4\pi R^2} \quad (W/m^2) \tag{1.40}$$

If the antenna is directional and has gain G_t, then for any receiving point at distance R on the boresight of the transmit antenna

$$W_d = \frac{P_t G_t}{4\pi R^2} \tag{1.41}$$

The numerator of equation (1.41) is a useful figure of merit for a transmitter. It is known as the *Effective Isotropic Radiated Power* (EIRP) where

$$EIRP = P_t G_t \quad (W) \tag{1.42}$$

If we had an ideal receiving antenna with an aperture area of $A_r (m^2)$, we would receive (collect) power P_r where

$$P_r = W_d A_r \tag{1.43}$$

For a practical antenna with an aperture area of A_r, some of the incident energy on the antenna is reflected away from the aperture or absorbed by lossy components. This reduction in efficiency is described by using an effective aperture area A_e, where

$$A_e = \eta_a A_r \qquad (1.44)$$

and η_a is called the *aperture efficiency* of the antenna. For antennas with paraboloidal reflectors, η_a is typically in the range 0.5 to 0.75, being lower for small antennas and higher for large Cassegrain antennas [1]. Sometimes if the precise value of η_a is unknown for a particular antenna, a figure of $\eta_a = 0.6$ is assumed. Horn antennas can have higher efficiencies with values of η_a approaching 0.9.

The power received by a practical antenna with effective aperture area A_e will be

$$P_r = \frac{P_t G_t A_e}{4\pi R^2} \qquad (1.45)$$

From a fundamental relationship in antenna theory [1], the (maximum) gain of the receiving antenna and its effective area are related by

$$G_r = \frac{4\pi A_e}{\lambda^2} \qquad (1.46)$$

where λ is the carrier signal wavelength.

Substituting for A_e in equation (1.45), gives the power output from the receiving antenna as

$$P_r = P_t G_t G_r \left[\frac{\lambda}{4\pi R}\right]^2 \qquad (1.47)$$

This equation, known as the *Friis transmission equation*, is important in the analysis of any radio or satellite link.

Path Loss: In relation to the last term in Equation (1.47), the path loss L_p is defined as

$$L_p = \left[\frac{4\pi R}{\lambda}\right]^2 \qquad (1.48)$$

It is not strictly a loss in the sense of power being absorbed. Rather it accounts for the way energy spreads out as an electromagnetic field travels away from a transmitting source.

In communications, the decibel is often used to facilitate calculations of the type given by equation (1.47). Hence we express the transmit power (P_t) in decibel form by using $10 \log P_t$ (dBW). Note that the expression dBW means decibels either greater than or less than 1W. The dBm unit is also sometimes used where the reference power is 1mW. Likewise the EIRPis expressed in decibel form by using $10\log(P_tG_t)$ and it also has the units dBW. Furthermore, once terms are expressed in dB ratio form, then equations such as Equation (1.47) can be evaluated by addition and subtraction. That is, the downlink *receiver input power* (in dBW) is given by

$$P_r = EIRP + G_r - L_p \quad (dBW) \tag{1.49}$$

where all terms in the equation are in decibel form.

In general, we choose not to use different forms for such terms to differentiate between their original form and their decibel form. Instead, it is assumed that this will be indicated by showing the units in parentheses as shown in Equation (1.49). Equation (1.49) represents the idealized case where there are no additional losses in the link. In practice, it is necessary to consider other losses, including

L_a = loss in the atmosphere due to attenuation by rain or dust
L_{ta} = losses associated with the transmitting antenna, such as feeder losses and loss of gain due to pointing error
L_{ra} = losses associated with the receiving antenna.

Including all these, we obtain the *link power budget equation*

$$P_r = EIRP + G_r - L_p - L_a - L_{ta} - L_{ra} \quad (dBW) \tag{1.50}$$

The received power P_r obtained from Equation (1.50) is commonly referred to as the *carrier power*, C. For most digital modulation types (including M-PSK), the received power is equal to the unmodulated carrier power.

Equation (1.50) is usually calculated by setting the various terms out in tabular form. This tabular identification of transmit power, losses and received power is referred to as a *link power budget*. To see how this is done, we will consider an example of a link power budget for a typical downlink used for broadcasting television to a certain zone on the earth. However before we can undertake this, it is first necessary to explain what is meant by a satellite's earth coverage zone.

A satellite transmitting antenna is designed to provide coverage of a certain *zone*, or area on the earth's surface. One way this antenna coverage area is sometimes specified is in terms of the *3dB beamwidth* of the antenna. This is the angle between the directions in which the radiated (or received) power falls to half the power in the direction of boresight. This angle is designated θ_{3dB} as shown in Figure 1.24.

The earth coverage zone for a satellite antenna may then be defined as that area within the 3dB beamwidth of the antenna radiation pattern as it points to the earth.

The 3dB beamwidth of an antenna is related to the antenna aperture width in the plane in which the radiation pattern is measured. A useful rule of thumb is that the 3dB beamwidth (sometimes just referred to as the *antenna beamwidth*), in a given plane for an antenna with width D in that plane is

$$\theta_{3dB} \cong \frac{72\lambda}{D} \quad \text{degrees} \tag{1.51}$$

It is also useful to note that from Equations (1.46) and (1.51) that the gain and beamwidth of an antenna are related and for a circular aperture antenna with $\eta_a = 0.55$ the relationship becomes

$$G \cong \frac{30,000}{(\theta_{3dB})^2} \tag{1.52}$$

Exercise 1.5

Consider a satellite antenna with a circular aperture with diameter D and aperture efficiency $\eta_a = 0.55$. The antenna is required to provide global coverage from geostationary orbit. If the earth subtends an angle of $17°$ when viewed from that orbit, what are the required diameter and gain of the antenna if it is required to operate at 4GHz?

Solution

At 4 GHz, the wavelength is $\lambda = 0.075$ m. If the antenna is designed to have a 3dB beamwidth of $17°$, then from Equation (1.51) the required antenna diameter is

$$D = \frac{75 \times 0.075}{17} = 0.33 \ m \tag{1.53}$$

The antenna gain is obtained from Equations (1.44) and (1.46) as

$$G = \frac{4\pi \times 0.55 \times \frac{\pi}{4} \times 0.33^2}{(0.075)^2} = 105 \equiv 20.2 \ dB \tag{1.54}$$

Note that although the gain of the antenna in Exercise 1.5 is 20.2 dB at the centre of the beam, it will be 3dB less at the edge of the earth coverage zone. Hence in designing the satellite communication system, it is necessary to use the edge-of-beam gain figure of 17.2 dB. This is to ensure acceptable performance to those earth stations which are close to the earth's horizon, as viewed from the satellite.

Sometimes a satellite's earth coverage zone may be described in terms of *EIRP Contours.* That is, rather than simply specifying the coverage zone to be within the 3 dB beamwidth of the satellite antenna, it can be described by means of a set of contours of EIRP values subtended by the satellite on the earth.

As an illustration, Figure 1.25 illustrates a typical example of EIRP contours subtended on the continent of Australia by a 30W transponder and spot beam antenna located on one of Australia's domestic satellites. The Figure shows that over much of the continent, the EIRP is greater than or equal to 42 dBW. However for earth stations in the vicinity of Brisbane or Hobart, the EIRP values are in the range 39-40 dBW.

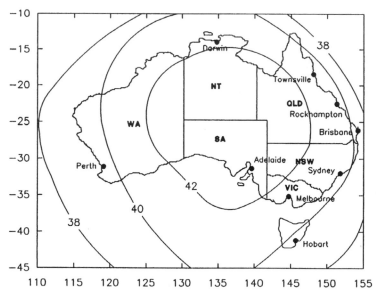

Figure 1.25 Typical EIRP contours (Australia demestic satellite).

Now let us return to an illustrative example showing how a link power budget is calculated.

Exercise 1.6: Broadcast Television

A direct broadcast satellite (DBS) television service is to be provided by a satellite with transponders capable of transmitting 200W per TV channel in a band near 12 GHz. The transmitting antenna on the satellite has a gain of

27 dB (peak). At the earth station receivers, the minimum acceptable power level for good reception of any channel is −115 dBW. The atmospheric loss is $L_a = 0.5$ dB for clear air and $L_a = 3$ dB for periods of heavy rain (0.5% of year). The receiver feeder losses are 0.5 dB. A loss of 1 dB is to be allowed for pointing error in the receiving antenna.

DBS Satellite Transmitter		
Transmit power per channel (P_t) *200W*		*23.0 dBW*
Transmit antenna gain (G_t)		*27.0 dB*
EIRP per channel		*50.0 dBW*
Transmit frequency (f_c)	*12GHz*	
Transmit wavelength (λ)	*0.025 m*	
Path length (typical) (R)	*38,000 $\times 10^3$ m*	
Losses		
Path loss $= 10 \log \left[\frac{4\pi R}{\lambda} \right]^2$		*205.6 dB*
Atmospheric loss (clean air)		*0.5 dB*
Losses at receive antenna		*1.5 dB*
Loss to receiver on boresight		*207.6 dB*
Earth Station Receiver		
Actual antenna area $A_r = 3.14m^2$		
Antenna gain $G_r = 4\pi A_r / \lambda^2$ *45.8 dB*		
Max. actual received power (P_r)		*-111.8 dBW*
Min. acceptable received power		*-115.0 dBW*
Clear air margin (max.)		*3.2 dB*
Margin (3dB contour)		*0.2 dB*

Table 1.1: Downlink power budget for direct broadcast TV system.

- *Determine a power link budget for the system if it is assumed that the receiving antenna is a parabolic reflector type with a diameter of 2 m.*

- *What is the power margin for people living in line with the satellite beam and for those on a 3dB contour of the satellite's transmit antenna pattern, respectively?*

- *If the satellite is directly overhead the coverage zone, find the diameter of the coverage zone.*

Solution

The downlink power budget is given in Table 1.1. Note that for periods of heavy rain, an acceptable signal will be received at the centre of the coverage zone. However at the edges (on the 3dB contour), the received signal will be 2.8 dB below the acceptable minimum. (This occurs for 0.5% of the year). The transmit antenna beamwidth, from Equation (1.51), is approximately $\theta_{3dB} = 8$ degrees so that the diameter of the coverage zone is

$$D_r = 2 \times 38,000 \times \tan 4° = 5300 \ km.$$

Note that in the above example, the coverage would extend over the whole of the specified coverage zone. On the other hand, if separate spot beams of approximately 4 degree beamwidth each were used to provide coverage at the east and west coast regions of the zone, respectively, then the EIRP per channel would be increased by 6 dB. For the same power margins as before, the diameter of the receiving antennas could be reduced by half to 1 m.

1.3.6   System Noise Temperature

A primary objective in the design of receiver systems is to keep the system noise as low as possible. Consider a typical antenna and receiver system as illustrated in Figure 1.26.

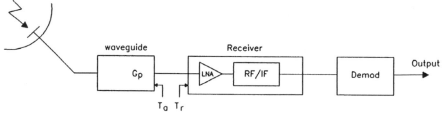

Figure 1.26   Receiving antenna and receiver system.

The receiving antenna is connected to the receiver by a waveguide. Let the waveguide loss (in dB) be equal to $-10 \log G_l$ (dB) where G_l is the ratio of waveguide output power to input power, and hence $G_l \leq 1$. The total system *noise temperature*, T_S, of the antenna and receiver referred to the receiver input, is given by

$$T_S = T_a + T_r \quad (K) \tag{1.55}$$

where T_a is the antenna noise temperature including waveguide losses and T_r the receiver noise temperature.

The *antenna noise temperature* is a measure of the sky noise at the antenna input and of any antenna or feeder losses. Sky noise is noise at the antenna input which is radiated from bodies at temperatures above 0 *K*. This may include atmospheric particles, rain, snow, the sun and other planets. Other sources of sky noise include storms and galactic noise. Sky noise is not a constant but it depends on the antenna elevation angle and its radiation pattern. It also varies with frequency. It is usually calculated from graphs under clear-sky conditions as illustrated in Figure 1.27.

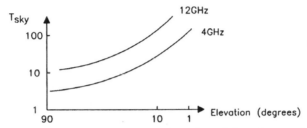

Figure 1.27   Sky noise temperature as a function of elevation angle (clear sky).

Figure 1.27 shows how the sky noise temperature T_{sky} varies with antenna elevation angle for two frequencies, 4 GHz and 12 GHz. Sky noise increases at low elevation angles because of the radiation contribution from the relatively "hot" earth.

Any losses between the antenna aperture and the receiver input affect the antenna system noise temperature (T_a). If a high gain antenna is connected to the receiver via a waveguide with loss $1/G_l$, then the antenna noise temperature referred to the receiver input is given to a good approximation by

$$T_a = T_{\text{sky}} + (1 - G_l)T_l \qquad (1.56)$$

where T_l is the physical temperature of the waveguide (normally 290K). See Reference [1] for more precise methods of accounting for reflection spillover and other factors affecting calculations for large antennas.

Exercise 1.7: Calculation of Antenna Noise Temperature

Consider an antenna operating at a frequency of 4 GHz and pointing at 5° elevation angle. The associated sky noise temperature is 32K. A waveguide with a loss of 0.8 dB is connected between the antenna and the receiver. Find the antenna temperature T_a.

Solution

Since the waveguide loss is $-10 \log_{10} G_l = 0.8$ *dB, then* $G_l = 10^{-0.08} = 0.83$. *Then from Equation (1.56) we obtain*

$$T_a = 32 + 48.3 = 80.8 \quad K \tag{1.57}$$

Note that the effect of the lossy waveguide is very significant. If the noise temperature is low, then for each 0.1 dB addition of waveguide loss, the antenna noise temperature will increase by approximately 7K. One way to keep the waveguide loss to a minimum is to physically locate the receiver LNA as close as possible to the antenna output.

The above calculations were based on clear-sky conditions. *Rain* also increases the antenna noise temperature. See Reference [1] for details.

The *receiver noise temperature*, T_r, is the noise temperature of an equivalent noise source, located at the input of a noiseless receiver, which results in the same noise power as the original receiver, measured at its output. Let the equivalent noise source be matched to the receiver input, replacing the antenna and feeder. Then the available noise power from the equivalent noise source is

$$P_n = kT_r B \tag{1.58}$$

where k is Boltzmann's constant, and B is the equivalent noise bandwidth of the receiver [1]. Note from equation (1.35) that

$$P_n = N_o B \tag{1.59}$$

where N_o is the one-sided input noise spectral density. Note that the total noise power at the receiver input is

$$N = kT_S B = k(T_a + T_r)B \tag{1.60}$$

The receiver noise temperature T_r is a function of all sources of noise in the receiver. In general, the noise temperature of the first stage of the receiver (the RF stage), is the most important [1]. One way of obtaining an RF amplifier with a very low noise temperature is to immerse the amplifier in a cryogenic bath of liquid helium. If this is done, then the receiver noise temperature may be as low as 20-50K. It is also vital to keep feeder losses ahead of the RF amplifier to a minimum as these affect the antenna noise temperature.

1.3.7 Link Signal to Noise Evaluation

As discussed in Section 1.3.4 a digital satellite link must be designed to ensure that the E_b/N_o ratio at the output of the receiver IF stages is sufficiently large to ensure the BER performance criteria are met.

From equations(1.47) and (1.60), the link equation written in terms of the carrier to noise power ratio (C/N) at the receiver is

$$C/N = \frac{P_t G_t G_r}{kT_S B} \left[\frac{\lambda}{4\pi R} \right]^2 \tag{1.61}$$

where losses other than path losses are neglected. We can express the *carrier to noise power ratio* in decibels as

$$C/N = EIRP - L_p + 10\log_{10}(G_r/T_S) - 10\log_{10}(kB) \tag{1.62}$$

where $EIRP$ and L_p are also in decibels. The third term on the right hand side is a *figure of merit* for a receiver system. It is the G_r/T_S ratio, usually shortened to G/T *ratio*. It has the units of dB/K or dBK^{-1}.

Earth stations must be designed to have certain minimum G/T ratios. For example, in the Intelsat network, an earth station is referred to as a Standard A earth station if it has a G/T ratio of at least 40.7 dB/K at 4.0GHz and for a 5° elevation angle. For small earth stations and for some satellite receivers, if G_r (dB) is less than T_S (dBK), then the G/T ratio (in dB) may be negative.

Other ratios that are sometimes useful include the *carrier to noise density ratio*, given in decibels by

$$C/N_o = C/N + 10\log B \tag{1.63}$$

Also the *carrier to noise temperature ratio* in decibels is

$$C/T_s = EIRP - L_p + G/T \tag{1.64}$$

Finally, from Equation (1.39) , the *energy per bit to noise density ratio* for an M-ary digital modulation scheme, in decibels, is

$$E_b/N_o = C/N - \log_{10}(\log_2 M) \tag{1.65}$$

Cascaded Uplink and Downlink

Each single-hop earth station to earth station link via a satellite consists of an uplink and a downlink in cascade. We can calculate the receiver C/N ratio for each link in turn using Equation (1.61). The question is, how do we compute the combined effect of the two links in cascade?

To answer this question, consider the noise model for the cascaded uplink and downlink shown in Figure 1.28.

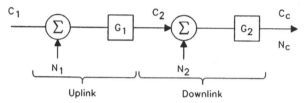

Figure 1.28 Noise model for cascaded uplink and downlink.

In Figure 1.28, G_1 is a ratio which represents the product of the transponder gain and the downlink loss (as a ratio). G_2 represents the receiver gain at the receiving earth station.

The satellite transponder input carrier to noise ratio is C_1/N_1. Since N_1, includes any noise power generated in the transponder, then the transponder output carrier to noise power ratio is also C_1/N_1. The input power to the receiving earth station is C_2 and the equivalent additional noise due to that receiver (and any additional downlink noise effects) is N_2.

Then the carrier to noise power ratio at the IF (demodulator input) output of the receiver at the output of the cascaded system is C_c/N_c where

$$N_c = N_1 G_1 G_2 + N_2 G_2 \qquad\qquad W \qquad\qquad (1.66)$$

and

$$C_c = C_1 G_1 G_2 \qquad\qquad W \qquad\qquad (1.67)$$

Hence we obtain the power ratio

$$\frac{N_c}{C_c} = \frac{N_1}{C_1} + \frac{N_2}{C_1 G_1} = \frac{N_1}{C_1} + \frac{N_2}{C_2} \qquad\qquad (1.68)$$

That is, the overall carrier to noise power ratio is

$$C_c/N_c = \frac{1}{(C_1/N_1)^{-1} + (C_2/N_2)^{-1}} \qquad\qquad (1.69)$$

Exercise 1.8: Cascaded Link C/N Ratio

An uplink is designed so that the transponder C/N ratio in decibels in 23 dB. This is cascaded with a downlink for which the receiver C/N ratio is 18 dB. Find the overall C/N ratio at the output of the cascade link.

Solution

In power ratio form, we obtain

$C_1/N_1 = 10^{2.3} = 199.6$ *and* $C_2/N_2 = 10^{1.8} = 63.1.$

Then from Equation (1.69)

$$C_c/N_c = \frac{1}{(199.6)^{-1} + (63.1)^{-1}} = 47.9 \ or \ 16.8dB$$

1.4 PROBLEMS

1. A BPSK satellite link utilizes a carrier frequency of 4 GHz. The input binary sequence consists of an alternating sequence of 0's and 1's at a bit rate of 8 Mbit/s. No Nyquist spectral filtering is used.

   Find algebraic expressions for and sketch the following:

   (a) the transmitted waveform

   (b) the transmitted signal spectrum

2. Consider the PSK signal referred to in Problem 1 but now consider the input to be a random binary sequence. If Nyquist filtering is used with roll-off factor $\alpha = 0.5$, sketch the transmitted signal spectrum.

3. The input to a satellite link with carrier frequency of 6 GHz is a 2 Mbit/s random binary input sequence. Sketch the transmitted signal spectrum for the following types of modulation. (For each case, plot spectra for no Nyquist filtering and for $\alpha = 0.5$ roll off, respectively.)

   (a) 4-PSK

   (b) 8-PSK

   (c) 16-PSK

4. Compute the spectral efficiency for each of the modulation types in Questions 2 and 3. Assume Nyquist filtering with $\alpha = 0$.

5. An input binary bit stream at a rate of 90 Mbit/s is to be transmitted in a channel bandwidth of 30 MHz using Nyquist filtering with $\alpha = 0.3$. Determine which M-PSK modulation schemes could be used.

6. A satellite carrying an 12.0 GHz continuous wave (CW) beacon transmitter is located in geostationary orbit 30,000 km from an earth station. The beacon's output power is 200 mW, and is the input to an antenna with an 19 dB gain toward the earth station. The earth station receiving antenna is 4 m in diameter and has an aperture efficiency of 60 percent.

   (a) Calculate the satellite EIRP in W, dBW, and dBm.

   (b) Calculate the receiving antenna gain in dB.

   (c) Calculate the path loss in dB.

   (d) Calculate the recieved signal power in W, mW, and dBm.

7. If the overall system noise temperature of the earth station in Problem 6 is 1050 K, determine

   (a) The earth station G/T in dBK^{-1}.

   (b) The receiver noise power in a 100-Hz noise bandwidth in W and in dBm.

   (c) The receiver carrier-to-noise ratio in dB in a 100-Hz noise bandwidth.

8. Consider a satellite communications link with uplink E_b/N_0 of 11 dB and downlink E_b/N_0 of 12 dB. The link employs uncoded QPSK. Ideal coherent demodulation is assumed.

   (a) Determine the overall E_b/N_0 ratio and BER if the satellite is a transparent repeater type.

   (b) Find the BER at the receiving earth station if the satellite is a regenerative repeater type.

9. Consider a satellite communications system with the following specifications.

Satellite	Transponder bandwidth	=	36 MHz
	Transponder gain	=	90 dB (max)
	Input noise temp.	=	550 K
	Output power	=	6.3 W (max)
	4 GHz antenna gain	=	20 dB
	6 GHz antenna gain	=	22 dB
Earth Station	4 GHz antenna gain	=	60 dB
	6 GHz antenna gain	=	61 dB
	Receiver system		
	noise temp.	=	100 K
Path loss	At 4 GHz	=	196 dB
	At 6 GHz	=	200 dB

(a) If the system is operated in TDMA mode with the transponder saturated by each earth station in turn, find

- the required earth station power
- the downlink C/N ratio.

(b) Repeat (i) for the case where four FDMA carriers share the transponder power equally, given that 5 dB input and output backoff is used.

10. For a satellite with an uplink E_b/N_0 of 14 dB and a downlink E_b/N_0 of 18 dB, determine the overall E_b/N_0 ratio.

11. Consider the Intelsat SPADE system in which voice channels are provided on a single channel per carrier (SCPC) basis, each channel using QPSK at a symbol rate of 32 ksymbol/s in a bandwidth of B = 38 kHz. This achieves a bit error rate of 10^{-4} (without coding) at an E_b/N_0 of 9.4 dB.

A single carrier for this SPADE channel is radiated by an INTELSAT V transponder with an EIRP (for one carrier) of 0 dBW to an earth station 40,000 km away. The channel centre frequency is 4095 MHz.

(a) If the transponder's EIRP (total) is 29 dBW, estimate the total number of simultaneous channels in the system.

(b) What value of C/N must be present at the input to the earth station receiver? (Ignore any uplink or intermodulation effects).

(c) What is the minimum G/T that the earth station must have to achieve the specified bit error rate with a 3 dB margin?

REFERENCES

1. Pratt, T., and Bostian, C.W., *Satellite Communications*, Wily, 1986.

2. Li, T., *Topics in Lightwave Transmission Systems*, Academic Press, 1991.

3. Pelton, J.N. and Wu, W. W., "The Challenge of 21st Centory Satellite Communications: INTELSAT enters the Second Millenium", *IEEE Journal on Sel. Areas in Comm.*, Vol. SAC-5, No.4, pp571-591, May 1987.

4. Leopold, R.J., "The Iridium Communications System", *Proc. Mobile & Personal Comm. Systems Conf.*, Adelaide, Australia, Nov. 1992.

5. Jansky, D.M., and Jeruchim, M.C., *Communication Satellites in the Geostationary Orbit*, Artech House, 1987.

6. Korn, I., *Digital Communications*, Van Nostrand, 1985.

Chapter 2

PROPAGATION

by Branka Vucetic
School of Electrical Engineering
University of Sydney, Australia

In this Chapter we will analyze propagation phenomena in both fixed and mobile satellite communications. It is shown how statistical communication theory can be applied to predict signal variations in the presence of the observed transmission factors. The Chapter begins by consideration of propagation effects experienced for fixed services. The effects of rain attenuation on propagation at high frequencies is explored and a statistical model for signal attenuation is discussed.

Then propagation for mobile satellite services is considered. In mobile satellite services the most serious propagation problem is signal fading caused by shadowing and multipath propagation. For land mobile satellite services, shadowing from trees and other roadside obstructions will often be dominant. For maritime and aeronautical services, multipath effects may be more significant. Statistical methods for modelling the effects of the observed propagation properties on signal attenuation are discussed and quantified on the basis of experimental measurements. It is shown how the Rayleigh, Rician and log-normal distributions are useful for predicting different aspects of short-term signal variations. Markov processes are useful for long term signal modelling. The Chapter is concluded by an assessment of the validity of the channel models in comparison with data derived from experimental measurements on typical land mobile satellite propagation paths [1].

2.1   FIXED SATELLITE SERVICES

2.1.1   Free Space Loss

In fixed satellite services, an earth station and the associated satellite are reciprocally visible and propagation of radio waves takes place via an unobstructed line-of-sight path. Free space attenuation is the major factor in link design for that case.

The *free-space loss* on a line-of-sight path is due to spherical dispersion of the radio wave. As explained in Chapter 1, this loss depends on the distance R between the transmitting and receiving antennas and the signal wavelength λ. The free space loss is given by Equation (1.48), repeated here for convenience

$$L_p = \left(\frac{4\pi R}{\lambda}\right)^2 \tag{2.1}$$

If we express R in kilometers and f in gigahertz, the free space loss in dB can be written as

$$L_p\,(dB) = 92.4 + 20log_{10}f_c\,(GHz) + 20log_{10}R\,(km). \tag{2.2}$$

The loss in dB increases as the logarithm of path length. Thus, the loss increases by 6 dB for each doubling of the distance.

Exercise 2.1
A satellite communications link is illustrated in Figure 2.1.

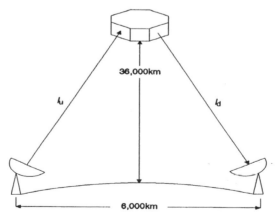

36,000km

l_u l_d

6,000km

Figure 2.1:   A satellite relay system.

It operates at carrier frequencies of 6 GHz for the uplink and 4 GHz for the downlink. The satellite is positioned 36000 km from the earth. Calculate the total free-space loss.

Solution

The satellite is about 36,000 km from either earth station. The downlink free-space loss is given by

$$L_{pd} = 92.4 + 20log_{10}4 + 20log_{10}36000 = 195.6 \ dB \qquad (2.3)$$

The total free-space loss is given by the sum

$$L_{pt} = L_{pu} + L_{pd} = 199.1 + 195.6 = 394.7 \ dB \qquad (2.4)$$

2.1.2 Rain Attenuation

Atmospheric conditions can introduce an additional loss in the direct signal resulting from such effects as atmospheric refraction, precipitation (rain, fog, snow) and cloud cover. Signal losses due to atmospheric effects are dependent upon carrier frequency. The growth of satellite communications has led to the use of higher and higher carrier frequencies, including those in the range 20-50 GHz.

Unfortunately, it turns out that propagation at frequencies above 10 GHz may be severely hampered by attenuation due to rain, fog, snow and clouds. In an early Telstar experiment [2], it was observed that the level of noise increased significantly when it was raining in the vicinity of the receiving earth station. In numerous later experiments involving rain propagation it has been found that rain produces some other degradations, such as signal attenuation, depolarization and interference.

The attenuation caused by rain, snow and fog. increases with the signal frequency and the amount of moisture on the propagation path. Figure 2.2 shows rain attenuation measured at 100 mm/hr rain rate, averaged over 1km, plotted as a function of frequency. The attenuation per km due to atmospheric loss is referred to as the *specific attenuation*. The attenuation on any given path depends on the values of specific attenuation, frequency, polarization, temperature, path length, and latitude.

A number of methods have been proposed for the prediction of rain attenuation [3], [5], [6]. A method recommended by the International Radio

Consultative Committee (CCIR) is used most often as the basis for international planning. In the CCIR method the specific attenuation is obtained as:

$$\gamma = aR_r^b \quad dB/km \tag{2.5}$$

where a and b are coefficients and R_r is the rainfall value occurring for 0.01% of the time in the region of interest. It is recommended that a representative value for R_r be obtained from local rain rate statistics measured by rain gauges with an integration time of 1min. If no such data exist, CCIR Report 563 [6] provides estimates of the value for R_r for 14 various climatic zones.

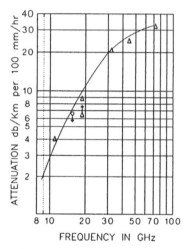

Figure 2.2:   Rain attenuation versus frequency.

The path rain attenuation in dB for 0.01% of the time is given by

$$A = \gamma \cdot L_r \cdot r \tag{2.6}$$

where L_r is the path length through rain and r is a reduction factor dependent on path length and elevation angle to account for the inhomogeneity of rainfall.

The path length in the CCIR method is given by:

$$L_r(km) = \begin{cases} 4 - 0.075(\phi - 36) & \phi > 36° \\ 4 & 0 \le \phi \le 36° \end{cases} \tag{2.7}$$

where $\phi(deg)$ is the latitude of the region of interest.

Figure 2.3 shows the CCIR estimates of the rain attenuation for circularly polarized transmission in a particular climate zone (zone K). In the Figure, the rainfall rate = 15 mm/hr for 0.01% of the time; latitude = 65°; elevation angle = 15°.

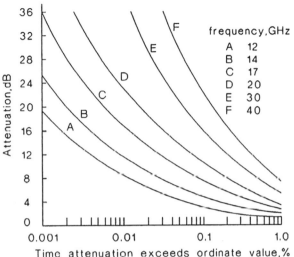

Figure 2.3: CCIR prediction of long term rain attenuation (After [34] with permission).

Exercise 2.2
Table 2.1 lists the required parameters for the CCIR rain attenuation method for satellite transmission at 12.5 GHz from three selected locations in Australia. Compute the rain attenuation for each location.

Location	Region	Latitude	a	b	R_r mm/h	r
Perth	K	31°57'S	0.048	1.299	34.9	0.392
Darwin	N	12°26'S	0.020	1.383	97.4	0.356
Melbourne	F	37°49'S	0.056	1.288	27.2	0.3049

Table 2.1: Rain attenuation parameters.

Solution
a) Perth: *The specific attenuation is computed by the CCIR method as*

$$\gamma = aR_r^b = 4.846 \ dB/km. \tag{2.8}$$

The path attenuation is given by

$$A = r \cdot L_r \cdot \gamma = 7.6 \ dB. \tag{2.9}$$

In comparison the CCIR "estimate" for region K and frequency 12.5 GHz is obtained from (Figure 2.3) as 9 dB. By applying the same computational procedure we obtain the attenuation for the other two regions.

b) **Darwin:** $\gamma = 11.25 \ dB/km$; $L_r = 4km$; $A = 16dB$

c) **Melbourne:** $\gamma = 3.944$; $L_r = 4 - 0.075 \ (37°49' - 36°) = 4.075km$; $A = 4.9 \ dB$

As the above Exercise illustrates, the rain attenuation is much lower for temperate climates, (typified by Melbourne), than for tropical areas, (like Darwin). Note that in tropical areas, rain attenuation is a major factor in designing satellite earth stations operating at Ku-band.

An earth station operating at higher frequencies must be designed to include a *rain attenuation margin* in its link budget design. The rain attenuation margin is defined as the attenuation caused by rain for a particular percentage of time. Note that rain causes two additional effects that must also be considered in order to determine a total effective rain margin. They are the effects of increased interference due to a reduction of cross-polar discrimination and the effect of a rain-induced higher system noise temperature in the earth station.

As explained in Chapter 1, a method of achieving efficient use of the radio spectrum which has been applied in operational satellite communication systems for frequencies above 10 GHz is based on the simultaneous operation of two users on adjacent carrier frequencies, but on orthogonal polarizations. Even though the adjacent frequency spectra may overlap, the resultant interference is maintained small through the use of orthogonal polarization. The major problem in implementation is to provide sufficient isolation between the orthogonally polarized transmissions in order to eliminate the interference. Let us denote as the *co-polar signal* the signal sought at the receiver on a particular polarization. We also denote as the *cross-polar signal* the adjacent unwanted signal of the other polarization type. The cross-polar signal can be regarded as a noise component to the co-polar signal. Then the *cross-polar discrimination* is defined as as the ratio in decibels of the power of the received co-polar signal to that of the cross-polar signal.

Degradation in cross-polar discrimination is caused by rain drops and ice particles. These are generally not spherical, and introduce differential attenuation and differential phase shift between signals on horizontal and vertical polarizations. Rain attenuation for the vertical polarization is smaller than that for horizontal polarization. Hence the cross-polar discrimination for transmitted vertical polarization can be expected to be better than for transmitted horizontal polarization. For more details on these effects, see Reference [7].

Figure 2.4 gives examples of rain margins for two earth stations operating at 12/14 GHz.

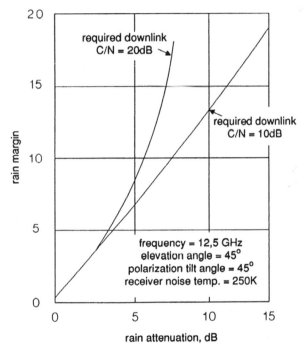

Figure 2.4: Typical down-link rain margins (After [5] with permission).

The small earth station has a figure of merit G/T without rain margin of 20 dB/K and a required downlink C/N of 10 dB. The medium size earth station has a G/T without rain margin of 30 dB/K and operates with a required downlink C/N of 20 dB. From Figure 2.4 one can see that the rain margin to compensate for increased system noise and cross-polarization interference is larger for the medium station. For more information on computing the rain margin refer to [5] and [7].

2.2   PROPAGATION
TO MOBILE RECEIVERS

2.2.1   Propagation Environment

The propagation conditions for a satellite mobile system are usually considerably more hostile than in fixed point-to-point systems. Link losses for mobile services are often influenced by the small antenna size on the mobile vehicles, and by fading due to multipath effects and signal shadowing. It is important to understand the mechanisms that give rise to multipath and shadowing, respectively.

In land mobile satellite systems the vehicle antenna may be typically of the order of 1m above ground level. The signals passing between a land mobile vehicle and a geostationary satellite will often be obstructed by roadside trees and by buildings which considerably attenuate the direct line-of-sight wave. The resultant signal losses are referred to as *shadowing*.

By contrast, for a maritime satellite environment, a shipborn earth station may not suffer shadowing. However it will be surrounded by moving sea waves and ship structures which act as signal scatterers. The resultant effect is referred to as *multipath propagation*. Note that the mobile antenna will usually exhibit a much lower gain than that of a large directional antenna such as would be used for fixed services. As a result, it will have a relatively broad beamwidth and is more likely to pick up multipath reflections of the satellite signal.

2.2.2   Signal Fading

Shadowing and multipath reflections give rise to *signal fading*, that is, variations in the amplitude and phase of the received signal as the vehicle moves. Fades of 20dB or more below the mean signal level may be quite common for land mobile satellite systems. These signal variations have the potential to introduce bursts of transmission errors, resulting in intolerable distortion and noise for speech transmission, or in high error rates for data transmission. In these circumstances, increasing the transmitter power may not necessarily improve the reception quality. Hence, special signal processing techniques are needed to alleviate fading effects.

In order to develop these techniques and assess their performance, a detailed knowledge of the statistics of the propagation phenomena is required. The nature of signal fading is closely related to the propagation environment.

Example 1: Single carrier wave

Consider a situation when a received signal consists of only one carrier wave with frequency f_c. Let us assume that the wave is coming in at an azimuthal angle θ with respect to the vehicle motion. The signal is described by a sinusoid in both the time and spatial domains. The variations in the spatial domain can be expressed as

$$r(d) = R_o \cos\left(2\pi f_c \frac{d}{v} - \frac{2\pi}{\lambda} d \cos\theta\right) \tag{2.10}$$

where R_o is a constant amplitude, d is the mobile receiver displacement from a reference point, v is the vehicle speed and λ is the signal wavelength.

The time domain signal variations can be written as

$$r(t) = R_o \cos\left[2\pi t(f_c - \frac{v f_c}{c} \cos\theta)\right] \tag{2.11}$$

The envelope R_o of the received signal is constant while its frequency f_r is given by

$$f_r = f_c - \frac{v f_c}{c} \cos\theta \tag{2.12}$$

where c is the speed of light ($c = 3 \cdot 10^8 m/s$).

The received frequency differs from the transmitted frequency f_c by

$$f_D = \frac{v f_c \cos\theta}{c} \tag{2.13}$$

This displacement in frequency is called the *Doppler shift*.

Example 2: Two carrier waves

Let us now consider two waves, each with the same amplitude R_o and frequency f_c, arriving at the receiver at different azimuthal angles $\theta_1 = 0^o$ and $\theta_2 = \theta$ as illustrated in Figure 2.5.

The receiver perceives the resulting signal as the sum of the two waves

$$r(t) = R_o \cos\left[2\pi t(f_c - \frac{v f_c}{c})\right] + R_o \cos\left[2\pi t(f_c - \frac{v f_c}{c} \cos\theta)\right] \tag{2.14}$$

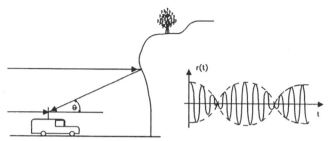

Figure 2.5: Signal fading caused by one scatterer.

The received signal can be expressed in terms of its in-phase $R_I(t)$ and quadrature $R_Q(t)$ components as

$$r(t) = R_I(t)\cos 2\pi f_c t + R_Q(t)\sin 2\pi f_c t \tag{2.15}$$

where

$$R_I(t) = R_o \left[\cos(2\pi \frac{v f_c}{c} t) + \cos(2\pi \frac{v f_c}{c} t \cos\theta) \right] \tag{2.16}$$

and

$$R_Q(t) = R_o \left[\sin(2\pi \frac{v f_c}{c} t) + \sin(2\pi \frac{v f_c}{c} t \cos\theta) \right] \tag{2.17}$$

The received envelope is

$$R(t) = \sqrt{R_I^2(t) + R_Q^2(t)} \tag{2.18}$$

which can be written as

$$R(t) = 2R_o \cos[2\pi \frac{v f_c}{2c} t(1 - \cos\theta)] \tag{2.19}$$

The received envelope pattern described by Equation 2.19 represents a standing wave. The envelope varies at frequency

$$f_d = \frac{f_D}{2}(1 - \cos\theta) \tag{2.20}$$

For this particular case the Doppler shift is equal to the maximum envelope frequency, obtained for $\theta = 180°$. Note that in this example, the effect of noise is ignored. Noise is likely to be much less significant than fading.

Example 3: N reflected waves

Let us now assume that the received signal consists of N reflected waves of the same frequency f_c with random amplitudes R_i and random azimuthal angles θ_i taking values between $0°$ and $360.°$ This is illustrated in Figure 2.6.

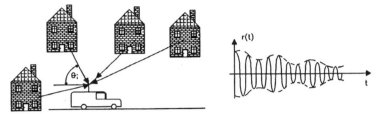

Figure 2.6: Signal fading caused by N reflected waves.

The resulting signal is given by the sum

$$r(t) = \sum_{i=1}^{N} R_i \cos(2\pi f_c t - 2\pi f_{Di} t) \qquad (2.21)$$

where

$$f_{Di} = \frac{v f_c}{c} \cos \theta i \qquad (2.22)$$

is the Doppler shift of wave i. In the in-phase and quadrature representation the received signal can be written as

$$r(t) = R_I(t) \cos 2\pi f_c t + R_Q(t) \sin 2\pi f_c l \qquad (2.23)$$

where

$$R_I(t) = \sum_{i=1}^{N} R_i \cos 2\pi f_{Di} t \qquad (2.24)$$

and

$$R_Q(t) = \sum_{i=1}^{N} R_i \sin 2\pi f_{Di} t \qquad (2.25)$$

For N large, according to the central limit theorem, both $R_I(t)$ and $R_Q(t)$ will be zero mean Gaussian variables. The signal envelope

$$R(t) = \sqrt{R_I^2(t) + R_Q^2(t)} \qquad (2.26)$$

will then be a Rayleigh random variable. Its probability density function will be given by

$$p(R) = \begin{cases} \frac{R}{\sigma} e^{\frac{-R^2}{2\sigma_r^2}} & 0 \leq R \leq \infty \\ 0 & \text{otherwise} \end{cases} \qquad (2.27)$$

where $2\sigma_r^2$ is the mean signal power.

The envelope signal spectrum consists of the set of sinusoids at frequencies f_{Di}. Therefore, the envelope signal bandwidth is identical to the span between maximum Doppler shift limits of the received signal.

Example 4: Direct and multipath signals

As another example, consider a maritime propagation scenario, as illustrated in Figure 2.7.

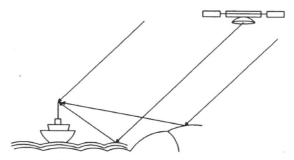

Figure 2.7: Propagation in a maritime mobile system.

The receiver antenna is assumed to pick up a direct signal s(t) and multiple propagation reflections. Many multipath reflections can be described by one signal d(t) with random Rayleigh distributed envelope D(t). The composite received signal is given by the sum

$$r(t) = s(t) + d(t) \tag{2.28}$$

Using an I and Q representation, the received signal is

$$r(t) = v_I(t) \cos 2\pi f_c t + v_Q(t) \sin 2\pi f_c t + D_I(t) \sin 2\pi f_c t + D_Q(t) \sin 2\pi f_c t \tag{2.29}$$

where $v_I(t)$ and $v_Q(t)$ are the in-phase and the quadrature components of the direct signal, and $D_I(t)$ and $D_Q(t)$ are the in-phase and the quadrature components of the reflected signal, respectively.

The received envelope

$$R(t) = \sqrt{[v_I(t) + D_I(t)]^2 + [v_Q(t) + D_Q(t)]^2} \tag{2.30}$$

has a Rice distribution. Its probability density function will be given by

$$p(R) = 2R\sqrt{(1+K)}e^{-K-(1+K)R^2} \; I_o(2R\sqrt{K(K+1)}) \tag{2.31}$$

where K is the ratio of the direct ray to multipath phasor powers, and I_o is a modified Bessel function

$$I_o(t) = \sum_{n=0}^{\infty} \frac{t^{2n}}{2^{2n}n!n!}o$$

This type of channel is referred to as a Rician fading channel.

Example 5: Two-path propagation – frequency selective fading

The mobile receiver is standing still. The received signal consists of a direct and a reflected wave. The reflected wave has a propagation delay that is larger than that of the direct wave due to a longer propagation path. The situation is illustrated in Figure 2.8.

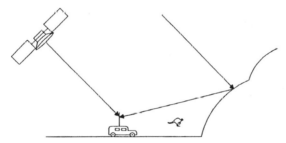

Figure 2.8: Two-path propagation with frequency selective fading.

The received composite signal is given by the sum of the direct signal and a delayed attenuated replica of the direct signal. That is

$$r(t) = s(t) + bs(t - \tau) \qquad (2.32)$$

where τ is the relative delay of the reflected signal and b is the relative amplitude of the reflected signal.

In in-phase and quadrature representation the received signal can be expressed as

$$
\begin{aligned}
r(t) &= v_I(t)\cos 2\pi f_c t + v_Q(t)\sin 2\pi f_c t + v_I(t - \tau)\cos 2\pi f_c(t - \tau) \\
&\quad + v_Q(t - \tau)\sin 2\pi f_c(t - \tau)
\end{aligned} \qquad (2.33)
$$

Consider ideal coherent demodulation of the in-phase component with a carrier

$$v_c(t) = 2\cos 2\pi f_c t \qquad (2.34)$$

Then the demodulated in-phase signal is given by

$$R_I(t) = LPF\{2r(t)\cos 2\pi f_c t\} \tag{2.35}$$

where LPF[·] represents low pass filtering. After some trigonometric manipulations we obtain for the I-channel output

$$R_I(t) = v_I(t) - bv_I(t - \tau)\cos 2\pi f_o \tau + bv_Q(t - \tau)\sin 2\pi f_o \tau \tag{2.36}$$

where the frequency

$$f_o = f_c - f_n \tag{2.37}$$

is defined as the difference between the carrier frequency and the frequency

$$f_n = \frac{2n+1}{2\tau} \qquad n = 1, 2, 3... \tag{2.38}$$

The frequency f_n is called the *notch frequency*.

In Equation(2.36), the first term represents the desired transmitted in-phase signal, and the remaining two terms are intersymbol interference (ISI). The first ISI term is fading induced interference from the in-phase channel and the second term represents co-symbol interference from the quadrature channel.

The transfer function of the channel with the response given by Equation(2.32) is illustrated by Figure 2.9 and can be written in the form

$$H_c(f) = 1 + b(t)e^{-j2\pi f\tau} \tag{2.39}$$

where b(t) is the time varying amplitude of the multipath signal relative to the direct signal.

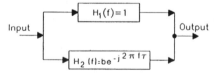

Figure 2.9:   Transfer function model for two-path propagation.

The amplitude frequency response of the channel is given by

$$\mid H_c(f) \mid = 2b(t)\mid \cos \pi f\tau \mid +1 - b(t) \tag{2.40}$$

and is shown in Figure 2.10.

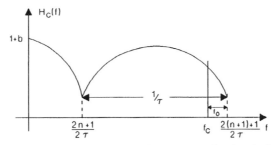

Figure 2.10: Amplitude response of frequency-selective fading model.

As Figure 2.10 shows, the resultant amplitude-frequency response exhibits "notches" spaced $1/\tau$ Hz apart.

The signal is therefore subject to frequency dependent, fading induced interference. This is referred to as *frequency selective* or *dispersive* fading. Note that the type of fading described in previous examples is known as *flat* or *nondispersive* fading. As a rule of thumb, a flat fading assumption is accurate for cases where the signal bandwidth is at least an order of magnitude less than the separation between two notch frequencies $1/\tau$.

2.2.3 Reflection from Flat Terrain

In a satellite radio link, energy can be transmitted from the transmitter to the receiver by the direct wave and also by reflection from surfaces separating media having different constitutive parameters. For example, for land mobile channels, the two media of interest are the air and the surface of the earth.

When the surface is sufficiently smooth and flat it may be regarded as a plane interface between two media. Examples of a smooth flat surface are a snow covered flat terrain or a calm sea.

A simple model of signal reflection is illustrated in Figure 2.11.

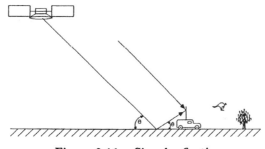

Figure 2.11: Signal reflection.

The received signal consists of two components, the direct wave and the reflected wave. The amplitude of the reflected wave is measured by the *reflection coefficient* R of the surface. The reflection depends on the incident angle, the polarization of the wave and the surface electrical characteristics. The wave is said to be horizontally polarized if the electric field is perpendicular to the plane of incidence and parallel to the reflecting surface and vertically polarized when the magnetic field is parallel to the reflecting surface.

If θ is the angle of incidence of the wave to the surface, measured from the surface, then for horizontal polarization the reflection coefficient (ratio of reflected to direct wave amplitude), is given by

$$R = \frac{\sin\theta - \sqrt{\frac{\varepsilon'}{\varepsilon_o} - \cos^2\theta}}{\sin\theta + \sqrt{\frac{\varepsilon'}{\varepsilon_o} - \cos^2\theta}} \qquad (2.41)$$

and for vertical polarization

$$R = \frac{(\frac{\varepsilon'}{\varepsilon_o})\sin\theta - \sqrt{\frac{\varepsilon'}{\varepsilon_o} - \cos^2\theta}}{(\frac{\varepsilon'}{\varepsilon_o})\sin\theta + \sqrt{\frac{\varepsilon'}{\varepsilon_o} - \cos^2\theta}} \qquad (2.42)$$

where ε_o is the permittivity of free space and ε' is the complex permittivity of the surface.

As

$$\varepsilon' = \varepsilon_o\varepsilon_r + \frac{\sigma}{j2\pi f} \qquad (2.43)$$

the ratio $\frac{\varepsilon'}{\varepsilon_o}$ may be expressed as

$$\frac{\varepsilon'}{\varepsilon_o} = \varepsilon_r - \frac{j\sigma}{2\pi f\varepsilon_o} \qquad (2.44)$$

where ε_r is the relative permittivity of the surface and σ is the surface conductivity.

Note that the reflection coefficient is complex. The values of its magnitude and phase for the sea are shown in Figure 2.12 as a function of incident angle for horizontal and vertical polarization and for two radio frequencies of 130 MHz and 1.6 GHz. For a vertically polarized wave the reflection coefficient is a minimum at a low incident angle called the Brewster angle. At this angle the sign of the reflection coefficient changes, indicating a 180° phase reversal in the reflected wave.

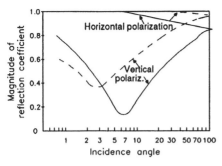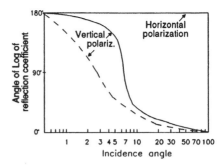

Figure 2.12: Reflection coefficient of smooth sea versus angle of incidence, left: Magnitude, right: Angle.

For horizontal polarization the magnitude of the reflection coefficient decreases regularly with increasing angle θ and the phase of the reflection coefficient is essentially 180°.

As the phase difference due to reflection is not the same for both horizontal and vertical components, an incident wave with inclined linear polarization gives rise to a reflected wave with elliptical polarization.

2.2.4 Reflection from Rough Terrain

The previous expressions for the reflection coefficient have been computed assuming that transmission is taking place over a smooth surface. If the surface between the two propagation media is not smooth, the signal is scattered in many directions, though for a surface with random roughness the maximum energy reflection will occur in the *specular direction*. The specular direction is defined as the direction that a reflected wave would take if the incident wave was reflected from a mirror positioned at the mean surface height.

The surface is regarded as rough if the irregularities of the surface result in propagation path variations of more than an eighth of the carrier wave length. The surface roughness is measured by the *terrain roughness coefficient*

$$\rho_c = e^{-\frac{c^2}{2}} \tag{2.45}$$

where c is the Rayleigh criterion which is equal to the phase difference between two waves caused by surface irregularities, given by

$$c = \frac{4\pi h \sin \theta}{\lambda} \tag{2.46}$$

and h is the mean height of the surface and λ is the wavelength. Figure 2.13 illustrates this. Empirical results indicate that the surface can be considered as smooth when c is well below $\pm \frac{\pi}{2}$. If c is higher then $\pi/4$ the reflection coefficient decreases rapidly and tends toward the value 0.2. At lower frequencies both land and sea will behave as smooth surfaces under most conditions. However, at higher frequencies (above 1GHz) the land is unlikely to appear smooth while the sea will be so only under very calm conditions. For a rough ground the effect of the ground reflected wave becomes small as signal power is scattered in all directions and the power reflected in the specular direction becomes negligible.

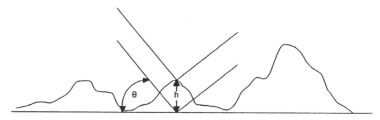

Figure 2.13:   Reflection from rough surface.

Due to the large distances involved in satellite communications, the curvature of the earth's surface will alter the effective reflection coefficient. This effect may be taken into account by a divergence factor D. Tables are given for estimating the divergence factor. See for example Kerr [13]. The magnitude of the ground reflected specular wave is given by

$$G = \rho_c D R S \tag{2.47}$$

where S is the amplitude of the direct wave, and R is the reflection coefficient of the smooth surface.

2.3   LAND MOBILE SATELLITE PROPAGATION

2.3.1   Shadowing and Multipath

In transmission between a moving vehicle and a satellite, propagation often takes place via many paths as illustrated in Figure 2.14. Experimental results indicate that a significant fraction of the total energy typically arrives at the receiver by way of a *direct wave*. The remaining power is received by way of a *specular ground reflected wave* and the many randomly scattered rays which form a *diffuse wave*. The direct wave arrives at the receiver site via a line-of-sight path without reflection. The propagation of the direct wave may be affected by free space attenuation, Faraday rotation, ionospheric scintillation and shadowing. For frequencies up to about 10 GHz, the propagation is not significantly affected by rain. Note that it is convenient to use the direct wave as a delay, Doppler and amplitude reference for other reflected and diffuse waves.

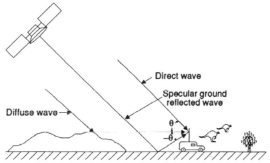

Figure 2.14: Multipath propagation.

Faraday Rotation is rotation of the polarization axis of a non-circularly polarized wave caused by the ionosphere. This can be a cause of additional propagation loss for a direct wave. Extensive experimental measurements show that the loss introduced by Faraday rotation at L-band for linearly polarized antennas for the worst 1% of a year under worst conditions (low elevation angles of 20° and in a period of solar maximum (eg. 1979-1980) [14] is of the order of 3dB. The loss can be eliminated by using circular polarization.

Ionospheric scintillation is produced by irregularities in the horizontal distribution of free electrons in the ionosphere. It is least severe for mid-latitude with respect to the geomagnetic equator. It has been observed [14] that the loss caused by scintillation at L- band is less than 1 dB for 99.9% of the time for most locations. Therefore, scintillation does not have a significant effect on satellite mobile communication performance.

Shadowing is the attenuation of the direct wave caused by roadside trees, buildings, hills and mountains. Experimental tests [15] indicate that shadowing caused by trees is the most dominant factor determining signal degradation for land mobile satellite systems. There are a great many factors affecting shadowing. Its effect strongly depends on the elevation angle from the mobile terminal to the satellite and on the type of surrounding vegetation and buildings.

Fading due to shadowing attenuation increases as the elevation angle decreases. It is more severe at low elevation angles where the obstacle projected shadow is high. It has been observed that if one assumes a 10 dB fade at 30°, then a 5 dB fade reduction exists at 45° and 7 dB reduction exists at 60° under otherwise similar propagation conditions [15].

The extent of fading induced by vegetation is also dependent on the frequency of tree interception, the path length through the trees and the density of branches and foliage. Single trees in full foliage have been found to cause transmission attenuation valves of between 10 and 20 dB at frequencies between 800 and 900 MHz and elevation angles 10° to 40°[16]. The seasonal variation in density of leaves has been found to give rise to fade variation of about 40% [15].

For a given tree density and elevation angle, shadowing attenuation increases with increasing frequency. At 100 MHz the average attenuation from moderately thick trees may be 5 to 10 dB for vertical polarization and 2 to 3 dB for horizontal polarization. At frequencies higher than 300 MHz the tree attenuation tends to become independent of polarization. It has been observed that fades at L-band (1.5 GHz) exceed those at UHF band (870MHz) by 1 to 2 dB [17].

Buildings are mostly rather opaque at L-band frequencies. Typical shadowing attenuation of a brick wall is in the order of 2 to 5dB at 30MHz and 10 to 40dB at 3GHz, depending on whether the wall is dry or wet [18].

Changing the mobile vehicle between lanes on a road might result in considerable variations in the tree attenuation. This is attributed to the fact that the probability of intersecting trees, the path length through trees and the elevation angle vary with position on the road.

Accurate estimates of shadowing are difficult to obtain due to the randomly varying parameters affecting propagation. If the number of these parameters is very large, then as a consequence of the central limit theorem, the amplitude R of the direct wave follows a log-normal distribution with probability density function (with parameters σ, and μ)

$$p(R) = \begin{cases} \frac{1}{\sqrt{2\pi}\sigma_1 R} exp\left(-\frac{(ln R - \mu)^2}{2\sigma_1^2}\right) & 0 \le R \le \infty \\ 0 & \text{otherwise} \end{cases} \qquad (2.48)$$

The amplitude mean and the variance strongly depend on the type of the shadowing environment. The amplitude variations due to shadowing occur at a slower rate than for multipath fading. The phase distribution of the shadowed direct wave is Gaussian.

A land mobile satellite system should be designed to have a sufficient link margin for shadowing. Typical shadowing margins for systems to operate satisfactorily for at least 90% of the time are estimated at 10% cumulative probability distribution of signal attenuation. These margins are shown in Table 2.2 for a value of 35% tree density and in Table 2.3 for 85% tree density. The tree density is defined as the percentage of optical shadowing caused by roadside trees at an elevation angle of 45°. For example, operating at L-band at 30° elevation angle requires a shadowing margin of 10dB to provide satisfactory service in an 85% tree density area [20].

Frequency (MHz)	Elevation angle (degrees)		
	30	45	60
893	5.0 dB	4.2 dB	3.1 dB
1550	5.6 dB	4.9 dB	4.0 dB
2660	8.6 dB	7.7 dB	5.2 dB

Table 2.2: Shadowing margin for 35% tree density.

Frequency (MHz)	Elevation angle (degrees)		
	30	45	60
893	9.3 dB	7.0 dB	5.5 dB
1550	10.8 dB	8.6 dB	6.9 dB
2660	15.4 dB	10.6 dB	9.2 dB

Table 2.3: Shadowing margin for 85% tree density.

The signal power received at a mobile terminal via *the specular wave* is produced by signal reflection from the ground in the direction of the satellite. This wave arrives at -θ, where θ is the elevation angle of the satellite, as seen from the vehicle. The magnitude of the ground reflected wave is proportional to the terrain roughness factor. The effect of the specular wave on the direct wave at the receiver site is usually significantly reduced by the vehicle antenna directivity and its impact can be neglected.

Power from *the diffuse wave* results from various reflections from the surrounding terrain. This component varies randomly in amplitude and phase. This form of multipath propagation does not cause significant losses for land mobiles. Even on roads with snow, for elevation angles above 24°, the ratio of the direct to multipath component is higher than 10 dB [21].

2.3.2   Modelling of Satellite Mobile Channels

Recently a number of stochastic models have been proposed to characterize and simulate propagation in satellite mobile communication systems. Some of them are coding channel models. That is, they have been developed to represent digital error sequences encountered in real digital communication links [22]. Digital error sequences in these models are characterized by M-state Markov chains. The error sequence models based on Markov chains describe channels with binary inputs and outputs. Such models are useful in voice codec and hard decision error control codec testing. We will examine two such models later in this Section.

However, it is often preferable to represent the channel by an analogue model of the radio frequency fading. Such a requirement arises, for example, in systems with forward error correction, where to obtain the optimum use of the received signal, soft decision decoding algorithms are used. In the simulation of such communication systems, an analog model, described by amplitude and phase channel distortion, is required.

As indicated in the previous Section, a wide range of experiments have been performed on satellite mobile communication systems to obtain fading channel characteristics including signal attenuation statistics, phase variations, fading rate and Doppler spread [26], [23],[21],[24]. The results obtained from the propagation experiments indicate that the statistical character of signal variations strongly depends on the type of environment in which the vehicle is located. Three representative types of environment can be distinguished according to the degree of the line-of-sight obstruction. These are

- urban areas with almost complete obstruction of the direct wave,

- open areas with no obstruction of the direct wave, and

- suburban and rural areas with partial obstruction of the direct wave.

Let us now consider the development of fading channel models for each of these environments in turn.

2.3.3   Channel Models for Urban Areas

In urban areas the direct line between the mobile unit and the satellite may be almost completely obstructed by tall buildings and multistorey residences. Therefore, electromagnetic energy propagation in urban areas may

occur largely by way of scattering. In such a case, the vehicle may pick up
reflected signals from all directions in the horizontal plane. The received
signal consists of many independent components of random phase. These
signals collectively add to give a nett signal at the receiver that varies randomly in amplitude and phase. Then as discussed in Example (3) above, the
amplitude of the received signal envelope undergoes fading with a Rayleigh
statistical distribution given by Equation 2.27. where $2\sigma_r^2$ is the mean signal power. This depends on the properties of the surrounding terrain. The
phase of the received signal is uniformly distributed from zero to 2π.

If an unmodulated frequency f_c is transmitted, the ith component has
Doppler shift of $\frac{vf_c}{c}\cos\theta_i$. Since $-1 < \cos\theta_i < 1$, all the frequencies of the
N components will be shifted into the range $\pm vf_c/c = \pm f_d$. Therefore, the
received signal experiences a form of frequency spreading and occupies a
band between $f_c \pm f_d$. The frequency f_d is called the *fade rate*.

Assuming N to be very large, and that uniform power is received for all
θ, then the power spectrum of the received unmodulated signal is:

$$S(f) = \begin{cases} \frac{2\sigma_d^2}{2\pi f_d}\left[1 - \left(\frac{f-f_c}{f_d}\right)^2\right]^{-1/2} & for \ f \leq \mid f_c \pm f_d \mid \\ 0 & for \ f > \mid f_c \pm f_d \mid \end{cases} \qquad (2.49)$$

For example, typical received signal variations recorded from the Japanese ETS-V satellite at a land mobile terminal in an urban area in Sydney
[27] are illustrated in Figure 2.15.

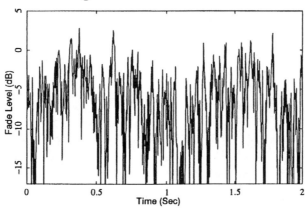

Figure 2.15: Measured signal envelope for an urban area.

The elevation angle was 51°, the signal frequency was 1.54515 GHz and
the measurements were taken at a mobile speed of 60 km/h.

A simulated channel waveform for the same channel can be generated using the Rayleigh model of Equation (2.27). A fading pattern simulated by such a model is shown in Figure 2.16. The experimental and simulated probability density functions for the same channel are depicted in Figure 2.17.

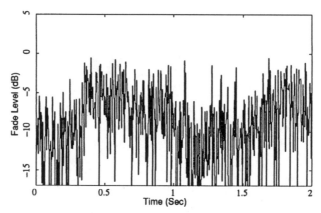

Figure 2.16:   Simulated signal envelope for an urban area.

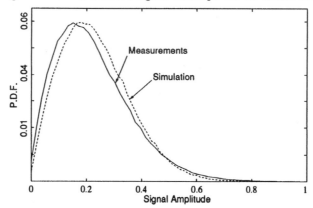

Figure 2.17:   Signal pdf's for an urban area.

2.3.4   Channel Models for Open Areas

In open areas such as in farm land or open fields, there are assumed to be essentially no obstacles on the line-of-sight path. The distortion-free direct wave s(t) is received together with a diffuse wave d(t). The resulting receiver signal is given by the sum

$$r(t) = s(t) + d(t) + n(t) \tag{2.50}$$

where $n(t)$ represents additive white Gaussian noise (AWGN). The diffuse component consists of many reflections from the nearby surface, each of them being independent and randomly phased. Hence, as described in Example (4) above, the probability density function of the sum of a constant envelope direct wave and a Rayleigh distributed diffuse wave results in a signal with Rician envelope. Its pdf is given by Equation 2.31.

Experimental propagation measurements show that the Rician factor K, being dependent on elevation angle, varies typically between 10 and 20 dB, while the diffuse component bandwidth is less than 200 Hz for vehicle speeds of less than 90 km/h [24], [26]. A typical field recording of the received envelope in an open area is shown in Figure 2.18.

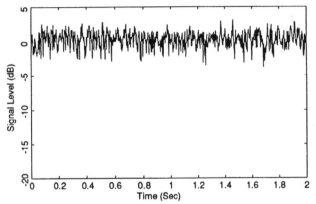

Figure 2.18:   Measured signal envelope for an open area.

For comparison, Figure 2.19 shows a best fit simulated Rician envelope for K= 14dB.

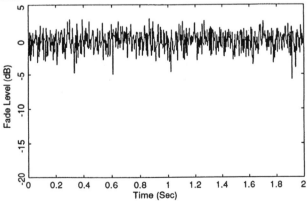

Figure 2.19:   Simulated signal envelope for an open area.

Figure 2.20 shows the measured probability density function along with that of the best fit simulated Rician distribution for the same channel [1].

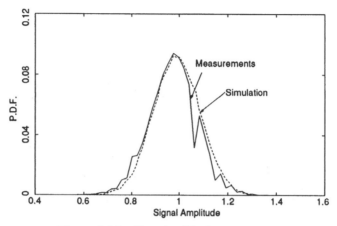

Figure 2.20: Signal pdf's for an open area.

2.3.5   Channel Models for Suburban and Rural Areas

In rural and suburban areas, roads may be surrounded by trees, houses or small buildings. These obstacles cause signal shadowing, manifested as an attenuation of the direct wave. As discussed previously, the attenuation depends on the signal path length through the obstacle, the type of obstacle, the elevation angle, the direction of travel and the carrier frequency. As the vehicle moves along the road the attenuation of the direct signal varies. The attenuation variations of the direct wave can be modelled by a log-normal distribution [23].

In areas where statistically, the environmental properties remain stationary, the channel can be adequately modelled by assuming that the received signal is a linear combination of a direct and a diffuse scatter component received in the presence of white additive Gaussian noise $n(t)$. The resulting signal can be represented by the sum

$$r(t) = s(t) + d(t) + n(t) \qquad (2.51)$$

The fading rate of the direct component, however, is significantly less than that of the diffuse component [23], [24], [21]. Since the variations of the direct signal envelope are slow, we can assume that its value remains constant for vehicle movement over small areas (typically several tens of carrier wavelengths). The conditional probability density function of the received signal envelope for a given envelope value of the direct wave S, is Rician, with

$$p(R \mid S) = \frac{R}{\sigma_d^2} e^{-\frac{R^2+S^2}{2\sigma_d^2}} I_0\left(\frac{RS}{\sigma_d^2}\right) \tag{2.52}$$

where $2\sigma_d^2$ denotes the average scattered power due to multipath propagation.

The statistics of the direct signal envelope are described by the log-normal probability density function

$$p(S) = \frac{1}{\sqrt{2\pi}\sigma_1 S} e^{\frac{-(\ln S - \mu)^2}{2\sigma_1^2}} \quad 0 \le S \le \infty \tag{2.53}$$

with mean

$$m_s = e^{\mu + \frac{\sigma_1^2}{2}} \tag{2.54}$$

and variance

$$\sigma_s^2 = e^{(2\mu + \sigma_1^2)}(e^{\sigma_1^2} - 1) \tag{2.55}$$

where μ *and* σ_1 are parameters.

The total probability is given by

$$p(R) = \int_0^\infty p(R \mid S)p(S)dS \tag{2.56}$$

The probability density function of the received direct signal phase can be approximated by a Gaussian distribution [27]

$$p(\phi) = \frac{1}{\sqrt{2\pi}\sigma_\phi} e^{-\frac{(\phi - m_\phi)^2}{2\sigma_\phi^2}} \tag{2.57}$$

where m_ϕ *and* σ_ϕ are the mean and the standard deviation of the received signal phase, respectively.

The phase distribution of the diffuse wave is uniform between 0 and 2π. From Equations (2.52)-(2.57), the generation of a stochastic model for this type of propagation requires a knowledge of the following parameters

1. Average attenuation of the direct wave m_s.

2. The variance of the direct wave σ_s^2

3. The power of the diffuse component $2\sigma_d^2$.

4. The mean value of the received signal phase m_ϕ.

5. The variance of the received signal phase σ_ϕ^2.

The fact that the fade rate of the direct wave f_s (usually several Hz) is significantly lower than the fade rate of the diffuse wave f_d (100-200 Hz) can be used to compute the statistical parameters of the model [27].

The computation method is based on filtering the baseband received signal by a low-pass filter with cut-off frequency f_s to obtain the direct wave fading. This signal is subtracted from the total received baseband signal to obtain the diffuse multipath fading. The direct wave fading signal is then processed to compute the envelope mean m_s, the envelope standard deviation σ_s, the phase mean m_ϕ and the phase standard deviation σ_ϕ.

The diffuse fading signal is used to compute the power of the diffuse multipath fading $2\sigma_d^2$. Channel parameters for a typical suburban areas are given in Table 2.4. These values refer to a normalized transmitted signal with the average power of 1. The channel parameters are assumed to be constant over large areas with constant environment attributes, such as the type of trees and buildings, elevation angles and vehicle speed.

Type of shadowing	Length of run	m_s	σ_s	σ_d	m_ϕ (deg)	σ_ϕ (deg)
Light	222 sec	0.787	0.259	0.151	0.086	9.680
Heavy	106 sec	0.670	0.30	0.154	0.0115	30.325

Table 2.4: Channel parameters for rural or suburban areas.

The direct signal attenuation in rural timbered areas can be computed by using an empirical formula [21]

$$\alpha_s = 0.45 f_c^{0.284} d \quad dB \quad 0 \le d \le 14 \tag{2.58}$$

where d is the signal path through the tree in metres and f_c is the carrier frequency in GHz.

A typical field recording of the signal amplitude in a suburban area is shown in Figure 2.21. Simulated variations for that channel using the above statistical modelling procedure are shown in Figure 2.22.

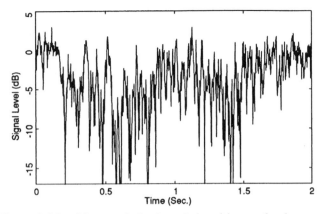

Figure 2.21: Measured shadowed signal in a suburban area.

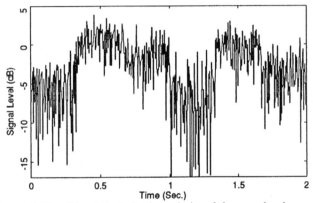

Figure 2.22: Simulated shadowed signal for a suburban area.

Typical measured and simulated pdfs for a shadowed signal in a suburban area in Sydney [27] are shown in Figure 2.23.

The fade duration is defined as the length of time over which the signal attenuation is higher than a specified value in dB. It has been found [23] that fade durations in an environmentally homogeneous area typically follow a log-normal distribution given by

$$p(x) = \frac{1}{x\sqrt{2\pi\sigma_x^2}}e^{-\frac{\ln x - \mu_x}{2\sigma_x^2}} \qquad x \geq 0 \qquad (2.59)$$

where σ_x is the standard deviation of $\ln x$ which has the normal distribution.

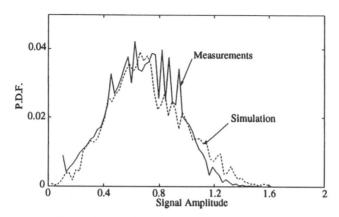

Figure 2.23:   Signal pdfs for a suburban area.

2.3.6   Channel Models for Very Large Areas

As the vehicle moves from one location to another, the environment properties change. This results in variations of the statistical character of the
received signal. That is, the received signal cannot be represented by a
model with constant parameters. Such channel models are referred to as
nonstationary. Although statistical channel characteristics can significantly
vary over very large areas, propagation experiments indicate that they remain constant over areas with constant environment attributes. Therefore,
a land satellite mobile channel can be modelled as "stationary" over these
areas.

For very large areas, the channel can be modelled by a finite state *Markov
model*. In order to derive a general model for any area, the whole area
of interest can be divided into M areas each with constant environment
properties. Each of the M areas is modelled by one of the three models
considered in the previous section. A nonstationary channel can then be
represented by M stationary channel models.

From a range of experimental recordings it has been found that it is
usually sufficiently accurate to use a Markov model with a maximum of four
states [27]. Such a model is illustrated in Figure 2.24. Each of the particular
channel states is described by a Rician, a Rayleigh or a shadowed short-term
model. Rician and Rayleigh channel models can be viewed as special cases
of the shadowed model. (A Rayleigh channel is obtained if the log-normal
component is zero, whereas a Rician channel is generated if the log-normal
component is constant).

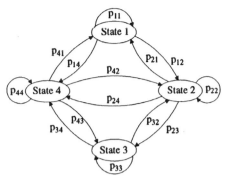

Figure 2.24: Four state Markov propagation model

For the four-state Markov model, the matrix of transition probabilities may be given in the form

$$
\mathbf{P} = \left[
\begin{array}{cccc}
p_{11} & p_{12} & \cdots & p_{1M} \\
p_{21} & p_{22} & \cdots & p_{2M} \\
\cdots & \cdots & \cdots & \cdots \\
p_{M1} & p_{M2} & \cdots & p_{MM}
\end{array}
\right] \tag{2.60}
$$

The matrix of the state probabilities is defined as

$$
\mathbf{W} = [W_1, W_2, ..., W_M] \tag{2.61}
$$

It can be computed from the matrix equations

$$
\mathbf{WP} = \mathbf{W}
$$
$$
\mathbf{WE} = \mathbf{I}
$$

where $\mathbf{E}$ is a column matrix whose entries are 1's and $\mathbf{I}$ is the identity matrix.

Example: The four-state channel model can be applied to an experimental data file recorded in Australian land mobile satellite propagation experiments [1]. The experiment was conducted using the Japanese ETS-V satellite. This particular experimental record was measured between Sydney and Coolangatta, Queensland, over a distance of 110 km with varying vehicle speeds and with environment conditions ranging from urban to rural. The speed of the mobile receiving van varied between 50 and 100 km/h. A left-hand circularly polarized unmodulated carrier at 1545.15 MHz was received at elevation angles ranging from 51° to 56°. A low-gain omni-directional antenna with 4dB gain was utilized.

The channel was modelled by four states. Two of the states represent Rician fading with ratios of direct to diffuse signal component powers of $K = 14dB$ and $K = 18dB$, respectively. The third and fourth states are modelled by a combination of Rayleigh and log-normal fading with the parameters given in Table 2.5.

State	m_s	σ_s	σ_d	m_ϕ (deg)	σ_ϕ (deg)
1	0.944	0.956	0.097	$-9.74 \cdot 10^{-4}$	7.395
2	0.978	0.912	0.063	$-3.15 \cdot 10^{-3}$	4.930
3	0.655	0.273	0.194	-0.063	22.758
4	0.474	0.201	0.205	0.132	37.204

Table 2.5: Four-state model parameters.

It was observed experimentally that the rate of variation of the log-normal component are two orders of magnitude slower than that of the Rayleigh component. The bandwidth of the Rayleigh component in this particular experiment was 100Hz while the bandwidth of the log-normal component was 3Hz. This fact allowed separation of the low and high frequency signal components and facilitated computation of the mean values and variances for the log-normal and Rayleigh signal components.

The matrix of the state probabilities was obtained as follows

$$[W] = [0.169, 0.461, 0.312, 0.058]$$

The transition probabilities for the experimental channel were computed as

$$[P] = \begin{bmatrix} 0.679 & 0.179 & 0.129 & 0.013 \\ 0.052 & 0.925 & 0.023 & 0.000 \\ 0.104 & 0.007 & 0.750 & 0.139 \\ 0.000 & 0.000 & 0.778 & 0.222 \end{bmatrix}$$

2.3.7   Probability of Error for QPSK

In experimental measurements of signal fading to obtain analogue fading models for mobile satellite channels, it is often useful to also measure transmission error statistics. These can subsequently be compared against the error statistics which can be predicted from the fading models. To see how these error statistics can be derived from the fading model, let us assume that v_i represents a QPSK symbol sample transmitted at time i.

The corresponding received sample at the input of the coherent demodulator is

$$r_i = a_i v_i + n_i \qquad (2.62)$$

where a_i is a real random variable equal to the envelope of the channel attenuation normalized to the value of the transmitted signal, and n_i is a sample of zero mean complex additive white Gaussian noise. We assume that the signal phase changes due to fading are fully compensated in the receiver.

The conditional bit error probability for QPSK modulation, with a Gray mapping, given the fading attenuation due to shadowing a_i in the i-th channel state is given by

$$P_{bci} = Q\left(\frac{a_i}{\sigma}\right) \qquad (2.63)$$

where σ is the variance of the Gaussian noise in each signal space coordinate.

The average bit error probability in the i-th channel state is given by

$$P_{bi} = \int_0^{\infty} Q\left(\frac{a_i}{\sigma}\right) p(a_i))da_i \qquad (2.64)$$

where $p(a_i)$ is the probability distribution of fading attenuation in the i-th state. Figure 2.25 gives typical values for the four-state model [27].

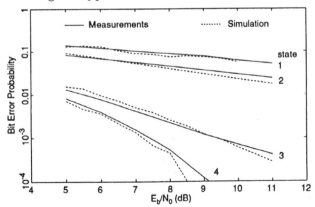

Figure 2.25: BER for each state in four-state Markov model for a land mobile satellite channel.

The average bit error probability at the output of the demodulator for an M-state Markov channel model is given by

$$P_b = \sum_{i=1}^{M} P_{bi} W_i \qquad (2.65)$$

The block error probability is defined as the probability that at least one bit is in error within a block of a given length. The average block error probability at the output of the demodulator for an M-state Markov channel model is given by

$$P_e = \sum_{i=1}^{M} P_{bli} W_i \qquad (2.66)$$

where P_{bli} is the block error probability in state i.

Plots of average bit error and block error rates for the 4-state model is shown in Figure 2.26. Simulated values are for a four-state model with parameters given in Table 2.5.

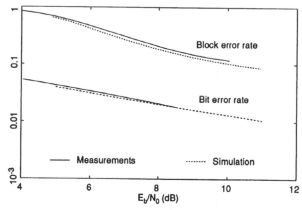

Figure 2.26: Average bit and block error rates for a land mobile system.

2.3.8   Coding Channel Models

Coding channel models for land mobile satellite systems are useful for evaluating the effectiveness of error correcting codes and for predicting the effects of channel errors on digital speech transmission. Errors on land mobile satellite channels occur in bursts. The channel models must represent as accurately as possible the statistical distribution of errors as they occur on real channels.

On-Off Model

A two-state Markov model known as an *On-Off model* has been specified by Inmarsat for modelling error patterns [33]. The model has been found useful in comparative evaluation of alternative speech codec algorithms for use on land mobile satellite channels. The two states of the model characterize the channel as being either in a period of a fade, or else being in a non-fade period, respectively. During a fade, it is assumed that the channel conditions have deteriorated to the point where the channel capacity has become zero, corresponding to a BER=0.5. During a non-fade interval the channel is considered to be ideal with BER=0.

For the On-Off model, the time spent in each visit to the non-fade state is usually fixed (at say, 10ms). Then the time assigned to the stay in the fade state (that is, the *fade duration*) is a random variable with probability density function (pdf) which is obtained from actual propagation measurements. Then the state transition probabilities are chosen so that they, in conjunction with the pdf of the fade duration, result in a given value of *link availability*. The link availability is defined as the fraction of the total time during which the channel is in the good (non-fade) state. A typical value for link availability is in the range 0.7-0.9. For more details, see Reference [33].

Other Markov Models

Although the On-Off model is simple to implement, unfortunately it does not necessarily generate error sequences which are typical of land mobile satellite channels. A more accurate representation has been obtained through the use of Markov models with higher numbers of states [22]. Table 2.6 describes the probability values associated with a four-state model that has been found to accurately represent error patterns measured on a land mobile satellite channel.

Transition Probabilities			
0.3897	0.0462	0.4000	0.1641
0.1467	0.0533	0.5333	0.2667
0.0495	0.0281	0.4345	0.4880
0.0042	0.0026	0.0938	0.8993
State Probabilities			
0.832	0.028	0.101	0.038
Bit Error Probabilities in Each State			
0.00022	0.0025	0.0140	0.190
Total bit error probability: 0.0090			

Table 2.6: Coding channel model parameters

The four-state model is one in which each state represents a binary symmetric channel of a given BER. The selection of the BER values together with the transition probabilities can define a model which produces error statistics that closely match measured results.

As an illustration, a four-state Markov model has been developed to represent propagation involving 9600 bit/s transmission through road-side tree foliage of 35% density [22]. The measurements were performed on an L-band channel for which the non-faded $E_b/N_0 = 9dB$. The resultant model values are given in Table 2.6. From the model values, it can be seen that the system is characterized by one bursty state with BER=0.19, and three random error states. The system stays in the burst state for only 3.8% of the total time and yet generates approximately 80% of the total errors.

2.4   MARITIME MOBILE CHANNEL

The received signal in maritime satellite communications systems consists of the sum of a direct component and reflected waves. Reflected waves coming from the sea surface are composed of a specular coherent component and a diffuse incoherent component. The coherent component is produced by specular reflection from a very calm, mirror-like sea surface and it exists for low elevation angles only. The diffuse incoherent component consists of reflections from many statistically independent points on the surface. The coherent reflected component is phase coherent with the direct wave while the phase of the incoherent diffuse component varies randomly with the motions of the sea waves.

In current Inmarsat communication systems, since the elevation angle is larger than 10° and the sea surface is usually rough enough, there is no significant specular component.

The magnitude of the diffuse component depends on the sea surface roughness and the elevation angle. It increases with decreasing elevation angles because of increasing reflections from the sea as well as the fact that the main lobe of the satellite-directed antenna also picks up more reflected waves. An antenna with a broad beam tends to receive diffuse signals even at high elevation angles.

The diffuse signal will seriously affect the line-of-sight transmission when the elevation angle goes below 5° [28]. The diffuse component will have its largest magnitude under rough sea conditions. However, experimental measurements show that the influence of the sea surface state on the diffuse signal is small compared to the elevation angle and the antenna beamwidth.

Large vessels usually employ highly directive antennas while small vessels use more compact antennas with broader beamwidths. Therefore, small vessels will receive more significant diffuse components. Typical fading depths of the received signal for 99% of the time have been measured to be 4.5 to 6.5dB for antennas with a gain of 20dBi and 8 to 10.5dB for antennas with a gain of 10dBi at the elevation angle of 5° [29]. For L-band frequencies employed in maritime communications, the received signal fading is usually essentially frequency non-selective.

In general, a maritime satellite mobile channel can be modelled by a three-component model which includes the direct signal $s(t)$, the specular reflection $g(t)$ and the diffuse signal $d(t)$. The received signal $r(t)$ is given by the sum

$$r(t) = s(t) + g(t) + d(t) + n(t) \tag{2.67}$$

where n(t) is Gaussian noise. The direct, specular anf diffuse waves can be represented in the form

$$s(t) = Re\left[Se^{j(\omega_c t + \phi)}\right] \tag{2.68}$$

$$g(t) = Re\left[Ge^{j(\omega_c t + \psi_G)}\right] \tag{2.69}$$

$$d(t) = Re\left[\sum_k D_k e^{j(\omega_c t + \phi_l)}\right] \tag{2.70}$$

where ω_c is the carrier frequency, S and G are the amplitudes of the direct and the specular components and D_k arc the amplitudes of the diffuse components. The individual components can be expressed in terms of in-phase and quadrature signals as

$$r(t) = r_i(t)\cos\omega_c t \ - \ r_q(t)\sin\omega_c t \tag{2.71}$$

$$s(t) = s_i(t)\cos\omega_c t \ - \ s_q(t)\sin\omega_c t \tag{2.72}$$

$$g(t) = g_i(t)\cos\omega_c t \ - \ g_q(t)\sin\omega_c t \tag{2.73}$$

$$d(t) = d_i(t)\cos\omega_c t \ - \ d_q(t)\sin\omega_c t \tag{2.74}$$

where subscript i denotes the in-phase and q denotes the quadrature components. For a large number of diffuse components, as a consequence of the central limit theorem, d_i and d_q are Gaussian random variables. We assume that these two variables have zero mean and equal variance σ_d.

The in-phase and quadrature components of the received signal can be expressed as

$$r_i(t) = s_i(t) + g_i(t) + d_i(t) \qquad (2.75)$$

$$r_q(t) = s_q(t) + g_q(t) + d_q(t) \qquad (2.76)$$

or equivalently

$$r_i(t) = S\cos\phi + G\cos\phi_G + d_i(t) \qquad (2.77)$$

$$r_q(t) = S\sin\phi + s_o\sin\phi_G + d_q(t) \qquad (2.78)$$

The distributions of ϕ and ϕ_G are assumed to be uniform over the interval 0 to 2π. The joint probability density function of r_i, r_q, ϕ_G and ϕ can be expressed as

$$p(r_i, r_q, \phi, \phi_G) = \frac{1}{8\pi^3\sigma^2} e^{\frac{-(r_i - S\cos\phi - G\cos\phi_G)^2}{2\sigma_d^2}} e^{\frac{-(r_q - S\sin\phi - G\sin\phi_G)^2}{2\sigma_d^2}} \qquad (2.79)$$

Noting that the envelope and the phase of the received signal are

$$R(t) = (r_i^2 + r_q^2)^{1/2} \qquad (2.80)$$

$$\psi(t) = tan^{-1}\frac{r_i}{r_q} \qquad (2.81)$$

the joint probability density of the envelope and phase can be derived from Equation (2.79). The probability density of the envelope is given by

$$
\begin{aligned}
p(R) &= \int_o^{2\pi} p(R, \psi) d\psi \\
&= \frac{R}{\sigma_d^2} c^{\frac{-R^2 + S^2 + G^2}{2\sigma_d^2}} \sum_o^\infty I_m\left(\frac{RS}{\sigma_d^2}\right) I_m\left(\frac{RG}{\sigma_d^2}\right) I_m\left(\frac{SG}{\sigma_d^2}\right)(-1)^m
\end{aligned}
\qquad (2.82)
$$

where I_m is the modified Bessel function of the first kind and order m.

Special cases of this three-component model distribution are

1. Rayleigh distribution corresponding to a one-component model with $S = 0$ and $G = 0$

$$p(R) = \frac{R}{\sigma_d^2} e^{\frac{-R^2}{2\sigma_d^2}} \qquad (2.83)$$

2. Rician distribution corresponding to a two-component model with $S = 0$

$$p(R) = \frac{R}{\sigma_d^2} e^{\frac{-R^2 + S^2}{2\sigma_d^2}} I_o\left(\frac{RS}{\sigma_d^2}\right) \qquad (2.84)$$

The distribution given by Equation (2.82) applies only to a smooth sea surface at low elevation angles.

For most situations, the sea surface is sufficiently rough to suppress the specular component. Therefore, most maritime channels can be modelled by the Rician distribution, given by Equation (2.84). If the power ratio of the direct signal to the diffuse signal is denoted by K where

$$K = \frac{S^2/2}{\sigma_d^2} \qquad (2.85)$$

the envelope distribution is given by Equation (2.31) . Statistical properties of maritime channels have been verified by analyzing the envelope and the phase of the recorded baseband signal. For more details, see Reference [21].

2.5 AERONAUTICAL MOBILE CHANNELS

A typical aeronautical satellite mobile situation is illustrated in Figure 2.27. The propagation between the satellite and the aircraft takes place via the direct path, the specularly reflected diffuse path and the diffuse path. Thus the signal received by the aircraft can be expressed by the sum

$$r(t) = s(t) + g(t) + d(t) + n(t) \qquad (2.86)$$

where the same notation as in the previous section is used.

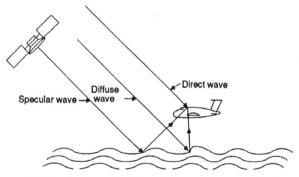

Figure 2.27:   Aeronautical mobile satellite propagation.

The direct signal s(t) is subject to free-space propagation loss, antenna gains and Doppler shift. We will consider the direct signal to be the reference for reflected signals in terms of their envelope value, Doppler shift and propagation delay.

The specular component g(t) is obtained as a mirror reflection in the specular direction from the surface. It has been reported [30] that the roughness of the surface at L-band is large enough to make the specular reflection negligible most of the time.

The diffuse multipath component consists of many reflected waves in directions other than the specular one. The diffuse reflections during flight over land are generally less significant then those over sea, because land surfaces are less reflective than water surfaces. The diffuse component at low elevation angles (below 30°) is suppressed by directive antennas.

In land mobile and maritime channels, there is no propagation delay between the direct and reflected waves. Their interference results in instantaneous variations of the received signal amplitude and phase. That is, all frequencies across the signal spectrum are attenuated equally. In the aeronautical channel the diffuse component experiences a propagation delay that is greater than that of the direct wave. In fact, each of the many diffuse reflected waves will have a slightly different propagation delay, but for the typical data rates used in this communication service this time spread can be ignored. If the differential delay between the direct and the diffuse component makes a significant proportion of the symbol period, the diffuse wave will introduce severe intersymbol interference. This type of signal distortion is known as frequency selective fading. On real channels, if the differential time delay is an order of magnitude less than the symbol period it is accurate enough to apply a frequency non-selective channel model. The differential propagation delay depends on the elevation angle and the altitude of the aircraft, as shown in Figure 2.28.

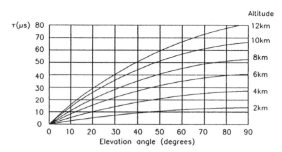

Figure 2.28: Differential delay as a function of elevation angle and altitude. (After [31] with permission).

Since the diffuse component is significant only for elevation angles below 30°, the delay is usually less than $40\mu s$. For this value of delay, the symbol rate should be less than 2400 baud to prevent frequency selective fading. It has been shown that for low symbol rates the Rician frequency non-selective model (Equation(2.31)) is very appropriate for the aeronautical channel [21].

The diffuse component is also subject to a differential Doppler shift relative to the direct signal. The diffuse component Doppler bandwidth B is related to the system parameters by

$$B = 2f_c \frac{v\alpha}{c} \sin \theta \qquad (2.87)$$

where f_c is the carrier frequency, v the aircraft ground speed, c is the speed of light, α is the root mean square sea slope and θ is the incidence angle. It has been reported [31] that typical values for two sided fading bandwidth is 200 Hz with the maximum fading bandwidth expected not to be higher than 600 Hz.

2.6 PROBLEMS

1. An aeronautical satellite mobile channel has the differential propagation delay of $30\mu s$ and the Doppler spread of 120 Hz. The ratio of the direct to the diffuse component powers is 20 dB. The transmitted symbol interval is 4.167 ms.

    (a) Is the channel frequency selective?

    (b) Is the channel fading slow or fast?

    (c) For QPSK modulation what approximate SNR is required to achieve a bit error probability of 10^{-3}?

2. A land mobile satellite system operates at 1.55 GHz. Compute the Doppler shift of the unmodulated carrier received by a vehicle travelling at 100 km/h if the elevation signal propagation angle is 30°.

3. A mobile unit is travelling at a constant speed of 60 km/h and communicates via a satellite at a carrier frequency of 1.55 GHz. Plot the Doppler power spectrum and determine the Doppler spread of the multipath fading.

4. Assuming free space propagation compute the receiver carrier power in a fixed satellite system operating at 14/12 GHz with transmitting and receiving antenna gain of 50 dB and the satellite repeater amplifier

with the gain of 80 dB. Assume that the transmitter power output is 100 W.

5. Compute the rain attenuation on a satellite channel for an area with a 0.01% rain rate of 32 mm/h, a latitude of 45° and the CCIR model parameters a = 0.03, b = 1.15 and r = 0.32.

6. Compute the attenuation of the direct signal due to tree shadowing for a land mobile satellite system operating at 1.55 GHz in a wooded area if the depth of the foliage traversed by the direct wave is 10 m.

7. A ship communicates at the carrier frequency of 1.542 GHz and at an elevation angle of 45°. The root mean square (rms) height of the sea surface is 0.25m.

   (a) Compute the Rayleigh roughness factor. On the basis of its value estimate whether the specular component is significant.

   (b) Find the rms of the specular component if the rms value of the direct component is 1mV, the magnitude of the reflection factor for a flat area is 1 and the divergence factor is also 1.

   (c) Compute the Rician factor K in dB if the diffuse component rms value is 0.2mV.

REFERENCES

1. Data Files with I & Q data, *Texas University, Houston, USA* (supplied by AUSSAT, Australia).

2. "The Telstar Experiment", *Bell Syst. Tech.* J.,Vol. 42, pp.739-1908, July 1963.

3. Hogg, D.C., "Statistics of Microwavews by Intense Rain", *The BSTJ*, pp. 2949-2962, Nov. 1969.

4. Lin, S.H., "Empirical Rain Attenuation Model for Earth-Satellite Paths", *IEEE Trans. Commun.* Vol. COM-27, No. 5, pp. 812-817, May 1979.

5. Flavin, R.K., "Rain Attenuation Considerations for Satellite Paths in Australia", *A.T.R.*, Vol. 16, No. 2, pp. 11-24, 1982.

6. "Radiometeorological Data", *CCIR Report 563-2, Recommendations and Reports of the CCIR, International Telecommunication Union*, Geneva, 1982.

7. "Special Issue on the COST 205 Project on Earth-Satellite Radio Propagation above 10 GHz", *Alta Freq.*, 1985, LIV, No. 3

8. Skellern D., Nicolas J., Vucetic B. and Jeffrey B., "A Comparison of Performance of a Solid State and a TWT Amplifier in a Digital Satellite Link", *Proc. IREECON'87*, Sydney, pp. 257-259, Sept. 1987.

9. Hetrakul P. and Taylor D.P., "The Effects of Transponder Nonlinearity on Binary CPSK Signal Transmission", *IEEE Trans. Commun.*, Vol. COM-24, pp. 546-553, May 1976.

10. Shimbo O., "Effects of Intermodulation, AM-PM Conversion, and Additive Noise in Multicarrier TWT Systems", *Proc. IEEE*, Vol. 59, pp. 230-238, Feb. 1971.

11. Saleh, A.A.M., "Frequency Independent and Frequency-Dependent Nonlinear Models of TWT Amplifiers", *IEEE Trans. Commun.*, Vol. COM-29, No. 11, pp. 1715-1720, Nov. 1981.

12. Satoh, G. and Mizuno, T., "Impact of a New TWTA Linearizer Upon QPSK/TDMA Transmission Performance", *IEEE J. Select. Areas in Commun.*, Vol. SAC-1, pp. 39-45, Jan. 1983.

13. Kerr, D.E., "Propagation of Short Radio Waves", *McGraw Hill Book Company*, 1951.

14. CCIR Report 263-5, "Ionospheric Effects upon Earth-Space Propagation", Geneva, 1982.

15. Goldhirsh, J., and Vogel, W.J., "Roadside Tree Attenuation Measurements at UHF for Land-Mobile Satellite Systems", *IEEE Trans. on Antennas and Propagation*, Vol. AP-35, pp. 589-596, May 1987.

16. Vogel, W.J. and Goldhirsh, J., "Tree Attenuation at 869 MHz Derived from Remotely Piloted Aircraft Measurements, *IEEE Trans. on Antennas and Propagation*, Vol. AP-34, pp. 1460-1464, Dec. 1986.

17. Goldhirsh, J., and Vogel, W.J., "Mobile Satellite System Fade Statistics for Shadowing and Multipath from Roadside Trees at UHF and

L-Band", *IEEE Trans. on Antennas and Propagation*, Vol. AP-37, pp. 489-498, April 1989.

18. Bullington, K. "Radio Propagation for Vehicular Communications", *IEEE Trans.* Vehic. Technol., Vol. VT-26, pp. 295-308, Nov. 1977.

19. "Properties of Mobile radio Propagation Above 400 MHz", *IEEE Trans. Vehic. Technol.*, Vol. VT-23, pp. 143-160, Nov. 1974.

20. Rahman, M. and Bundrock, A.J. "Propagation Characteristics of Mobile Satellite Links", *Telecom Research Report*, Telecom Research Laboratories, Melbourne, Australia,

21. Jongejans, A., et al, Prosat Phase I Report, ESA STR-216, May 1986.

22. Rahman, M. and Bulmer, M.,"Error Models for Land Mobile Satellite Channels", *Proc. International Conference on Mobile Satellite Communications*, Adelaide, Australia, August 1990.

23. Vogel, W. J. and Smith E. K., "Theory and Measurements of Propagation for Satellite to Land Mobile Communication at UHF", *Proc. IEEE 35th Vehic. Technol.*, Conf., Boulder, CO, pp. 218-223, 1985.

24. Davarian F., "Channel Simulation to Facilitate Mobile-Satellite Communications Research", *IEEE Trans. Commun.*, Vol. COM-35, No. 1, pp. 47-56, Jan. 1987.

25. Beckman, P., "Probability in Communication Engineering", Harcourt, Brace & World, Inc., 1967.

26. Vogel W.J., Goldhirsh, J. and Hase, Y., "Land-Mobile-Satellite Propagation Measurements in Australia Using ETS-V and INMARSAT-POR", University of Texas Report.

27. Vucetic, B. and Du, J., "Channel Modeling and Simulation in Satellite Mobile Communication Systems", *Proc. International Conference on Satellite Mobile Communications*, pp. 1-6, Adelaide, Australia, August, 1990.

28. Fang, D.J., Tseng, F.T. and Calvit, T., "A Low Elevation Angle Propagation Measurement of 1.5 GHz Satellite Signals in the Gulf of Mexico", *IEEE Trans. on Antennas and Propagation*, Vol. AP-30, pp. 10-15, Jan. 1982.

29. Karasawa, Y. and Takayasu, S., "Characteristics of L-Band Multipath Fading due to Sea Surface Reflection", *IEEE Trans. on Antennas and Propagation*, Vol. AP-32, pp. 618-623, June 1984.

30. Bello, P. "Aeronautical Channel Characterization", *IEEE Trans. Commun.*, Vol. COM-21, No. 5, May 1973.

31. Lodge, J., "Modulation for Aeronautical Mobile Channels", *FANS WG/B Report*, November 1986.

32. Sutton, R.W., Schroeder, E.H., Thompson, A.D. and Wilson, S.G., "Satellite- Aircraft Multipath and Ranging Experiment Results at L Band", *IEEE Trans. Commun.*, Vol COM-21, No. 5, p. 639-647, May 1973.

33. Inmarsat, ICEP/WK/1, *"Codec Evaluation Procedure"*, London, August 1989.

34. Davies, P.G. and Lane, J.A., "A review of propagation characteristics and prediction for satellite links at frequencies of 10–40GHz", *IEEE Proc.*, Vol.133, Pt. F, No. 4, pp.420-427, July 1986.

Chapter 3

MOBILE SATELLITE SYSTEM DESIGN

by Nick Hart

Project 21 Division, Inmarsat, London, UK.

This Chapter examines satellite-based mobile networks from a system designer's point of view. It first describes how mobile satellite communication networks are evolving. Systems such as those operated by Inmarsat offer global coverage. Some countries such as the U.S., Canada, and Australia, are developing their own national or regional mobile satellite systems.

Mobile satellite systems enable mobile users to communicate via the public switched telephone networks in much the same way as cellular radio telephone users are able to do. However unlike terrestrial cellular systems, mobile satellite users are not tied to operating in the vicinity of local radio base stations. They can utilize mobile satellite systems in remote areas and for land air and sea applications.

This Chapter begins by examining the range of mobile satellite services either currently available or in the planning stages. Then from a system designer's point of view, it identifies the important network performance objectives that must be met. The process of system design for a typical mobile satellite system is then illustrated by a case study. The case study is based on the Australian L-band Mobilesat system, the first domestic mobile satellite system to be placed in operation, offering telephone, facsimile and data services.

3.1   MOBILE SATELLITE SERVICES

3.1.1   Introduction

The number of new and proposed mobile satellite systems is growing rapidly as system designers seek to utilise the key satellite advantages of global earth coverage, network flexibility and broadcasting capabilities. These new systems are designed to support large numbers of mobile terminals. They will herald the introduction of personal satellite communication equipment to the general public.

Mobile satellite communication networks will provide users with global roaming capabilities and "Mobile ISDN" features. Mobile terminal prices are eventually expected to be comparable to current cellular radio equipment. This remarkable capability for low cost mobile communications via satellite can only be achieved by using state of the art technologies in voice signal processing, modulation and coding techniques and terminal equipment design. Each of these key technologies is considered in this Chapter. They are also studied in greater depth in other Chapters of this text.

The mobile satellite systems engineer faces a tremendous challenge in the design of intelligent networks and signalling systems which provide advanced user features and capabilities. This Chapter seeks to give the reader an insight into the complex system engineering challenges in designing such mobile satellite systems and networks. We concentrate on geostationary satellite systems. Later in this text, Chapter 9 provides an interesting contrast with its discussion of future low earth orbit (LEO) systems.

The systems engineering principles and challenges described here will generally apply to any satellite or even to some radio networks. They involve taking the broad service requirements and the engineering design constraints and developing from them communication systems which are both technically and commercially viable.

3.1.2   Global Mobile Satellite Systems

The opportunities that satellites offer for mobile communications were first recognised in the early 1970s by the International Maritime Organisation (IMO), then known as the Inter Governmental Maritime Consultative Organisation (IMCO). In particular the global coverage and high reliability of satellite communications were seen to be important for relieving the hazards and isolation of maritime travel.

The IMCO convened a conference in 1973 to decide on the principles of establishing an international maritime satellite system. This culminated in September 1976 in the adoption of what became the Inmarsat Convention and its complementary operating Agreement, which came into force in July 1979.

Inmarsat (International Maritime Satellite Organisation) was formed by a consortium of 28 founder countries. It was charged with providing a world-wide communications service which would, as stated in the convention; "... make provision for the space segment necessary for improving distress and safety of life at sea, communications, efficiency and management of ships, maritime public correspondence services and radio determination capabilities."

Keen to offer a service as soon as possible, Inmarsat took advantage of existing satellite developments by leasing three Marisat satellites from Comsat General, and subsequently, two Marecs spacecraft from the European Space Agency (ESA).

The Marisat consortium had been set up in the U.S. in 1976, to provide a maritime satellite capability using spare capacity on a series of satellites launched primarily for the U.S. Navy. A commercial system based on the use of military satellites which were outside international control was not acceptable to the world community in the long term, and in 1982 the Inmarsat global communication system was established.

The technical standards established by Marisat were perpetuated by Inmarsat to maintain continuity of service, and these became the Inmarsat Standard A specifications (later simply Inmarsat A). The Inmarsat A system provides speech, data and telex circuits on demand directly between ships at sea and the international telecommunications network via a Coast Earth Station (CES).

Maritime communications was one of the first satellite communications service to benefit from the application of *Demand Assigned Multiple Access* (DAMA) assignment of satellite capacity. In a DAMA scheme, satellite channels are allocated upon the user's demand. That is, the DAMA technique assigns satellite link capacity on an "as needed" basis to terminals. This dynamic assignment of channels enables the network efficiency to be maximized and the user charges to be minimizeded. It is described in more detail in Chapter 4. All mobile satellite communications systems currently under development base their design on this DAMA principle. It is one of the key methods by which a large terminal population can be supported by a relatively small number of satellite channels.

Inmarsat currently (1993) operates its own four second-generation satellites and provides mobile satellite services to over 14,000 ships at sea. Figure 3.1 illustrates how the current Inmarsat operational satellites provide global coverage over four ocean regions. The number of countries which are Inmarsat signatories has now grown to more than 60.

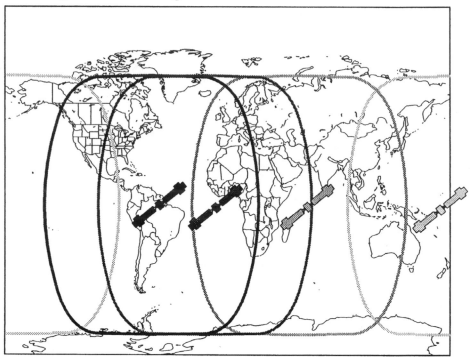

Figure 3.1:   Inmarsat global L-band coverage (Permission of Inmarsat).

In 1985 the Inmarsat Convention was extended to include providing Aeronautical services and this was further amended in 1989 to also support land mobile services. In order to provide services to these additional markets which are expected to grow rapidly, Inmarsat has placed a contract for at least four additional third-generation satellites. These will, in addition to providing global coverage, incorporate powerful spot beam antennas.

Inmarsat has introduced or is currently planning a number of new services to cover a range of different market segments. These all utilize links operating at L-band frequencies (near 1.5 GHz) for transmission between Mobile Earth Stations (MES) and satellites, and links at C-band (near 4 GHz and 6 GHz) for transmission between satellites and Coast Earth Stations (CES).

The types of Inmarsat services which are either operational or planned arc as follows:

1. Inmarsat A : This analogue FM system provides a telephone, telex, facsimile or data circuit between an MES and a CES with data rates of up to 56 kbit/s. The MES are relatively complex and expensive ($25,000 to $50,000 each). They can usually only be afforded by larger ocean-going vessels. More recently, transportable "suitcase"Inmarsat A terminals, with unfurlable antennas (with gains greater than 20 dBi), were introduced particularly for land mobile applications.

2. Inmarsat B: This system which will be available worldwide in 1993, will provide a digital replacement for Inmarsat A. It will utilise a 16 kbit/s digital voice codec.

3. Inmarsat C: This is based on a low cost MES and was introduced in 1991 to provide a two way data messaging service. The primary objective was to reduce the cost, size and weight of the terminals. The MES uses a small, omnidirectional, low gain antenna, to support a 600 bit/s data channel. The smallest available MES weighs around 4 kg. The terminals currently cost $5,000 to $8,000 each. This cost is expected to be reduced in the future.

4. Aeronautical System: This was introduced in 1991 and is designed to provide a digital voice and data service directly between jet aircraft and Land Earth Stations (LES). It provides interconnection into the Public Switched Telephone Network (PSTN) It will provide voice transmission (using a 9600 bit/s voice codec) and data services up to 9600 bit/s, using high gain (12 dBi) steerable antennas on the aircraft. The use of a small omni-directional antenna supports a low bit rate (600 bit/s) data service which is used for aircraft fleet management and in the future, air traffic control (ATC) purposes.

5. Inmarsat M: Introduced towards the end of 1992, this provides low cost digital voice, facsimile and data services for both maritime and land mobile applications. The service is aimed at extending Inmarsat services to a greater market segment by reducing the MES costs and introducing cheaper call charges. This is achieved by using a 6.4 kbit/s digital voice codec to provide voice transmission and by only support-ing 2400 bit/s data services.

The MES will use medium gain 12 dBi steerable antennas. The terminals are envisaged to cost between $10,000 and $15,000 when they are first introduced. For transportable applications, "briefcase" terminals will be introduced which will initially weigh around 10 kg. These will offer far greater mobility than the current Inmarsat A "suitcase" units.

In addition, Inmarsat is currently developing a satellite paging service due for introduction in 1994 and planning lower cost handheld mobile terminals. These are envisaged to be handheld units with omnidirectional antennas, which will offer voice and data services at very attractive tariffs, by year 2000.

3.1.3   National Mobile Satellite Systems

Canada was one of the first countries to appreciate that satellite communications could offer tremendous opportunities to improve domestic communications in rural and remote areas. In the early 1980s a programme was initiated by the Canadian Government to explore how a two-way mobile satellite voice and data service could be provided within Canada. Therefore much of the original research work on mobile satellite systems, particularly those designed to operate in a land mobile environment, was carried out in Canada. This early work assumed a UHF mobile satellite frequency allocation would be made available.

Unfortunately a battle for UHF spectrum allocation between the terrestrial and satellite mobile communities delayed the development programme. It was finally agreed in North America that the UHF frequency bands would be used solely for terrestrial mobile applications and that mobile satellite systems would operate at frequencies above 1 GHz.

This spectrum dispute provided an opprtunity for the Geostar Corporation to provide the United States with the first national land mobile commercial satellite service. Initially it was implemented as a one-way inbound data reporting service at L-band (1.6 GHz) using spread spectrum techniques. However it was designed to support the introduction of two way messaging at a later time. The two-way messaging service was never implemented as Qualcomm developed shortly afterwards a competitive two way Ku-band (11.0 and 13.0 GHz) data/messaging service, called Omnitracs. This system also utilised spread spectrum techniques to minimise interference with other transponders. Both these data only services were aimed at providing position reporting and messaging services for the trucking community.

By early 1992 Qualcomm had over 25,000 terminals in use on trucks across North America and were successfully marketing their services world-wide.

In 1988 the American Mobile Satellite Consortium (AMSC) was licensed by the Federal Communications Committee (FCC) to be the provider of L-band mobile satellite capacity in the United States. At the same time, a subsidiary of Telesat was formed called Telesat Mobile Incorporated (TMI). Its goal was to provide mobile satellite services in Canada. The AMSC and TMI organisations jointly contracted to procure two multibeam L-band spacecraft, which will provide voice and data services from the end of 1994. Each satellite will support five spot beams covering the North American continent. The service coverage area is illustrated in Figure 3.2.

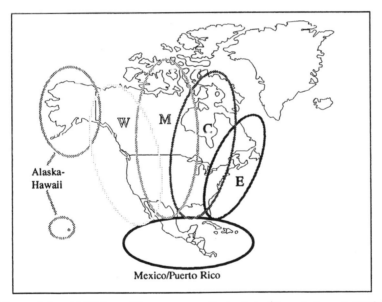

Figure 3.2: AMSC satellite coverage footprints (Permission of AMSC).

In order to meet the challenge from the Omnitracs service, both the AMSC and TMI introduced interim two way data messaging services in the early 1990's, by leasing capacity from Inmarsat and using Inmarsat C type terminal equipment. A number of other countries are also either considering or committed to implementing national mobile satellite services. These include Mexico, Japan, Europe (through the Eutelsat organisation), and Australia.

In this Chapter we will, as an illustrative case study, concentrate on the new Australian mobile satellite service called Mobilesat. This will be the first domestic mobile satellite system to offer full voice and data services in the world. The service is planned to be introduced by Optus Communications in mid 1993, and will be based on an integrated service concept which will offer users "Mobile ISDN" features.

Optus Communications represents a consortium comprising Bell South, Cable and Wireless, and a number of Australian-based companies In early 1992, a Memorandum of Understanding (MOU) was signed between Optus Communications, the AMSC, TMI and Industrias S.A de CV (IUSA) of Mexico, to develop common technical standards for their respective regional mobile satellite systems. This MOU was established in recognition of the fact that technically, each of these regional mobile satellite networks will be providing very similar services and will operate under similar design constraints.

The Mobilesat system design and specifications are in a relatively mature form. Thus Mobilesat provides a very informative example of future regional mobile satellite systems which will be implemented world-wide.

We will examine a number of the key design issues and trade-offs in the development of a system like the Mobilesat system. We examine differences between a national mobile satellite system and a global mobile satellite system such as the Inmarsat M system. Specifically, we concentrate on describing the required service and networking features, and the important system engineering goals. Then it is possible to determine appropriate values for satellite and mobile terminal parameters. An important element of the system design process is the selection of the modulation type to be used and the error control strategies to be adopted. We examine these issues in some detail.

3.2   MOBILE SATELLITE SYSTEM PARAMETERS

3.2.1   Service Characteristics

The design of a new mobile satellite network must recognize that internationally, telecommunications is currently undergoing dynamic change in the increasingly deregulated and competitive business environment. The introduction of new communication networks , such as the Integrated Services Digital Network (ISDN), have allowed a sophisticated and versatile range of services to serve the communications and information needs of business.

In the mobile communications field, cellular telephony is enjoying rapid growth in the urban areas and the key trunk routes. In some countries, cellular radio users are offered a national roaming service and a diverse range of network facilities. Similarly, users of private mobile networks now have access to the latest trunked mobile radio schemes offering efficient, integrated, voice/data networks.

Within certain sparsely populated countries such as Australia and Canada, the cellular radio network will only reach a limited percentage of the population (perhaps 70-80%) and a small fraction of the land areas. This is due to the vast distances and relatively sparse distribution of population outside of the metropolitain cities.

Against this background, it was determined that within Australia for example, a national mobile satellite communication system offering nationwide coverage could expect around 50,000 customers by the year 2000. So in 1988, the Australian national satellite system operator Aussat, (now fully owned by Optus Communications) contracted for L-band capacity to be provided on two new satellites it was planning. These so-called B Series satellites were to replace older satellites currently in use and near the end of their life. The new mobile satellite system, called Mobilesat, is targeted at providing public and private telephony services to the rural regions of the country.

There are a number of key issues to be faced in implementing a national mobile satellite service which must coexist with existing terrestrial mobile networks. In order to compete with the expanding range of terrestrial radio systems it is important to provide in the mobile satellite network, integrated voice and data services which offer good service quality and reliability. National mobile satellite systems will face increasing competition from current analogue and future digital cellular radio services and private mobile radio systems.

Terrestrial mobile radio networks offer some distinct advantages over geostationary satellite systems. These advantages follow from the much shorter distances between terminals and terrestrial base stations compared to distances of the order of 36,000 km over which satellite mobile terminals must operate. The relative advantages that potentially hold for terrestrial mobile networks include the following.

- cheaper and smaller terminals and lower call charges;

- superior radio coverage in urban and in-building areas;

- minimal signal propagation delay;

- greater network capacity (number of mobile users per given area).

However, mobile satellite systems can exploit three advantages of satellite communications to the maximum:

- wide area coverage

- network flexibility

- broadcast capability

Wide area coverage enables satellite networks to offer services nationally to customers who operate in rural areas. This may be at the expense of greater user call charges and more expensive radio equipment than for conventional terrestrial services. However, it is possible to offer a range of diverse services, using the inherent flexibility of satellite systems. This has the potential to increase the terminals' usefulness and effectively minimise the user charges. It enables the satellite service to not only complement but also to compete against terrestrial radio networks.

For example, Mobilesat will offer integrated circuit-switched full duplex voice, data and facsimile, as well as messaging services from mobile terminals into the public switched networks. Reliable error free data and facsimile services with standard facsimile and data equipment will enable the user to pass information in a compact and efficient form. The use of a cheaper data messaging service by using short messaging packets to transfer routine data, will minimise the customer charges. These data and messaging types of services are currently not reliably supported with analogue cellular systems. They will provide a clear differentiation between mobile satellite and cellular radio services. In addition to these basic services, the Mobilesat network will have the option of being configured to operate in a number of broadcast call modes, as required by private network operators who may be currently using terrestrial mobile radio services.

In general, customers will naturally seek the cheapest, most reliable mobile communication system that meets their requirements. In high density population areas, such as urban areas, terrestrial networks offer advantages. Therefore mobile satellite services can be expected to be always complementary to these terrestrial services.

Ideally, from the user's viewpoint, it would be desirable to see the integration of the cellular and satellite mobile networks into a single service. This could even lead to the use of a single customer mobile telephone number and the need for only one item of radio equipment in the vehicle or a single handheld unit to access both services. We may well see this in the future.

Any mobile communication network must provide a range of user services. Advanced networking features, such as call diversion and the use of common cellular/mobile satellite handsets are compromise first generation solutions which will initially meet the customer needs. In addition to services such as call diversion, the more advanced networking features of, for example, Calling Line Identification, can be provided to enable the called party to determine who is calling them. These and other features allow network customers to more effectively benefit from the use of their mobile terminals.

3.2.2 A Typical National Mobile Satellite Network

As a case study of a typical national mobile satellite network, let us review the main features of the first such network, namely the Australian Mobile-sat network. The Mobilesat system is based on the use of two spacecraft, each with a single L-band (1.5/1.6 GHz) transponder of bandwidth 14 MHz for communications with mobile terminals. Ku-band (12/14 GHz) communications is used for communications between the satellites and fixed earth stations. The system is illustrated in Figure 3.3.

The satellites will each provide at L-band an approximate usable EIRP = 48 dBW. A single beam will cover Australia and its surrounding waters. The satellite L-band receivers are designed to have a minimum G/T=-1.5 dB/K. It is estimated that the two satellites will be required to provide approximately 1000 channels in order to service the forecast mobile terminal population.

The Mobilesat network illustrated in Figure 3.3 will be implemented with two fully redundant fixed earth stations at separate locations. Each of these incorporates a Network Management Station (NMS). One NMS site will be configured as the Primary NMS. It will control and monitor access to the network. The other NMS site will act as a backup. On request from users, the Primary NMS will allocate communications channels between the mobile terminals and base stations on a per-call basis using a DAMA technique.

It is necessary to distinguish between the different types of fixed earth stations illustrated in Figure 3.3. They can be described as follows:

- *Network Management Stations* which monitor and control access to the network and allocate channels to mobile terminals.

- *Gateway Stations* which provide an interface to the public switched telephone network and are usually connected to an ISDN exchange. An ISDN interface will allow the satellite system to support the advanced user call facilities normally associated with ISDN. The Gateway Station is often co-located with an NMS, but not necessarily so.

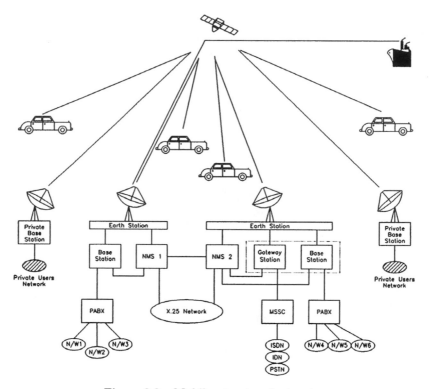

Figure 3.3: Mobilesat network structure.

- *Major Base Stations* which act as a 'hub' shared between multiple private user groups and provide terrestrial interface to private networks.

- *Minor Base Stations* which are private non-shared stations, each serving a particular closed user group. They not only provide communications to mobile terminals but also must be able to communicate with the NMS via the satellite.

The mobile terminals will transmit and receive signals at L-band. These will be frequency translated on board the spacecraft to Ku-band for transmission to the base stations. Thus a mobile to mobile call will have to be routed via a base station, and will suffer a "double hop". That is, the users will experience twice the normal satellite delay.

The communication channels between the mobile terminals and the Gateways will support voice, facsimile and data traffic. They may be referred to as *communication channels*. The Mobilesat system also incorporates separate *signalling channels*. A *common channel signalling* protocol will be used to provide a mechanism for mobile terminals to communicate with the NMS to establish and clear down calls. Common channel signalling systems are based on the provision of a signalling channel separate from the users' speech/data circuits, the signalling channel being used to carry the signalling and control information associated with all users in the network. More details will be given in Section 3.3, and in Chapter 4.

Each Gateway station will typically support up to 900 traffic channels, split equally between the two spacecraft, and thus will require two separate Ku-band antennas. Each antenna is expected to have a diameter of between 7 and 13 metres. Gateway Stations will also include an ISDN switch which will act as a Mobile Satellite Switching Centre (MSSC). The MSSC will support the billing and a number of supplementary services which will be offered, as well acting as a traffic switch providing alternative routing capabilities into the public telephone network.

In addition to the public services, the Mobilesat system will support a Mobile Radio Service optimised for private network user requirements. The Mobile Radio Service will enable all mobile terminals within a closed user group to use common communication channels. A base station operator could manually control access to the network. Access to this service is via either a Major Base Station or a Minor Base Station. The use of a privately owned Minor Base Station eliminates the terrestrial tail costs and will allow more network flexibility. A Minor Base Station will only support a maximum of 6 channels, accessing a single satellite and will have a single Ku-band antenna of 3 metre diameter. Minor Base Stations will communicate to the NMS using the standard common channel signalling system, and will therefore also require an L-band antenna.

A Messaging Service will also be provided. This service will route short, pre-formatted ("canned"), messages from terminals via the NMS, using reserve capacity on the signalling channels, into the terrestrial packet data X.25 networks. The NMS simply acts as a message switch between the customers' Despatch Centre and the mobile terminals, with an End-to-End protocol ensuring message delivery. The processing in the NMS will allow a poll/response message transfer time of around 3 seconds. This messaging service is particularly aimed at supporting position location and general data reporting applications.

3.2.3   Global and National Networks

Having provided the reader with an overview of a national (or regional) mobile satellite system it is worth considering some of the key technical differences between national and global mobile satellite network services. The Inmarsat global network provides a similar set of voice, facsimile and data services to those described for Mobilesat through the Inmarsat M system (voice, facsimile and data) and the Inmarsat C service (messaging).

Currently the most significant difference between national and global services are with respect to the spacecraft. All national satellite systems will be based on the use of spot beams. These offer improved performance compared to the Inmarsat global beams but with limited regional coverage. The higher satellite EIRP obtained from the use of spot beams enables the antenna gain of the mobile terminals to be reduced compared to the Inmarsat service. With the introduction in the mid-1990s of the Inmarsat third-generation spacecraft , each supporting a number of spot beams, these differences will be less significant.

The various Inmarsat services are supported by different Network Control Stations (NCS) and Land Earth Stations (Gateways) for each of the services. These networks operate as separate systems even though they use the same satellites and for which the base stations may actually coexist at the same physical ground station. This is in contrast to national mobile satellite systems where there is expected to be a high level of integration between the different services and the ground infrastructure.

The Gateways in a national system also support a limited number of user interfaces, whilst the global satellite network has to support a range of network interfaces which vary from country to country. Therefore it is difficult for the global system to provide a uniform level of service. Indeed, signatory countries compete with each other to provide customers with access to the Inmarsat system, by offering different services and tariffs.

Inmarsat signatories currently use large C-band (4 and 6 GHz) earth stations to access satellites which actually support relatively few communication circuits. This is in contrast to regional mobile satellite operators who use Ku-band (12 and 14 GHz) earth stations to access regional satellites which will support a larger number of channels.

3.3 SYSTEM DESIGN OBJECTIVES

Let us review of some of the major objectives which must be incorporated into the engineering design of a mobile satellite system. These can be briefly listed as follows:

- To provide bandwidth efficient modulation techniques. This is essential as there is limited spectrum available for mobile satellite services;

- To utilize techniques which minimise the carrier power required from the satellite. Current geostationary satellite systems are power limited;

- To provide for a range of evolving mobile terminal specifications. For example, the network should permit usage of mobile or transportable (eg. suitcase) user terminals;

- To provide support for different service types, qualities, and priorities (data rates, link margins, and service priorities);

- To provide the capability to support growing numbers of mobile terminals. This requires a modular system design so as to facilitate increasing the number of signalling and communication channels. It is also desirable to provide for the addition of different types of channels as technology and service requirements evolve;

- To provide compatibility with the possible introduction of new spacecraft including the possible use of multiple spot beam satellites.

An important issue is to decide which multiple access strategy to adopt. At the present time, all mobile satellite systems which offer voice services have adopted a Frequency Division Multiple Access (FDMA) approach. That is, different users are provided channels on a single transponder by frequency channelisation. This technique is discussed further in Chapter 4.

The use of FDMA in conjunction with common channel signalling, allows future developments in modulation and processing technologies to be easily incorporated into the system and allows a variety of communication standards to be supported. Also, as growth in the mobile terminal population occurs, incremental increases in the required system spectrum allocation can be easily assigned.

The approach to the design of any satellite network must also consider a number of issues that are vital to the successful operation of the network, both in technical and marketing terms. Some of these key issues are as follows:

- To maximise *network availability*, that is, the probability that the network can provide service when it is required. The network availability of a mobile satellite network should ideally, be comparable to the public telephony network availability;

- To maximise *network capacity*, that is, the number of users able to be serviced by the system. Network capacity should be maximised, while maintaining quality, response times and acceptable level of service;

- To minimise *terminal price*. Terminal prices must be kept to a minimum (even if this means that the satellite must be more complex to achieve this). This is an acknowledgement of the sensitivity of market penetration to the mobile terminal price.

The remainder of this Chapter provides an insight into some of the complex trade-offs involved in trying to meet these design objectives.

3.4   NETWORK AVAILABILITY

As indicated in the previous Section, the term *network availability* describes the probability of being able to access the network service when it is required. In a conventional public switched telephone network (PSTN) the network availability is a function of the number of telephone lines that are available and the reliability of the network elements. Similarly in a mobile satellite system, network availability is dependent on the total number of available communication channels, the signalling channel capacity, and the reliability of the Network Management Station and the Gateways.

However, in the PSTN the actual subscriber telephone units are often connected by telephone cables which have a very high reliability. In mobile satellite networks the signals from the satellite are often subject to severe fading as described in Chapter 2. This not only complicates the design of the common channel signalling system and the users' communications systems. It also has the effect of dramatically reducing the service availability, particularly in urban areas. However, as the mobile satellite service is aimed mainly at rural and remote areas, it is usually more important to concentrate on providing acceptable service availability in rural environments.

In this Section, design issues relating to the reliability of the network elements and the common channel signalling system will be reviewed. We will also consider how the system design is influenced by the service coverage requirements. Chapters 4 and 5 concentrate on more detailed signalling channel and network performance issues. As in earlier Sections, we use the Mobilesat system as an illustrative network architecture since its features are likely to have much in common with any other national mobile satellite networks.

3.4.1 System Reliability

A high degree of system reliability is considered essential to gain and retain customer confidence in mobile satellite services. The mobile satellite network architecture must reflect this need for reliability by incorporting a high level of redundancy in key system elements.

The Network Management Station (NMS) is vital to the successful operation of the system. A minimum of two NMS installations is necessary to provide redundancy. These should not be located at the same site. This is because the Ku-band satellite link section from the satellite to the NMS may be subject to severe signal fades due to rain attenuation during storms. If the two NMS sites are at different geographical locations, then this provides "rain diversity" protection .

At the main NMS site, it would be desirable to provide duplicate network management computers to provide a high level of NMS reliability. In the Mobilesat system, the two NMS computers will be kept in event step on a call by call basis. A third NMS computer will be implemented at the second NMS site so as to offer site diversity. The second site will be connected to the main NMS site using terrestrial lines, so that if either the main NMS site fails or a rain fade occurs, a controlled switchover to the second site can be implemented. The availability design goal for each site is of the order 99.8%. Linking the two sites yields an NMS system availability of the order of 99.9991%. Similarly, coupling site diversity with terrestrial network call rerouting will ensure an overall high service availability.

3.4.2 Common Channel Signalling

As mentioned previously, a common channel signalling system is used by mobile terminals to communicate with the NMS for the establishment of calls. The performance of the signalling system is very important in ensuring a high network availability. The signalling system designer could begin by seeking to achieve performance compatible with the existing PSTN requirements. For example, this would specify a goal of 99% of calls being set up successfully in less than 5 seconds, whilst operating under a load of 13 call requests per second.

Let us briefly examine the essential elements of a signalling system for a mobile satellite network. (Chapter 8 explores in detail the operation of a typical signalling system, and provides an overview of the analysis and simulation tools that can be used to determine the likely performance of the signalling protocol). The signalling system consists of two parts, namely

- Outbound signalling channel (NMS to mobile terminal), which uses a time-division multiplexing (TDM) technique, and

- Inbound signalling channels (mobile terminal to NMS) using a time-division multiple access (TDMA) technique.

The use of the two terms TDM and TDMA may be confusing unless they are explained. A *TDM technique* is one which is used where there is a single transmitter communicating with many receivers. To do this, the transmitter sends frames of information, each frame being divided into a sequence of shorter units or sub-frames, each sub-frame containing information intended for a particular receiver. The intended receiver is indicated by inserting a binary address (receiver number) prior to the signalling information for that receiver.

On the other hand, a *TDMA technique* is one which is used where multiple transmitters wish to share a single channel to communicate with a given receiver. Each transmitter must transmit in short bursts, preferably timed and interleaved so as not to overlap each other. The bursts of each TDMA terminal may be controlled so that bursts are transmitted in assigned time slots with bursts arriving at the satellite one at a time in a closely spaced sequence that never overlaps. Alternatively, terminals may be permitted to operate in a random access mode, commonly referred to as an *Aloha* mode. This mode involves less complexity in the terminals but may result in overlap (collisions) between bursts causing errors. Because of the possibility of collisions and the need to repeat messages so affected, the Aloha TDMA procedure offers only typically 10% average channel utilisation.

For the TDM outbound signalling of the Mobilesat system, the NMS sends call signalling information to mobile terminals using frames, each frame containing a number of short 12-byte sequences known as Signal Units (SU's).

Each SU will contain the selected mobile terminal number as an identification code. The TDM signalling channel is assigned a fixed frequency allocation and will operate at a relatively high data rate (9600 bit/s) compared to the communication channels. In order to ensure a high integrity signalling channel, the TDM signalling channel should be designed so that it operates with a higher link margin than that used for the communication channels.

For the TDMA inbound signalling, TDMA techniques are used by the terminals to communicate with the NMS. Each mobile terminal first assembles and temporarily stores its user signalling information and then transmits this information in short TDMA bursts at 2400 bit/s. The signalling bursts will either be used for randomly transmitting call requests to the NMS, or for acknowledging signalling messages from the NMS. Mobile terminals can signal over one of 32 inbound frequencies. Of these, 24 are designated for use for inbound signalling using the Aloha technique. The remaining 8 frequencies are reserved for acknowledgement signals using assigned time slots.

A critical design challenge is to provide a robust signalling system capable of overcoming the signal fading commonly experienced in land mobile satellite systems. The signalling system must be capable of ensuring at least 99% throughput. That is, it must ensure that at least 99% of signalling packets are received without errors. This may require careful analysis of available measured propagation data for the system under consideration. Fading on heavily shadowed roads would cause severe errors and hence disruption of the signalling protocol.

For example, for the Mobilesat system, analysis indicated that the conventional use of FEC coding, combined with interleaving to spread the burst errors, would not provide the required 99% throughput. Instead it was found that the only way to provide protection against fading was to use a form of *time diversity*. Time diversity involves sending each SU a number of times, each transmission being spaced sufficiently apart in time to ensure that fading effects are uncorrelated. In the event, a protocol was developed in which all call signalling SU's are transmitted three times spaced apart prescribed intervals in time. Figure 3.4 shows the frame structures adopted for outbound signalling.

Each SU in Figure 3.4 is first encoded using a rate 3/4 error correcting convolutional code. Then a number of encoded SU's are formatted in a frame. Each frame is then transmitted three times from the NMS to the terminals. At the receiving end, each of the three copies of each SU unit is first decoded in an attempt to correct transmission errors. Then an error detection procedure is used to check whether any residual errors remain. If any one of the three SU transmissions is found to be error free, then that SU is accepted by the terminal. More detail is given in Chapter 8.

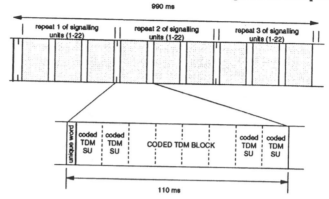

Figure 3.4: 9600bps TDM frame format.

One important design decision is to choose the appropriate length for a signalling frame. To minimise the call set up times, it would be desirable to make provision for the time diversity transmissions of each SU to be repeated rapidly. However, in order to gain maximum benefit from the time diversity scheme, it is important to try to ensure that the transmission error probabilities are independent. Extensive propagation data analysis proved that signal fades separated by more than 250 ms were largely uncorrelated. Therefore the time diversity repetition strategy was designed based on a 330 ms frame size, each frame being transmitted three times. The chosen frame size also allowed additional time for the demodulator to reacquire carrier synchronization following loss of synchronization due to a signal fade.

So far as outbound signalling is concerned, the use of a high outbound TDM transmission rate would give a short SU transmission time. This in turn could help to reduce the probability of an SU being lost by a signal fade. The transmission rate would need to be selected as a good compromise between

- providing a short SU transmission time,

- operating at a rate which can be readily demodulated in the mobile terminal, and

- a rate which would accommodate the expected signalling traffic load (refer Chapter 8 for details).

As a specific illustration, for the Mobilesat system a 9600 bit/s outbound signalling rate was selected for normal operation. This results in an SU transmission interval of 13 ms. (To provide additional system flexibility, it was decided to specify that the outbound TDM channel may be optionally set to 4800 bit/s, this mode being to support possible future standards for mobile terminals which may operate with low gain omnidirectional antennas).

This example of how a simple robust signalling protocol was developed illustrates clearly why careful system engineering is important. The conventional use of FEC and interleaving would not have met the system requirements. In the particular case studied it was found desirable to use a rate 3/4 convolutional code (with no interleaving) on both the outbound TDM and inbound TDMA channels. FEC provides improved performance with noise on the channel but the effect of FEC is much less significant than the time diversity (retransmission) strategy also adopted.

All common channel signalling systems being developed for land mobile satellite services world-wide have now adopted the use of packet repetition techniques. (More details of the signalling systems are provided in Chapter 8).

3.4.3   Service Coverage

The term service coverage is used to provide an indication of the geographical area over which a given availability requirement is met by the mobile satellite system. The service coverage requirements essentially define the link operating margins required for the system. For example, a decision to provide good quality mobile satellite services to vehicles which are operating in an urban environment would involve the provision of L-band link margins in excess of 15 dB (as discussed in Chapter 2). However, a typical mobile satellite user may not require service in a city environment, but will want access to the service in rural and remote areas.

Ideally the systems engineer needs to determine the exact areas where the mobile satellite potential customers will operate, and then perform extensive propagation measurements to determine the required link margins required in order to provide effective services into those areas.

This data, in conjunction with other information relating to the percentage of customers operating in any one area, would then enable the systems engineer to determine the exact link margin required (to provide say a 95% service coverage throughout a given country). In fact, at the expense of complexity, it would be possible to design a system to have a variable link margin which depends on the operating environment. The link margin could be either fixed for the duration of a call or dynamically variable during the progress of a call. In practice, it is neither feasible nor possible to determine precisely each of the various design parameters. Thus the systems engineer has to make a judgement based on incomplete propagation data and imprecise market research information. For example, in Figure 3.5 the L-band received signal probability density functions are given for a mobile receiver in a vehicle operating at different speeds on a number of rural and suburban roads.

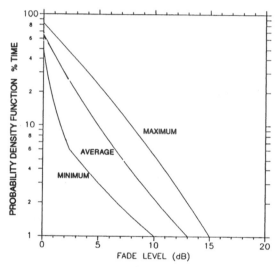

Figure 3.5 Fade distributions for rural and urban areas.

These statistics were derived from a series of detailed propagation measurements. From these measurements it was calculated that a 3 to 5 dB link margin would give a link availability of around 80%, even on the most heavily shadowed roads. As it transpired that only such a small link margin was required, it was not considered neccessary to build into the system the additional complexity of an adaptive link operating margin.

It should be noted that to ensure toll quality voice services along roads with only 80% link availability, requires the very latest techniques in voice signal processing and modem design. The development of these techniques will be briefly described in Section 3.4 which describes some of the key trade-offs. Chapters 5, 6, and 7 provide more details on the modulation, voice codec and FEC coding issues. Similarly, the provision of facsimile and data services along such roads is not a simple task. The low bit rate voice codecs which have been selected by mobile satellite network operators for the transmission of voice signals do not support the transmission of in-band data signals. That is, the voice channels cannot be used with conventional terrestrial data modems for data transmission. Therefore interface units are required at the mobile terminal and the Gateway stations to ensure all analogue data modulated signals from the PSTN are converted to baseband. These are then passed as a digital signal over the satellite link using FEC coding and QPSK modulation schemes.

Due to shadowing and other propagation irregularities on digital mobile satellite links, error rates on the links are often quite high, typical operating BER values being in the range 10^{-4} to 10^{-1}. Fortunately, some of the most recently developed low bit rate voice codecs can operate successfully at bit error rates in excess of 10^{-2}. However data services demand essentially error free transmission to maintain the integrity of the information. This can only be assured if an automatic repeat request (ARQ) retransmission scheme is used between the two ends of each data link.

For the Mobilesat network, the data channel will support information rates of either 2400 or 4800 bit/s. At the lower data rate, a rate 3/4 convolutional code will be inserted to reduce the required E_b/N_0 ratio at the receivers. The overall protocol is bit rate independent, and may support full duplex data transmission.

One application of the data channel is for facsimile transmission. In regard to facsimile transmission, a careful analysis of conventional terrestrial error protection schemes used for example, by Error Correction Mode (ECM) facsimile machines, has shown that they would not perform well over a shadowed mobile satellite link.

An additional problem is that typical terrestrial facsimile circuits are required by the facsimile machine standards to operate with tight command/response timing constraints. Analysis of the possible satellite link and interface delays indicates that the performance of a conventional synchronous facsimile protocol would be marginal on a mobile satellite circuit, even assuming a single satellite link. It is therefore necessary to develop a special protocol for facsimile transmission over mobile satellite networks. One such protocol is illustrated in Figure 3.6.

This system uses three separate synchronous links between the facsimile machines and the Facsimile Interface Units (FIU). This allows the decoupling of the two facsimile machines and the introduction of an ARQ protocol to ensure error free data transmission. Extensive experimental measurements were performed with prototype interface units to verify and optimise the overall protocol. The ARQ protocol ensures that the FIU's always provide error free facsimile reception, but the transmission times may become unacceptably long when channels are noisy.

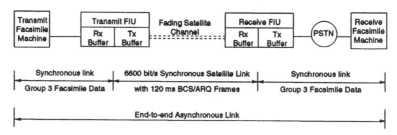

Figure 3.6: Mobilesat facsimile/data service schematic.

Experimental results have shown that reasonable transmission times are usually possible, even with heavy roadside shadowing. An illustration of the very high BER's to be expected is shown in Figure 3.7.

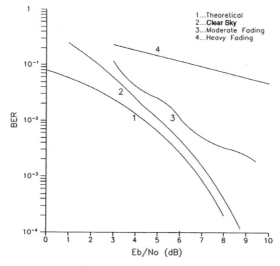

Figure 3.7: Bit error rate functions used for fax/data system testing.

These figures were obtained on a channel with parameters adjusted to give a nominal value of $E_b/N_0 = 6dB$ in clear sky conditions. Despite the high error rates, it has been shown that the facsimile system illustrated in Figure 3.6 can provide error free images at the receiving end.

This Section has sought to give the reader an insight into how a mobile satellite network can be designed to provide a high network availability. This requires not only the provision of a reliable ground segment, but also the development of signalling and communication channel protocols designed to operate in a particular radio propagation environment. Before commencing any system design it is important to determine and understand the fundamental operating conditions (propagation data) and the service requirements. Then a proper system design can commence based on a combination of theoretical calculations, computer simulation, and experimental measurements to optimise and verify the final design.

3.5 Network Capacity

For a mobile satellite service to offer a high network availability there must be a sufficient number of communication circuits available so that users can with a small probability of loss, access the network. For example, with the current terrestrial analogue cellular phone service there is sometimes a shortage of channels in high traffic city areas, which results in poor network availability during busy hour traffic times. On the other hand an excessive number of channels may result in unnecessary network operating costs and inefficient spectrum use.

The systems engineer has to therefore strike a careful balance between the expected traffic and the number of communication channels that the network is designed to support. For any radio system design, there is always the challenge of utilising the available signal power and bandwidth efficiently. This is why radio systems engineers must often consider tradeoffs between a variety of modulation and FEC coding schemes.

In this Section, based once again on the Mobilesat system design parameters, the reader is introduced to the trade-offs associated with selecting the optimum modulation scheme for a mobile satellite system given a specified network capacity requirement.

3.5.1 Power and Bandwidth Limits

Consider a mobile satellite network (such as Mobilesat) in which service will be based on the use of two satellites. Each satellite is assumed to have a single 14 MHz L-band transponder, providing 48 dBW of usable EIRP into a single beam covering the whole of the service area. Traffic studies estimate that to service the expected mobile terminal population will require a total of 1000-1500 voice communication channels. These are to be split equally between the two satellites.

Although each transponder will have a 14 MHz bandwidth, due to the close proximity of the two spacecraft, this spectrum must be shared between the two satellites. Not all channel frequencies may be usable, due to frequency co-ordination agreements with other operators.and for other technical reasons, so the usable spectrum may only be 7.5 MHz.

Exercise 3.1

Consider an outbound mobile satellite voice channel from a base station via the satellite to a mobile terminal. Determine the value of carrier to noise density ratio (C/N_0) expected at the mobile terminal receiver if the following parameters are assumed. (Note that these may not necessarily reflect the exact values for the Mobilesat or any other particular system, but are illustrative of typical system parameters).

Ku-band uplink:
The frequency is 14020 MHz, base-to-satellite-range is 37184 km, base EIRP is 49.5 dBW (clear sky), losses (atmospheric feeders) are 0.6 dB, satellite G/T equals -1 dB/K, and the intermodulation noise is negligible.

L-band downlink:
The frequency is 1550 MHz, satellite-to-terminal range is 38512 km, satellite EIRP is 24.5 dBW (single channel), Losses (feeders etc) are 24.5 dBW (single channel), mobile receiver G/T equals -13.0 dB/K, and the intermodulation noise is negligible.

Solution

The following link budget table summarizes the calculations required. Refer to Chapter 1 for details.

	Uplink	*Downlink*
EIRP	*49.5*	*24.5dBW*
Path loss	*206.8*	*188.0 dB*
Other losses	*0.6*	*0.4 dB*
Receiver G/T	*-1.0*	*-13.0 dB/K*
Boltzmann's constant	*-228.6*	*-228.6 dBW/K/Hz*
C/N_0	*69.7*	*51.7 dBHz*

Then using the complete cascaded link from Equation (1.69), we obtain the total carrier to noise power density ratio at the mobile receiver input as

$$C/N_0 = [(10^{6.97})^{-1} + (10^{5.17})^{-1}]^{-1}$$
$$= 1.46 \times 10^5 \quad (51.6 \ dBHz)$$

Let us now return to our two-satellite mobile satellite system. Using an approach similar to that illustrated in Exercise 3.1, it is possible to compute values for the required maximum satellite EIRP per carrier and the bandwidth available per carrier. The results for both 1,000 and 1,500 channel network capacities are summarised in Table 3.1. The required link C/N_0 values are also calculated for an assumed mobile terminal operating with an antenna of gain 8 dBi (decibels compared to an isotropic source).

No of channels	EIRP per carrier	C/N_0 with 8dBi mobile antenna	Channel spacing kHz
1000	25 dBW	47 dBHz	7.5 kHz
1500	23 dBW	45 dBHz	5.0 kHz

Table 3.1: Maximum available EIRP per carrier and channel spacings to support 1000 and 1500 voice activated channels.

The EIRP values and associated channel numbers in Table 3.1 assume a typical *voice activation factor* of 40%. Voice activation is based on the concept that in a normal telephone conversation it is very rare for both parties to be talking simultaneously. Therefore, on average at least half the time a one-way communication channel is not being actively used. In practice it is a slightly greater proportion than this, as there are always a number of pauses in a normal conversation. To reduce the total satellite EIRP required to support a given number of communication channels, voice activation can be introduced. That is, the transmitted carrier is turned off if there is no speech present on the channel. This voice activation factor is typically taken into account on power limited satellite systems which are used predominantly for voice traffic. Use of voice activation provides a "magical" two and half times increase in network capacity. The concept of voice activation and its resultant impact on network capacities is discussed more fully in Chapter 4.

3.5.2 Analogue vs Digital Modulation

In system design, a key issue is to choose between the use of analogue modulation (such as FM, SSB etc) or digital modulation (such as QPSK or 8PSK). For mobile satellite communications, near toll quality speech signals must often be carried in 5kHz channels and must survive the effects of signal fading due to shadowing and multipath.

The use of Amplitude Companded Single Side Band (ACSSB) was first proposed in Canada in the early 1980's as a suitable modulation scheme for mobile satellite services. It was expected to offer high quality voice service with a robustness to channel fading. Furthermore, to transmit a telephone channel band from 300-3300 Hz, ACSSB requires a signal bandwidth of less than 5 kHz. Experimental measurements on typical mobile satellite channels indicated that the receiver C/N_0 ratio required for high quality speech was at least 50 dBHz with unfaded links. This figure provided some margin for fading effects. Increasing the amount of companding of the ACSSB signal was found to reduce the demodulated background noise, but led to severe signal distortion.

The larger the required C/N_0 ratio, the greater is the value of the required mobile receiver G/T ratio. This will have a major impact on the cost and size of the mobile terminal antenna.

A low bit rate digital voice service using digital modulation might enable the required C/N_0 ratio to be significantly reduced. Digital voice codecs are not so subject to link noise effects providing sufficient C/N_0 is available on each link. In addition, powerful FEC coding techniques are then available to reduce the required carrier power. To illustrate this, consider the following Exercise.

Exercise 3.2

Consider a mobile satellite outbound link. It consists of a Ku-band uplink station to the satellite and an L-band downlink from the satellite to a mobile terminal. Assume that the uplink carrier to noise spectral density ratio at the satellite input is considerably greater than that on the downlink. Then, as shown in Chapter 1, the C/N_0 ratio at the mobile receiver input will be determined by the downlink values only. Calculate the value of C/N_0 required at the mobile receiver input if the following system parameters are assumed:

Modulation Type	=	*QPSK*
Demodulator Type	=	*Coherent with 1dB implementation loss*
Receiver noise bandwidth	=	*B=4kHz*
Maximum BER	=	10^{-2}
Fade margin	=	*5dB*

Solution

From Chapter 1, Figure 1.22 we obtain that for a value of $BER \leq 10^{-2}$, then

$$E_b/N_0 \geq 4.4 \ dB$$

Hence to provide a fade margin of 5 dB, we require that

$$E_b/N_0 = 9.4 \ dB$$

From Equations (1.63) and (1.65) it follows that for QPSK (M=4) we obtain

$$C/N_0 = 48.4 \ dBHz$$

Note that if FEC coding were used, the required C/N_0 would be reduced by an amount equal to the coding gain at $BER = 10^{-2}$.

During the development of mobile satellite systems in the late 1980's, there were two major problems that had to be overcome to allow the introduction of a digital voice service.

1. The development of low bit rate voice codecs!algorithms operating at around 4.8 kbit/s.

2. The selection of a modulation scheme and the development of modem techniques that allowed coherent demodulation of multiphase modulation schemes with shadowed land mobile satellite channels.

In 1990, over ten different 4.8 kbit/s voice codecs had been developed in a number of organizations, each codec having potential to meet the needs of the proposed Inmarsat M and Mobilesat systems. These were submitted for comparative evaluation to the Telecom Australia Research Laboratories. The overall voice quality of each codec was subjectively evaluated under a range of simulated mobile satellite channel conditions [1]. Finally, a codec utilizing the Improved Multiband Excitation (IMBE) algorithm was selected for both the new satellite services. The voice codec operates at 6.4 kbit/s (including FEC). The IMBE algorithm is described in Chapter 6.

Given the voice codec scheme to be used, the next design goal is to determine the most suitable modulation scheme. The modulation type must be such that it would provide acceptable BER values whilst operating with a low C/N_0 ratio. It also had to enable the use of channel spacings as low as 5.0 kHz.

As discussed in Chapter 1, QPSK modulation schemes are attractive digital modulation schemes for satellite communications due to their relatively high bandwidth and power efficiency. They have been used extensively in fixed satellite networks. The particular questions that remain about their application to the mobile satellite field concern the performance degradations likely to arise from the tough propagation conditions. It is also necessary to consider whether QPSK or one of its variants would be the best choice. That is, one is faced with the selection of a modulation scheme from QPSK, Offset QPSK (OQPSK), or $\pi/4$ QPSK. Trellis Coded Modulation (TCM) could also be considered as a suitable modulation scheme. TCM would provide the same spectral efficiency as QPSK schemes, but might possibly reduce the required C/N_0 ratio. However the use of trellis coded 8PSK modulation (a popular TCM scheme) raises issues related to the greater sensitivity of 8PSK to noise and phase jitter (since QPSK signals are spaced only half the angular distance apart compared with their QPSK counterparts).

The performance of different modulation types will be influenced by the type of demodulators to be used. To minimise system implementation losses, the use of coherent rather than differential detection would be desirable. Furthermore it is necessary to minimise any carrier or frame synchronisation overheads to reduce the overall channel bit rate, in order to allow for 5 kHz channel spacing. This in turn would impact on the design of the demodulator synchronization circuits.

Terminals mounted in vehicles driving at 100 km/hr are subject to Doppler frequency offsets. At L-band, these are expected to be limited to values of less than 150 Hz. Mobile terminals employing coherent demodulation will use the received signal to correct for frequency offsets arising within the terminal. On the inbound (L to Ku-band) link, total frequency offsets may be expected to reach up to ± 400 Hz. Hence the selected modulation scheme must be such as to be unaffected by adjacent carriers offset by 4.2 kHz from the wanted signal, (in order to allow 5.0 kHz channel spacing).

Finally, the selected modulation scheme also needs to be easily realisable and tolerant to channel nonlinearities. The outbound (Ku to L-band) link will operate over a relatively linear channel since the Base Station HPA can be designed to be relatively linear and the satellite L-band transponder can be expected to have good linearity. The inbound link will be subject to nonlinearities caused by the mobile terminal HPA's. It is desirable to use nonlinear HPA's in the mobile terminals so as to increase the power efficiency of the amplifiers.

However, as discussed in Chapter 1, non-constant envelope modulated signals are subject to spectral regrowth on passing through nonlinear amplifiers. This can result in interference between two or more carriers which are assigned adjacent frequencies. In that case, the modulated signals may spread into overlapping frequency bands. This form of interference is termed *adjacent channel interference* (ACI). This contrasts with interference that might be caused by a transmitter on the same carrier frequency. That is called *co-channel interference* (CCI).

3.5.3 QPSK and OQPSK Modulation

In theory OQPSK and QPSK should provide the same BER performance when operated on linear satellite channel with the same link parameters. However, this is not true when consideration is taken of system nonlinearities and implementation issues. The performance of QPSK and OQPSK modulation schemes for transmitting data over satellite channels has been extensively analysed in the literature. In particular researchers have analysed the BER performance of both schemes, studying the effects of carrier phase noise and transmitter non-linearities. [2-5].

Reference [6] describes the results of extensive simulation studies of the performance of OQPSK versus QPSK in terms of BER performance with varying Nyquist filter bandwidth constraints, with joint carrier phase noise and non-linearities. The simulation results were aimed at determining the optimum modulation scheme with Nyquist filter bandwidths between 0.4 and 0.6 bit rate. These narrow filter bandwidths are required to meet channel spacing requirements.

Figure 3.8 illustrates the BER degradation with variable roll-off factors for a nonlinear channel and carrier phase jitter.

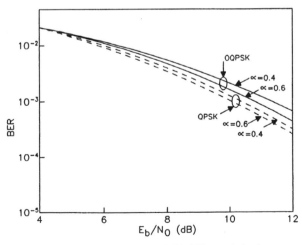

Figure 3.8 Error rates for QPSK and OQPSK modulation on a channel via Intelsat V with zero backoff.

Under tight Nyquist filtering constraints (40%), OQPSK performs up to 2.0 dB worse than QPSK, on nonlinear channels at a BER 10-4. Even under linear channel conditions OQPSK has a worse BER performance than QPSK with tight Nyquist filter bandwidths ($BT \leq 0.5$). The performance of QPSK and OQPSK with differing carrier phase noise levels, under linear conditions, is analysed in reference [7].

The difference in BER performance between QPSK and OQPSK, for different filter bandwidths,can be simply explained. With OQPSK the I and Q channel signals are staggered by a half symbol so that only one channel experiences a bit transition at a time. This results in minimal crosstalk between the channels in the demodulator, as at the time of sampling on one channel, the other channel will be passing through zero if a bit transition has occurred. Due to nonlinearities and recovered carrier phase errors the crosstalk significantly increases, resulting in the BER performance degradation. This occurs for two reasons. Firstly, a nonlinearity introduces amplitude errors on one channel as it is being sampled due to bit transitions on the other channel. Secondly, as the I and Q channels are offset by half a symbol, the Nyquist filtering inter symbol interference (ISI) criteria are no longer valid. In other words, the correct sampling points for bit decisions on one channel do not occur at the ISI nulls of the other channel. Therefore any crosstalk will result in interchannel ISI degradation. This is why the OQPSK BER performance worsens with reducing Nyquist filter bandwidths.

With QPSK modulation, the demodulator I and Q channel demodulation outputs are simultaneously sampled and therefore there are no crosstalk degradations due to bit transitions at the time of sampling with nonlinearities or interchannel ISI effects due to tight Nyquist filtering.

QPSK is the optimum modulation scheme for narrowband ($BT_s < 0.6$) signal transmission over linear and nonlinear channels. The only advantage OQPSK has over QPSK is in terms of reduced spectrum regrowth following transmission through a nonlinear transmitter.

3.5.4 $\pi/4$ QPSK Modulation

$\pi/4$ QPSK modulation is the modulation scheme favoured for some mobile satellite services, such as the North American Digital cellular (EIA) standard. As described in Chapter 1, $\pi/4$ QPSK is simply a QPSK modulation scheme with the addition of a 45^0 phase rotation for every alternate symbol. This ensures that there are no 180^0 phase changes, eliminating zero signal level crossings and reducing the peak signal amplitude fluctuations. For example, with 50% Nyquist filtering $\pi/4$ QPSK reduces the peak to average power ratio to 3.2 dB, from 4.0 dB for QPSK.

On analysis, $\pi/4$ QPSK modulation with carrier phase noise and tight Nyquist filtering offers the same BER performance advantages as QPSK, while still offering the possibility of improving the efficiency of the transmit amplifiers in the mobile terminals. That is, it permits the possibility of them using nonlinear amplifiers.

As discussed in Section 3.5, it is desirable in mobile terminals to use nonlinear (Class C) amplifiers as they offer the highest HPA efficiencies. The resultant spectral regrowth associated with using a Class C HPA is illustrated in Figure 3.9 for both $\pi/4$ QPSK and OQPSK modulation scheme with 40% Nyquist filtering.

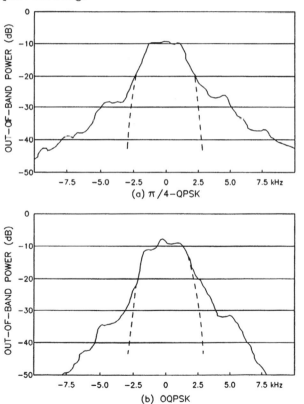

Figure 3.9: Spectra of $\pi/4$ QSPSK and OQPSK after nonlinear amplification.

Clearly, OQPSK offers significantly less spectral regrowth than $\pi/4$ QPSK due to its constant signal envelope characteristics. Unfortunately, even

OQPSK modulation will result in significant adjacent channel interference (ACI). This is such that it would not allow 5.0 kHz channel spacing and would only marginally allow 7.5 kHz channel spacing, after allowance is made for Doppler frequency offsets.

Therefore it is necessary for the mobile terminal to use quasi-linear HPAs, which may be either linearised Class C amplifiers or Class A/B amplifiers. This is necessary to reduce the ACI to acceptable limits. The use of quasi-linear amplifiers is currently being extensively studied for digital cellular applications.

Quasilinear amplifiers offer considerably less spectral regrowth than non-linear Class C amplifiers. For example, Reference [8] compares the relative performance of QPSK and $\pi/4$ QPSK modulation schemes with nonlinear Class B amplifiers. A summary of the resultant ACI levels with different modulation schemes using Class B and C amplifiers is given in Table 3.2.

Frequency offset	Class C and OQPSK	Class B and QPSK	Class B and $\Pi/4$ QPSK
3.0 kHz	15 dBc	27 dBc	29 dBc
5.0 kHz	25 dBc	32 dBc	38 dBc

Table 3.2:   Adjacent channel interference (ACI) levels with different non-linear amplifiers and 6.6 kbps modulation schemes.

A linear Class A amplifier operates with about 20 to 25% power efficiency, whilst a Class B amplifier would offer approximately 35 to 40% power efficiency and a Class C amplifier 45 to 50%. It is clear that the use of a Class B HPA would result in a significant decrease in the terminal power consumption.

In conclusion, simulation results indicates that either QPSK or $\pi/4$ QPSK would a superior BER performance over OQPSK with carrier phase noise and narrow filter bandwidths. However, $\pi/4$ QPSK modulation has an additional advantage as it allows the use of quasi linear Class A/B amplifiers which enables the efficiency of the transmit HPA in the mobile terminal to be improved, with no degradation in the system BER performance.

3.5.5   Trellis Coded Modulation

Trellis coded modulation (TCM) schemes also have some features which make them of interest for mobile satellite services. For example, by combining a trellis code and 8PSK modulation (for details, see Chapter 7), it is possible to achieve a spectral efficiency equal to that of uncoded QPSK, whilst at the same time have the added benefit of useful coding gain. The following outlines some studies undertaken to compare the performance of TCM and QPSK systems for a land mobile satellite channel.

It is advantageous to use satellite link simulation stored propagation!data, to evaluate the relative performance of uncoded QPSK and Trellis Coded Modulation (TCM) schemes with ideal coherent demodulation. Stored propagation data could be in the form of I and Q values, and therefore the effect of combined signal amplitude and phase variations can be evaluated. QPSK modulation was selected for the simulation experiments as the software module had already been developed. All the simulation results were obtained for linear channels.

The TCM scheme that was implemented was a 16-state TCM 8PSK scheme, studied with and without interleaving. A maximum likelihood Viterbi decoder was used after the demodulator to estimate the transmitted sequence. Grey mapping was selected as it offers 0.4 dB improvement in BER relative to Ungerboeck codes. Only limited interleaving was optionally used to randomise burst errors, as the maximum allowed interleaving delay was 60ms.

From the fade level pdf's, two fading sequences were selected representing average and poor channel conditions, respectively. Simulation results were then obtained and plotted in terms of both bit error rate (BER) and block error rate (BLER).

Generally, BER is used to compare the performance of different modulation schemes, but, in this case it is actually BLER which is more relevant to the operation of a digital voice codec. A 30ms block size was considered as a typical voice codecs!frame size used in Linear Predictive codecs (LPC). A block error was defined as any block with one or more bit errors occurring within a single block.

It is interesting to compare the BER results obtained for QPSK and TCM modulation in Figure 3.10 for the average channel, with the BLER results given in Figure 3.11.

Clearly in terms of BER, QPSK actually outperforms the TCM scheme at all E_b/N_0 values. However, when the BLER is plotted, TCM schemes are shown to offer a significant improvement in performance at low E_b/N_0 values. Similar results are obtained for poor channel conditions where the road is subject to severe shadowing. As the BER ans BLER are so high, the realtive difference between the two modulation schemes is smaller.

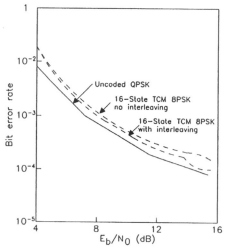

Figure 3.10: Bit error rates on a mobile satellite channel with typical shadowing from trees in a normal environment.

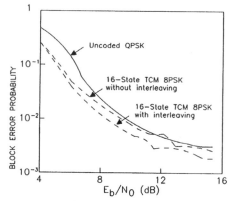

Figure 3.11: Block error rates on mobile channel with shadowing.

The simulation results support the following conclusions:

1. For typical land mobile channels, the uncoded QPSK modulation provides better BER performance than the TCM. The TCM can slightly improve the BLER performance at E_b/N_0 values below 7 dB. These diffrences arise as the coding can correct random errors and improve the BLER, but, when the channel is severly shadowed the TCM decoder will introduce burst errors causing more bit errors than with QPSK.

2. TCM modulation with limited interleaving offers only a small BER improvement relative to no interleaving. (Other simulation work indicated that interleaving can actually worsen the BER performance under severe shadowing.)

3. As SNR increases, the BER performance exhibits asymptotic behavior. Increasing the E_b/N_0 above 10 dB offers only a small improvement in either the BER or BLER.

4. Most importantly, the simulation results indicated that there is no severe phase jitter. Therefore coherent demodulation of both QPSK and TCM modulation schemes even under severe channel fading was feasible.

Although TCM theoretically offers a small reduction in BLER at low E_b/N_0 values, there are considerable concerns about realising such small improvements with real hardware. Therefore, $\pi/4$ QPSK modulation, using coherent demodulation, was selected as the optimum modulation technique for the Mobilesat system.

3.6 MOBILE TERMINAL PRICE

The mobile satellite service and terminal prices will naturally be compared to terrestrial cellular tariffs and equipment costs. The mobile satellite user will be prepared to pay a premium for the wide area coverage and extra service features, as discussed in Section 3.2.1. However, to ensure the widespread market acceptance of satellite services, it is essential that the mobile terminal purchase price is comparable to the cellular units. In this section a brief review of the existing mobile terminal architectures and their major component costs are provided, along with a description as to how new satellite services will allow the introduction of dramatically reduced terminal prices.

Mobile terminals generally comprise two distinct units as follows:

1. An Outdoor Unit (ODU) which includes a high/medium gain antenna, diplexer, Low Noise Amplifier (LNA) and High Power Amplifier (HPA);

2. An Indoor Unit (IDU) which contains the baseband functions and transmit/receive frequency conversion units. The IDU are generally designed to operate with different types ODUs.

Initially these two units were physically separated with for example the ODU mounted on a ship's superstructure whilst the IDU would be located in the radio room. The recent introduction of "suitcase" Inmarsat A satellite terminals has seen the physical integration of these two units. However, for convenience it is worth considering the technical requirements for these functional units separately.

3.6.1   Outdoor Unit (ODU) Requirements

With the current generation of mobile terminals the antenna, diplexer and High Power Amplifier (HPA) are the most expensive single items within the terminal. The introduction of new digital signal processing based on low bit rate voice codecs and digital modem technology will allow the mobile terminal G/T and EIRP parameters to be significantly reduced. For example, Table 3.3 provides a summary of the key radio parameters associated with different mobile terminals.

	Service date	EIRP (dBW)	G/T (dB/K)	Nominal antenna gain (dBi)	Threshold C/N_0 (dBHz)
Inmarsat A (global beam)	1980	36	-4	21–23	52
Inmarsat M (global beam)	1992	25	-12	12–13	42
Inmarsat M spot beam)	1995	19	-12	12–13	42
Mobilesat (spot beam)	1993	10	-17	7–8	42
Future hand-held Terminal (spot beam)	~ 2000	-1	-22	2–3	39–41

Table 3.3: Mobile Terminal performance characteristics.

Comparing the Inmarsat A to Inmarsat M specifications illustrates how reducing the antenna gain and HPA size will allow a large reduction in the terminal price.

However, the Inmarsat M terminal will still require a directional antenna when operating into a global hemispherical satellite antenna pattern. The introduction of spacecraft with high gain spot beams such as the Inmarsat 3 satellites and the various national satellite systems will allow the introduction of antennas which will not need to be steered in azimuth, and the use of lower power HPAs as shown in Table 3.3. The introduction of these new satellites will allow a dramatic reduction in the size and cost of the ODU.

3.6.2 Indoor Unit (IDU) Requirements

The current generation of mobile terminals are implemented with relatively bulky and expensive IDUs. The new generation of terminals will be considerably smaller and cheaper for the following reasons:

1. The use of digital signal processing techniques (based on low bit rate voice codecs) allowing the greater use digital technology;

2. The lowering of the ODU power consumption due to a reduction in the HPA size, which will allow a much smaller and reduced capacity power supply unit;

3. The new mobile terminals will be manufactured in larger numbers allowing greater allowing increased investments during the design and manufacturing processes so allowing reductions in the terminal manufacturing costs.

Finally, mobile satellite systems are becoming further integrated with terrestrial radio technologies as new frequency bands are opened up at L-band for terrestrial applications and digital mobile transmission schemes are introduced. In the future, terrestrial cellular radios and cordless telephones will be integrated into the mobile satellite terminal allowing for further reuse of the terrestrial radio technology. In addition to the technical advantages of further integration, there will be commercial advantages as the terrestrial radio distributors become involved in the retailing of mobile satellite terminals.

3.7 PROBLEMS

1. Consider a signalling scheme such as that illustrated in Figure 3.4. Calculate the number of repeated transmissions of a Signal Unit (SU) required to ensure a 99% probability of a call being established correctly (that is, the SU being received without errors), given that the

probability of an individual SU being received error free is 0.5. Assume independence of error events between repeated SU's. Repeat for the cases where the probability is 0.8 and 0.9, respectively.

2. Consider a mobile satellite system for speech, which uses $\pi/4$ QPSK modulation with mobile receiver operating point $C/N_0 = 49$ dBHz. The system provides a maximum of 400 speech channels. Calculate the number of channels that could be supported on the system, if, instead of $\pi/4$ QPSK, frequency modulation (FM) was used which required a receiver operating point of $C/N_0 = 55$ dBHz. Assume that in other respects, the mobile terminal and satellite specifications remain unchanged from those described in this Chapter.

3. Determine the threshold operating points C/N_0 for a voice codec which can operate at BER values equal to $1 \times 10^{-2}, 4 \times 10^{-2}$, and 8×10^{-2}, respectively. For each case, consider the use of QPSK modulation. Consider first that the channel can be modelled by an AWGN model. Then consider the case for a fading channel represented by the worst channel from those represented in Figure 3.5. Assume a 1% blocking probability is required.

4. Calculate the HPA power requirements for each of the mobile terminals defined in Table 3.3 assuming that the diplexer and antenna feed losses are no more than 1.5 dB.

REFERENCES

1. Bundrock, T., and Wilkinson, M., "Evaluation of Voice Codecs for the Australian Mobile Satellite System", *Proc. International Mobile Satellite Conference*, Ottawa, Canada, 1990.

2. Jeruchim, M., "Techniques for Estimating the Bit Error Rate in the Simulation of Digital Communication Systems", *IEEE Journal on Select. Areas in Comms.*, Vol. SAC-2, pp. 153-170, January 1984.

3. Benedetto, S., Bigilieri, E. and Daffara, R., "Performance Prediction for Digital Satellite Links - Volterra Series Approach", *Fourth International Conference on Digital Satellite Communications*, Montreal, Canada, October 1978.

4. Shimbo, O., "Effects of intermodulation, AM-PM conversion and additive noise in multicarrier TWT systems", *Proc IEEE* , Vol. 59, pp. 230-238, February 1971,

5. Hetrakul, P. and Taylor, D.P., "Nonlinear Quadrature Model for Travelling wave Tube Type Amplifiers", *Electronic Letters*, Vol. 11, January 1975.

6. Vucetic, B., "Effects of Phase Noise, Nonlinearities, and bandlimiting on QPSK and OQPSK modulation Schemes", *Journal of Elec. and Electronic Eng. of Aust.*, December 1990.

7. Fan, F. and Li, L.M., "Effect of Noisy Phase Reference on Coherent Detection of Band Limited Offset-QPSK Signals", *IEEE Trans. on Comms.*, Vol. 38, pp. 156-159, February 1990.

8. Ariyavisitakul, S. and Liu, T.P., "Characterising the Effects of Nonlinear Amplifiers on Linear Modulation for Digital Portable Radio Communications", *IEEE Trans. on Vehicular Technology*, Vol. 39, pp. 383-389, November 1990.

9. Akaiwa, Y. and Nagata, Y., "Highly Efficient Digital Mobile Communications with a Linear Modulation Method", *IEEE Journal on Selected Areas in Comms.*, Vol. SAC-5 pp. 890-895, June 1987.

Chapter 4

TRAFFIC CAPACITY AND ACCESS CONTROL

by Les Berry† and Sanjay Bose‡
† Royal Melbourne Institute of Technology
‡ Indian Institute of Technology, Kanpar.

Recent advances in technology have prompted increased interest in satellite communication systems using small earth stations either fixed directly at the user premises, as in very small aperture terminal (VSAT) systems, or mobile as in mobile satellite (MSAT) systems. Various countries and consortiums are in the process of developing such systems to provide a variety of communication services. Apart from providing telex-type low-rate data services, these MSAT and VSAT systems are now planned to support voice calls as well as higher-rate data communication applications.

In order for them to be economically viable, satellite communication systems should make efficient use of the satellite's limited resources of bandwidth and power. This is especially important in MSAT and VSAT systems where a large number of essentially uncoordinated and statistically bursty users are expected to share these resources in a mutually cooperative and efficient manner. Depending on system requirements and the nature of the applications supported by the system, it is important to choose the appropriate *multiple access* technique in the system design. Multiple access is a variant of multiplexing which is specific to satellite communications and describes the method to be used for sharing communication resources between a large number of users.

In this Chapter, we will first review the type of services required from satellite communications systems, with particular emphasis given to mobile satellite systems. Alternative multiple access protocols are then described. We examine methods for analysing the relative performance of some of these protocols. Details of a typical access protocol providing integrated voice and data services in a mobile satellite system are then presented.

4.1 SERVICE CHARACTERISTICS

The fundamental task of a typical satellite communication network is the requirement that a large number of earth stations are able to simultaneously interconnect their respective voice, data, teletype, facsimile, and television services through a satellite. As we have seen in Chapter 3, in system design it is important to first establish the network services required and the nature of the circuits needed to carry these services.

The types of information carried by a mobile satellite system, for example, may range from digital voice, of reasonable quality, to data at various speeds (depending on the requirements of the supported applications) or slow scan video for facsimile transfers. Teleconferencing applications may also be supported even though full motion video may be difficult to provide until a large amount of cheap bandwidth becomes available. A typical MSAT system [2] is expected to be able to provide at least three types of services to its users, namely

1. Continuous Voice Service

2. Continuous Data Service

3. Packet Data Service

The continuous services for voice or data are similar in the sense that both of them are based on *circuit switching*. A circuit switched connection is one for which a specific transmission path or circuit is established for dedicated use during the duration of each call, that is, from the time a call is set up to the time it is terminated.

The difference between voice and data calls is that a mobile terminal participating in a voice call will use a voice activity switch (to provide a time assigned speech interpolation (TASI) advantage in terms of the power-limited operation of the satellite), whereas no such activity switch is used for the data call (a circuit assigned to a data call is assumed to be always active). The voice activity switch turns off the transmitted signal from the mobile terminal during idle periods of voice. This reduces the power required by the satellite repeater to the level required by the active voice sources.

The packet data service is expected to carry low-rate, low-volume data for applications like messaging, paging or telex-type services. Closed user groups may also use the transmission capacity of these services to implement their own private data networks with their own base stations.

Systems required to carry data or voice, often have somewhat conflicting requirements. Data can be sent with random (and possibly large) delays between packets but must be sent in an error-free fashion. Systems carrying data must have built-in automatic repeat request (ARQ) techniques for the recovery (through retransmissions) of packets in error. In addition, buffering and sequencing may also be required to ensure that a receiver can reconstruct the transmitted source data stream in the correct order. On the other hand, because of its inherent redundancy, voice traffic can tolerate some errors without a significant loss in quality but circuit delay requirements are more stringent. The delay between successive voice packets must be below a certain system-specified limit and cannot be allowed to become arbitrarily large. (The need for delivery in proper sequence would still be there but one may not need to correct errors through ARQ.)

4.2 MULTIPLE ACCESS SCHEMES

The problem of multiple access is essentially one of allowing a number of uncoordinated users to share a common transmission resource. The multiple access protocol is fundamental to the performance of a satellite communications system since it affects all elements of the system, determines the system capacity and flexibilty, and has a major influence on costs. Numerous multiple access algorithms have been described in the literature. It is convenient to divide these into three basic types:

- Fixed Assignment,

- Random Access and

- Controlled Access techniques.

Fixed Assignment schemes, such as those using frequency division multiple access (FDMA), time division multiple access (TDMA), or code division

multiple access (CDMA), permanently assign a fraction of the system's resources to each user. (These schemes will be described in Section 4.2.1).
Such schemes are best suited to routes carrying large quantities of steady
traffic.

For an MSAT or VSAT system with a large population of bursty short
traffic sources, a permanent assignment scheme might be extremely inefficient. *Random Access* techniques like Pure or Slotted Aloha, described in
Section 4.2.2, may be better suited for such systems. This is despite the fact
that they have to be operated at low efficiencies in order to avoid problems
of instability.

For systems where the information generated by an active user tends to
be long (for example, in voice calls or long data transactions), *Controlled Access* techniques, described in Section 4.2.3, may be a better alternative. In
these schemes, a fraction of the system's resources (bandwidth or time) is set
aside to carry requests for (resource) assignment. Successful requests get an
appropriate assignment which can then be used for the actual message transmission. Circuits may be assigned in a random access mode. The division
of the system's resources may be done either over time, as in Reservation
Aloha, or over frequency. For telephone speech traffic on satellite mobile
systems and for other fixed thin route networks, a demand assignment multiple access (DAMA) scheme is often preferred in which a circuit-switched
FDMA, TDMA or CDMA channel is assigned.to the user only for the period
of the call for which it is needed.

Providing multiple access in a communication system essentially implies
the specification of a set of rules to be followed by each member of the user
population in order to cooperatively share the system's common transmission
resource (typically, bandwidth or time). In the following Sections we briefly
review some common fixed assignment and random access methods and then
describe in rather more detail a preferred controlled access protocol.

4.2.1   Fixed Assignment Schemes

For systems which generate traffic in a fairly regular fashion it would be
possible to permanently assign a part of the transmission resources to each
user. This might be done in a manner proportional to the requirements of
each station.The actual entity being shared could be either bandwidth or
time, giving rise to Frequency or Time-Division Multiple Access schemes,
respectively. Actually, one can also share the given transmission resource in
terms of a third variable, namely a set of mutually orthogonal codes. This

gives rise to Code Division Multiple Access schemes. We will briefly look at each of these schemes.

Frequency-Division Multiple Access (FDMA)

Consider a satellite system with a total satellite transponder bandwidth of B Hz serving N nodes. If the nodes generate equal amounts of traffic, a possible way of providing multiple access would be to divide the bandwidth into bands of B/N Hz and permanently assign one such band to each node for its transmissions. (If nodes generate unequal amounts of traffic, one can modify this to assign bandwidth in proportion to the traffic generated by each node.) This method of sharing is referred to as Frequency-Division Multiple Access and is illustrated in Figure. 4.1. The band assigned to a node is available to it for transmission over all time, or at least, for extended periods.

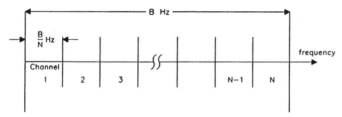

Figure 4.1: Frequency-division multiple-access.

Historically, FDMA was the first method of multiple access used for satellite communication systems. Examples include single-channel-per-carrier (SCPC) systems [24], [18], one of the best known being the Intelsat single channel per carrier pulse code modulation multiple-access demand-assigned equipment (SPADE) system [25]. SCPC systems are most suitable for use where each terminal is required to provide only a small number of telephone circuits. (This is referred to as a thin-route service).

For application in an Integrated Services Digital Network (ISDN), satellite circuits may be required to carry relatively large quantities of telephone and other traffic. In such systems, each telephone signal is first digitally encoded into a 64 kbit/s binary bit stream. Then several such bit streams representing multiple telephone signals are multiplexed in a time division mode to form higher rate bit streams. Common (approximate) rates used for higher order streams include 34 Mbit/s, 45 Mbit/s, and 140 Mbit/s. When these higher order bit streams are to be transmitted over a satellite communications link, each bit stream is usually modulated on a separate microwave carrier using QPSK or 8PSK modulation. Then one or more of these modulated signals are transmitted in an FDMA mode through the satellite transponder. For example, for transmission of a number of 45 Mbit/s signals (each equivalent to 672 telephone channels), it is common to multiplex pairs of such modulated signals side by side in the frequency band associated with single satellite transponders.

In some circumstances, FDMA schemes may be comparatively inefficient since they require guard bands between neighbouring bands to prevent the transmission in one band from interfering with those in neighbouring bands. These guard bands cannot be used for information transfer and represent a waste of the system resources. Given that the system requires guard bands of Δ Hz around each user frequency band, the bandwidth available to each station for information transfer is

$$B_T = \left[\frac{B - (N+1)\Delta}{N} \right] \tag{4.1}$$

Another problem with FDMA systems is that they require the satellite transponder to be linear in nature. The high power amplifier (HPA) on board the satellite is usually constructed from a travelling wave tube (TWT) device. An HPA works most efficiently when it is operated close to saturation but then, as described in Chapter 1, nonlinearities occur. Unfortunately, the linearity requirements of an FDMA system implies that the TWT has to be backed off substantially in order to operate it as a linear amplifier. This in turn, leads to inefficient usage of the available satellite power.

For this reason, fixed assignment strategies have tended to increasingly shift from FDMA to TDMA as certain technical problems associated with the latter were overcome.

Time-Division Multiple Access (TDMA)

Consider a system where the available system bandwidth is B Hz as before. Consider the total transmission time to be divided into frames of T seconds each. Let each frame be further subdivided into N slots of duration T/N seconds where each slot is permanently assigned to one of the N users for its transmissions.

Unlike the FDMA scheme where a user will continuously transmit in its assigned frequency channel, in a TDMA scheme each user will transmit over the full system bandwidth of B Hz but only during its own time-slot in each frame. This is illustrated in Figure 4.2.

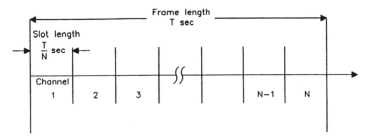

Figure 4.2: Time-division multiple-access.

Corresponding to guard bands in the FDMA scheme, TDMA will require guard times to separate one user's transmission from that of the next one in the frame. Since the transmission from each user is essentially done in the burst mode, additional preambles may also be required to enable carrier synchronization (for coherent demodulation) and clock synchronization (for bit timing in digital transmissions). Synchronization issues will be discussed in more detail in Chapter 5. These synchronization procedures are essential to allow the receiver to recover the information contained in the transmitted burst.

A TDMA system also requires overall system timing so that each user can unambiguously determine the frame boundaries and the position of its own slot within the frame. Any possibility of error in this determination will translate into a longer guard time requirement and will have a detrimental effect on the overall efficiency of the system. The receivers in a TDMA system must be capable of rapid burst synchronization, that is quickly acquiring carrier (and possibly, clock) at the beginning of the bursts they are intended to receive.

It is evident from the above that a TDMA system requires considerably more complex timing and synchronization than an equivalent FDMA system. Another disadvantage of TDMA is that the transmissions from each mobile station are in the form of bursts of T/N seconds with the station being idle for the remaining duration of the frame. This implies that the transmitters in a TDMA system would have a high peak power requirement even though their average transmitted power would be considerably lower. For VSAT or MSAT systems with small earth stations, this may be undesirable. Because of their hardware limitations, earth stations in such systems would prefer to transmit continuously at the average power level rather than in bursts of high power. In order to keep the power input to the satellite transponder within the required limit, the individual earth stations should use power control to limit their transmitted power.

In a TDMA system, at any given instant only one user will be in the transmitting mode. (In contrast to this, in an FDMA system, the time waveform of the total signal will be a combination of the instantaneous transmissions from each station.) This makes it easier to exercise power control in the system by requiring each user to adhere to a pre-specified transmission power limit. Because of this, the system may be operated such that the satellite transponder is close to its saturation limit. As discussed in Chapter 1, this ensures greater power efficiency in the system.

Code-Division Multiple Access

In FDMA and TDMA systems, different users are separated from each other in frequency and time, respectively. For digital transmission, it is also possible to have a number of users share the same bandwidth and transmission time, while keeping them separated by ensuring that they use different codes which are orthogonal to each other. This is referred to as Code-Division Multiple Access (CDMA).

For any given n, there exist special binary sequences, referred to as pseudo-random or pseudo-noise (PN) sequences. For large values of n, a number of such sequences will be available (for a particular choice of n). The sequences have the interesting property of being orthogonal to each other.

CDMA uses this property to separate the transmissions of different users even though they may overlap each other in frequency and/or time. Consider a situation where the choice of n is such that N such sequences are available where each sequence is of length M,

$$M = 2^n - 1 \qquad (4.2)$$

Each station is assigned a sequence for transmitting its message bits. The mode of transmission is to EX-OR (exclusive 'or') each data bit with the specified sequence and then transmit the resulting sequence. (Note that to support a data rate of D bits/sec, this would require transmissions to be at MD bits/sec. The bandwidth available to the system must be capable of transmissions at the latter rate.)

Knowing the sequence being used for transmission, the receivers can then extract the information being sent by essentially a process of correlation. The reader is referred to [9] for a more detailed description of the operation of the receivers and transmitters in such a system as well as for greater details on PN-sequences.

It is important to note that CDMA systems are inherently more inefficient than FDMA or TDMA schemes. This is primarily because the high effective transmission rate (compared to the actual bit rate) requires considerably more bandwidth for its operation. Acquiring and maintaining clock and carrier synchronization in such a system is also reasonably complex.

The main advantage of CDMA is that it inherently provides a processing gain of M and can be used in systems which must transmit at low power. It has indeed been successfully used for low data rate commercial VSAT operations as in [17].

In environments with noise, this processing gain may also be used to advantage. Moreover, the inherent anti-jam (AJ) and anti-interference (AI) properties of the PN-sequences provide a certain amount of security to data transmissions in a CDMA system when used in a military enviroment.

4.2.2 Random Access Schemes

Fixed assignment strategies tend to be inefficient when the user population consists of a large number of bursty users. Such low duty-factor users generate traffic in bursts with long inactive periods between bursts; hence, it will be inefficient to assign fixed transmission resources to each of them in a permanent fashion.

Random access schemes using contention based strategies for multiple access are more suitable for use in such an environment. In these schemes, a transmitter uses the entire transmission resource of the system only when it is needed. That is, there is no fixed (permanent) assignment of transmission capacity for it. Using statistical multiplexing techniques to share the transmission resource in this fashion allows a large number of bursty users to be supported by the system. In this section, we review some of these techniques which are suitable for use with satellite repeaters and hence can be used to support multiple access in VSAT and MSAT systems.

Two examples of random access schemes which may be used to provide communications to a large population of bursty users are the Aloha and Slotted Aloha schemes. Though they were initially proposed for terrestrial radio systems, they can be used in systems using a satellite repeater. Next, we describe these schemes and look at their throughput performance and stability behaviour. Note that for a system supporting communications between a large population of bursty users and a central station, the need to do statistical multiplexing only arises for transmissions from the users to the central station.

Pure Aloha System

The Pure Aloha system was the first example of a random multiple access scheme which used the concept of statistical multiplexing to provide data communications to a large population of bursty users. Such a system may be used to support communications between distributed users or between users and a common central station. In the following, we will examine these schemes for communications between users and a central station. Such schemes may be easily extended to support inter-user communications.

Consider the available bandwidth to be divided between a forward and a reverse channel where the forward channel carries data from the users to the central station and the reverse channel carries data in the opposite direction. Assume that the binary data to be transmitted from the users to the central station is packetized. Apart from the data contents, the packet will contain a synchronization header, the address of the source and a trailing CRC field for error detection.

The CRC field is generated by a Cyclic Redundancy Check (CRC) code. The transmitter uses an encoding algorithm to generate these bits depending on the contents of the packet. The receiver uses an appropriate error detection algorithm on the received bits of the packet and the received CRC bits to determine whether the packet has been received without error. (Note that packets from the central station must carry the destination address instead of the source address, but will otherwise be similarly formatted.) For ease of analysis and description, we assume constant length packets of duration T and a noise-free channel.

In the Aloha system, whenever a user has a packet to send, it sends it on the forward channel, that is from the user to the central station. If this packet reaches the central station without error, it is acknowledged and the user is now free to send its next packet, if it has any. On the other hand, if the packet is not acknowledged within a certain time-out period, it is repeated following a random rescheduling delay. This process is illustrated in Figure 4.3.

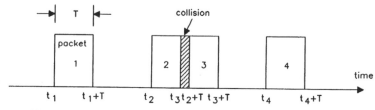

Figure 4.3: Forward channel of a pure Aloha system.

Here we have shown four packets labelled 1,2,3 and 4 from different users. Packets 1 and 4 are received correctly without errors. Packets 2 and 3 collide and neither of them will be received correctly. The central station will not be able to acknowledge them and hence they will both be repeated after a random rescheduling delay.

Retransmission of an unacknowledged packet is repeated following this procedure until it is finally acknowledged by the central station. In the following analysis the channel is assumed to be noise-less, thus the only source of error will be collisions between the transmissions of different users, when their packets overlap. The random rescheduling delay is needed to separate collisions in a probabilistic fashion. The time-out period should take into account the duration of the packet and its acknowledgement, the processing time and the round-trip propagation delay. If it is implemented in a satellite system, this round-trip delay will be a two-hop delay – one from the user to the central station and the other from the central station back to the user.

Assume that the arrival process of new packets is Poissonian (i.e. pure chance) with an average arrival rate λ per second. We also assume that the combined process of arrivals of new packets and retransmissions is also a Poisson distribution with an average arrival rate of λ_r. (Note that these arrival processes are really the net arrival processes from all the users in the system.)

Let P_S be the probability that a given packet transmission will be successful (that is it will not suffer from collisions or errors). The packet transmission process is illustrated in Figure 4.4.

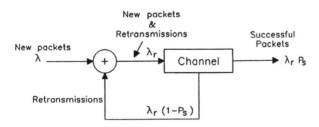

Figure 4.4: Operation of an Aloha system.

Since a given packet transmission starting at time t succeeds if there are no other packet arrivals in $(t-T, t+T)$, ignoring channel errors, the probability of success will be given by

$$P_S = e^{-2\lambda_r T} \qquad (4.3)$$

In equilibrium, the average arrival rate of new packets should be equal to the average rate of packets successfully carried by the system. This yields

$$\lambda = \lambda_r e^{-2\lambda_r T} \qquad (4.4)$$

Defining $S = \lambda T$, and $G = \lambda_r T$, Equation (3.4) is more typically written as

$$S = Ge^{-2G} \qquad (4.5)$$

where S represents the traffic carried by the system, and G is called the load offered to the system (in order for it to be able to carry the traffic S). Equation (4.5) is plotted in Figure 4.5.

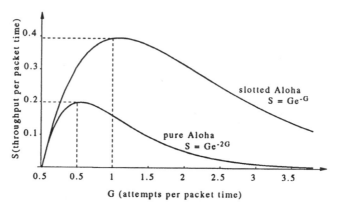

Figure 4.5: Throughput versus offered traffic for Aloha systems.

Examining Figure 4.5, one finds that the maximum achievable value of S is $1/(2e)$ corresponding to a load of $G = 1/2$. This is the capacity of the Pure Aloha system and it is only about 18%.

It is evident that even though this system is simple and can indeed support a large population of bursty users, its overall efficiency of bandwidth usage is rather low. As a matter of fact, from stability considerations described later, the system must operate considerably below its capacity, thereby making it even more inefficient. This is evident from the simplified analysis given above and is also borne out by a more detailed examination of its performance.

Exercise 4.1
The probability that a call request from a mobile vehicle through a satellite to a central station is successful is P_S. If unsuccessful, the mobile will repeat

the attempt after a random delay with mean $\bar{T}_r$. After transmitting each request there is a fixed timeout period, T_{out} (processing times and round trip delay). What is the average delay before a call begins?

Solution

The average delay incurred in carrying a packet using the Pure Aloha system is given by

$$\bar{D} = \left(\frac{1 - P_S}{P_S}\right)(T_{out} + \bar{T}_r) \tag{4.6}$$

where T_{out} is the time-out period (measured from the start of the packet) and $\bar{T}_r$ the average rescheduling delay.

The above follows from the observation that the distribution of the number of attempts before a success is a geometric distribution. The mean number of attempts before a success is $(1 - P_S)/P_S$.

Slotted Aloha System

Slotted Aloha [23] improves the performance of the Pure Aloha system by the incorporation of slotting. As shown in Figure 4.6, the time-axis is considered to be divided into slots equal to the packet duration T. All the users are assumed to be properly synchronized so that the slot definitions are common over the entire system.

Incorporation of slotting is the only added complexity of Slotted Aloha over the Pure Aloha scheme. (Compared to a terrestrial system, this common slot timing may be somewhat more difficult to achieve if a satellite repeater is being used.) In Slotted Aloha, arriving packets are constrained to wait till the next slot boundary before they can be transmitted.

As shown in Figure 4.6, this gives rise to a collision in slot i, only if two or more packets arrive for transmission in slot $(i - 1)$. Following the same terminology as in the previous section, for Slotted Aloha

$$P_S = e^{-\lambda_r T} \tag{4.7}$$

$$\lambda = \lambda_r e^{-\lambda_r T} \tag{4.8}$$

$$S = G e^{-G} \tag{4.9}$$

and hence the average delay in carrying a packet is

$$\bar{D} = \frac{1}{2}T + \left(\frac{1 - P_S}{P_S}\right)\left(T_{out} + \bar{T}_r + \frac{1}{2}T\right) \tag{4.10}$$

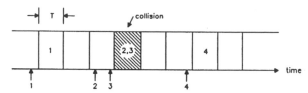

Figure 4.6:   The forward channel of a slotted Aloha system.

Note that from Equation (4.9), the capacity of the Slotted Aloha scheme is $1/e$ (36%) which occurs at a load of $G = 1$. This is twice the capacity of the Pure Aloha system. From considerations of stability, the system should however be operated well below its capacity.

Stability behaviour of (Slotted) Aloha

The Aloha systems - both pure and slotted- exhibit the undesirable property of instability. This means that, even when the offered traffic is below capacity, the system may at times reach a state where the load G increases to a large value (most of the users are backlogged) while the carried traffic decreases towards zero. Once this situation is reached, the system cannot return to its normal mode of operation without external assistance.

A more detailed analysis is needed as in [8], [11] and [3] to study the stability behaviour of Aloha-type systems. Such systems operate at equilibrium operating points which can be either desirable or undesirable. A desirable equilibrium point is one where the throughput is high and very few users are backlogged. An undesirable point, on the other hand, carries very little throughput even though most users are backlogged and the load is high.

The undesirable stability behaviour of Slotted Aloha stems from the fact that a system which starts from a desirable equilibrium point A will after a sufficiently long time transit to an undesirable equilibrium point B with a non-zero probability.

Moreover, once the system reaches B, it cannot recover from there (and return to A) unless some appropriate external action is taken. In other words, unless one can guarantee that the system operates in a way such that it does not have an undesirable equilibrium point , the system will eventually degenerate to a point where most users are backlogged, most transmissions collide, and the throughput is very small even though the load is high.

Finally, it should be noted that various distributed control algorithms have been proposed for such schemes so that transitions to the potentially damaging undesirable equilibrium point can be avoided. These control procedures require each user to estimate the system state based on the observed state of the previous slots and then appropriately modify the probability that a backlogged user transmits in a slot or the probability that a thinking user transmits in a slot, or both.

A particularly attractive version of such a control procedure is the Pseudo-Bayesian Control Algorithm where new arrivals are considered to be immediately backlogged. For the Slotted Aloha system this algorithm guarantees stability and that the theoretical capacity of $1/e$ can be reached given enough traffic. The reader is referred to [3] for a more detailed description of this algorithm. Slotted Aloha schemes may incorporate Collision Resolution Algorithms (CRAs) not only to improve system capacity (beyond the 36% value otherwise achievable) but also to avoid problems of instability as long as the schemes operate below their capacity limit. The added cost of achieving this is increased system complexity.

The first scheme of this type was proposed by Capetanakis [7] and could provide a capacity of 43%. A fairly straight-forward modification of this can improve the capacity to 46% [15] but leaves the system prone to dead-lock in case errors are made in determining the state of the channel. Gallager [10] proposes a further modification which increases the capacity to 0.4871 (but it is still liable to suffer from dead-lock). Without introducing special signalling or additional feedback mechanisms, the highest value of capacity reported for a scheme of this type is 0.4877 [16]. It should be noted that all of these schemes, except the first one proposed by Capetanakis [7], are liable to suffer from problems of dead-lock in case of erroneous channel feedback. However, additional control strategies can be used which eliminate dead-locks at some cost to the overall system capacity.

These schemes are stable if operated below their capacity and can be made to be first-come-first-served in nature, that is, (new) packets which arrive first are carried successfully before others which arrive later. (Note that the Aloha-type systems are not able to guarantee these properties.) We refer the reader to Massey's CRA [15] which is described in a terrestrial environment and point out that this and other schemes may be modified for use in a satellite system. The reader is referred to [14] for several other papers dealing with multiple access techniques using a collision resolution approach.

4.2.3 Controlled Access or Reservation Based Schemes

A large number of random access schemes have been proposed to allow a common transmission resource to be shared by a large population of bursty users. Unfortunately, most of the schemes of this type which can provide high capacity require channel sensing mechanisms which allow transmissions to be aborted if the channel is sensed to be busy. One example is the Carrier Sense Multiple Access (CSMA/CD) protocol used in the Ethernet system for LANs.

Since such sensing would not be practical in a satellite system, we cannot use these schemes for multiple access in a VSAT or MSAT environment. For these satellite systems, random access schemes which may be applied can provide only limited values of capacity. Systems such as Pure or Slotted Aloha have capacity values limited to (approximately) 18% and 36% respectively.

Higher capacity requires the use of more complex collision resolution schemes. Even these protocols are only able to increase capacity to about 48%. If bandwidth efficiency is a major criterion along with the ability to support a large and bursty user population, then a better option is to employ schemes with embedded reservations. In this section, we describe two such schemes which use reservations and requests for channel assignments which provide more efficient bandwidth utilization for bursty users.

Reservation Aloha

The Reservation Aloha system [22] improves the capacity of a Slotted Aloha system by incorporating a reservation procedure for the actual transmission of message packets. These reservation requests are made by special reservation packets which are considerably smaller than the actual message packets. These requests are made in an Aloha-mode of operation. Their smaller size ensures that their probability of failure due to collisions will be small. Once a reservation request is successful, the corresponding message packet(s) will be carried without collisions. The overhead of additional requests (for reservation) will typically be more than compensated for by the high efficiency of transmission of the message packets themselves.

A Reservation Aloha scheme is illustrated in Figure 4.7. As shown, the system uses a slotted time-axis with slots equal to the duration of a message packet. A frame consists of $N+1$ such slots where N should be such that for a slot duration of T seconds, NT is larger than the one-hop propagation delay. One slot in every frame is further subdivided into minislots of τ seconds each where τ is the duration of a request for reservation. The channel is assumed to be broadcast in nature so that the transmissions from any one of the users on the uplink (to the satellite) can be heard by that user and all other users on the downlink (after the one-hop propagation delay).

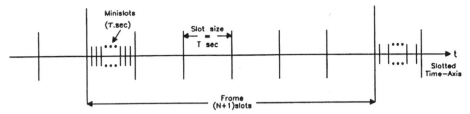

Figure 4.7: Organisation of frames in a reservation Aloha system.

In the Reservation Aloha system, a user can request a reservation for one or more slots for a multi-packet message. The reservation requests are essentially made to a single-server queue operated in a Blocked Customers Held mode. It is assumed that by keeping track of the successful reservations that have been made, each user knows the state of the queue at all times. The broadcast nature of the medium also allows the users to monitor the status (collision/no collision) of requests being made by the various users in the system.

A user with a message to transmit chooses one of the minislots at random in which to transmit a reservation request. The reservation request must identify the user and specify the number of message slots required for its transmission. The request packet also contains a synchronization header and a CRC field for error detection. The latter is used on reception to determine if the request packet has undergone a collision. If this request packet is received without errors on the downlink (by the source user and all other users in the system), then the reservation request has been successful. Otherwise, the requesting user repeats this request packet in a randomly chosen minislot of the next frame and continues this process until it finally succeeds.

Since everybody knows the global state of the queue, the corresponding message is assigned the requested number of slots and the global queue is increased by the appropriate amount. This information also helps the source user to identify which (consecutive) slots are to be used for this message transmission.

Note that the reservation requests are made in a Slotted Aloha mode to get slot assignments from a single server queue. Let λ_r be the average arrival rate of requests for new message transmissions. Then the sum of new and repeat request will have an effective average arrival rate λ'_r, where $\lambda'_r > \lambda_r$. Assuming both the arrival processes to be Poisson in nature, the effective arrival rate may be obtained by solving

$$\lambda_r = \lambda'_r e^{-(N+1)\lambda'_r \tau} \qquad (4.11)$$

The arrival process of (multi-packet) messages to the single-server queue will also have an average arrival rate of λ_r. Assuming this to be a Poisson process and the number of packets in a message to be a geometrically distributed random variable, one can use Erlang's Delay formula for the $M/M/1/\infty$ queue to obtain the average queueing delay encountered by a message.

The actual delay will consist of this delay, combined with the one-hop propagation delay and the (random access) delay encountered in successfully transmitting a reservation request. The reader is referred to [10] for a delay analysis of the Reservation Aloha system. Since the queue is of Blocked Customers Held type, it can be operated to its full capacity provided enough traffic is present. (Actually, the delays incurred in message transmission may become very large if the queue is operated close to its capacity.)

The corresponding request traffic must be such that the capacity limit of $1/e$ of the Slotted Aloha mode is not exceeded. In particular, for an average message length of T_m seconds, this implies that

$$(N+1)\lambda_r\tau \le \frac{1}{e} \tag{4.12a}$$

$$\lambda_r T_m \le \frac{N}{N+1} \tag{4.12b}$$

Note that Equation 4.12a arises from the capacity considerations of the imbedded Slotted Aloha system used for making the requests for reservations. On the other hand, Equation 4.12b arises from the queueing considerations and essentially indicates that one slot in every frame is "wasted" by the reservation mechanism.

In view of the stability problems of Slotted Aloha indicated previously, it should also be noted that the "equality condition" in Equation 4.12a should be avoided. As a matter of fact, assigning excess capacity to the reservation slots may be desirable so that the imbedded Slotted Aloha system never encounters problems of instability.

From the above description of Reservation Aloha, it can be seen that a user will encounter one single-hop delay before it can start transmission of its message. This will be true even when the queue is empty. This fixed delay will be present even when the system is lightly loaded. Since the slot with the minislots comes around only once in a frame, a new message arrival will have to wait half a frame, on the average, before it can transmit its request.

This delay will also be a fixed delay which will be present even under light loading.

The original Reservation Aloha scheme tried to alleviate both these problems by suggesting that when the global queue becomes empty, all the slots in the frame should be used in the minislotted mode. This should continue until the first successful reservation is made. Whenever such a reservation is heard on the downlink, the system should revert back to the frame and slot structure given in Figure 4.7. It also suggests that short messages should be directly transmitted on randomly chosen minislots instead of going through the reservation process. These modifications will improve the delay performance of the system when the traffic is low without affecting its capacity performance

DAMA Schemes with Random Access

It can be seen that the Reservation Aloha scheme is essentially one using Time Division Multiplexing to imbed a random access scheme in an essentially Demand Assigned approach. In this section, we briefly describe another approach using Frequency Division Multiplexing to achieve the same objective.

Consider the system illustrated in Figure 4.8, in which the available bandwidth is divided into N_r request channels and N_m message channels.

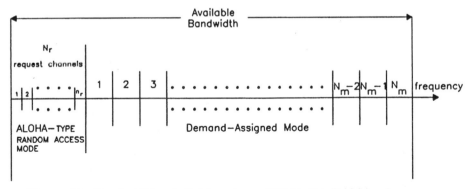

Figure 4.8 Bandwidth subdivisions in an FDMA RA-DAMA scheme.

The request channels are used in a random access mode (that is, Pure or Slotted Aloha) to carry requests for channel assignments to a Central Controller. The message channels are used to transmit the actual message. Note that the user can transmit on a message channel only after it gets an explicit assignment from the central station authorising it to do so. Once the message transmission is over, the station relinquishes this channel which can then be reassigned by the central station to another requesting user.

A user with a message to transmit first sends a request packet in an Aloha mode to the central station on a randomly chosen request channel. If there is no collision and this packet reaches the central station correctly, then channel assignment is done by the central controller using the particular queueing discipline being followed. The central controller informs the user of this using an appropriate frequency band in the reverse channel (not shown in Figure 4.8).

If a collision occurs, the request is lost; in this case, the requesting user waits for the assignment information from the central station until a pre-specified time-out period and then repeats the request following a random rescheduling delay. This process is continued until the request successfully reaches the central station.

This scheme is one of a class of random access, demand assignment multiple access (RA-DAMA) schemes. Various schemes of this type have been described in [6], [19] and [20]. The IA-MAP protocol of MSAT-X [13] as well as the protocol suggested by Aussat for its Mobilesat system are essentially variations of this approach with some added features.

Apart from the fact that the transmission resources are divided in frequency rather than time, an RA-DAMA scheme is reasonably similar to the Reservation Aloha protocol described in the earlier section. In both cases, the request for the actual transmission resource (frequency channel or time slots) is made using random access following an Aloha-approach. The actual resource assignment for the message transmission is done on a Demand Assigned Multiple Access (DAMA) basis. As noted earlier, the ability of the system to support a large, bursty user population is because of the random access method used for transmitting the request. The high efficiency of the scheme is a result of the DAMA mode of operation of the major part of the system's transmission resources (message channels or message slots).

Within the class of RA-DAMA schemes, a large number of choices can still be made. The request channels may be operated either in the Pure or Slotted Aloha mode. The former is simpler and does not require any timing information whereas the latter is more efficient and encounters fewer collisions.

The queueing discipline could either be Blocked Customers Cleared (BCC) or Blocked Customers Held (BCH) in nature. (Note that due to considerations of efficiency arising from its overall approach, Reservation Aloha is limited to the latter since it operates as a single server queue. The DAMA scheme presented here operates as an N_m-server queue and can use either discipline for handling requests which come when all the message channels are already assigned.)

If the BCC discipline is used, a user with a successful request may still not get a channel assignment if its request reaches the central station at a time when all channels are assigned. In such cases, one can either take the approach that the blocked call is discarded or consider the system to be such that the user tries again with a fresh request.

If the message length is random, then also two approaches are possible. One can consider a situation where the user announces the length of the message in its request and the central station accounts for that in its overall assignment strategy. Alternatively, the requesting user may be assigned a channel indefinitely and then send a special packet to the central station relinquishing the assigned channel once its requirements are over. This would effectively hold an assigned channel for an extra time equal to the one-hop propagation delay.

The former scheme would be more efficient in terms of channel usage than the latter. Unfortunately, it may not always be possible to announce beforehand the length of time for which the requested channel is to be held. This could be the case for a voice call or a remote transaction with a computer. Even for situations where a known amount of data is to be sent, the ARQ policy may make the actual transmission time random, depending on the error rate of the channel.

Note that the centralized assignment scheme introduces a minimum two-hop delay between the time a request is (first) made and the time when actual data transmission can start. This fixed delay would be present regardless of the actual load on the system. (The other component of delay would be the random access delay in getting the request through to the central station.)

This fixed delay component can be halved to a one hop delay if the assignment is done in a distributed fashion by all the users together, rather than by a central station. Such a distributed algorithm will obviously be more complex to implement. Note that the Reservation Aloha scheme is such a distributed algorithm.

The analysis of an RA-DAMAscheme would depend on which of the above options are actually chosen. In its simplest form, assuming the arrival processes of new requests and that of the new and repeat requests to be Poisson, the analysis would be fairly straight-forward. As an example, consider a case where Slotted Aloha is used for the request packets of length τ. Assume the queueing discipline to be BCC and that users who are blocked repeat their requests after a random rescheduling delay. If T_m is the mean message length, then the average arrival rate λ'_m of new and repeat requests required to support an average arrival rate λ_m of new requests is given by

$$\lambda_m = \lambda'_m e^{\frac{-\lambda'_m \tau}{N_r}} \left[1 - B(N_m, \lambda'_m T_m e^{\frac{-\lambda'_m \tau}{N_r}}) \right] \qquad (4.13)$$

where $B(s, a)$ is Erlang's Blocking Probability for a $M/G/s/s$ queue.

$$B(s, a) = \frac{a^s}{s!} \left/ \left[\sum_{i=0}^{s} \frac{a^i}{i!} \right] \right. \qquad (4.14)$$

Note that $\lambda_m T_m$ is the traffic carried by the system. Moreover, stability considerations of the imbedded Slotted Aloha scheme for requests imply that τ should be kept small enough or N_r should be made large enough so that the offered request traffic is well below its capacity of $1/e$.

As indicated earlier, the advantage of controlled access schemes with imbedded random access for channel resource assignment is that they can support a large population of bursty users with fairly high values of system capacity. The main disadvantage of this approach is the added delay involved in obtaining an assigned channel before actual message transmission can be started. In spite of the fact that this latter delay would be fairly large in satellite systems, the overall efficiency and simplicity (of operation) of this approach makes it a good candidate for VSAT and MSAT systems. To some extent, the performance of such schemes can be further improved by allowing random access on unassigned message channels [6]. This would improve the overall throughput of the system and, if properly used, can also help to reduce delay.

4.3   MSAT SYSTEM ACCESS PROTOCOLS

In Chapter 3, we described the overall architecture of a typical mobile satellite (MSAT) system. The services provided by systems such as the Australian Mobilesat system are predominantly targeted towards the requirements of land mobile and transportable communications in regional and remote areas. Apart from the mobile terminals (MT), such systems utilize one or more Network Management Stations (NMS), other Base Stations (BS) and Gateway Stations (GS) as shown in Figure 4.9.

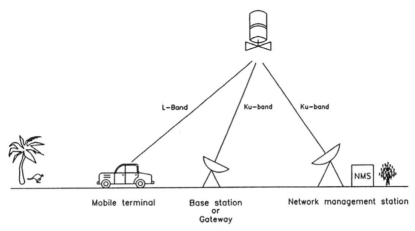

Figure 4.9: The Mobilesat network.

The NMS is responsible for the management of the overall system and provides appropriate control and signalling information to the other stations in the system. The communication services provided by the system allow information transfer between the mobile terminals and the base stations. Direct mobile to mobile communication is not supported (actually, it is physically prevented by the L/Ku-band mode of operation). If required, such communications will have to be carried out through an appropriate base station. The Gateway Stations are intended to be the primary interfaces of the MTs to the Public Switched Telephone Network (PSTN).

4.3.1 An Integrated Protocol Supporting Continuous Voice and Data Services

A satellite transponder for a typical satellite mobile service supports a total bandwidth of 14 MHz for mobile communications. The bandwidth is usually divided into 5 kHz channels providing a total of 2800 channels for operation. Linear operation over all channels requires that the total power received by the satellite be below a specified threshold. This power limited mode of operation implies that only a fraction of these 2800 channels may be active at any given instant.

In this Section, we describe a typical access protocol for a satellite mobile network. It is assumed here that the system supports either continuous voice or data services. We refer to the two modes of operation as voice calls or data calls respectively. For both types of calls, a request is sent from a terminal for channel assignment to the NMS. If the call is received successfully, the

NMS assigns a frequency channel for the call (provided such a channel can be assigned). The assigned channel is released at the conclusion of the call.

For providing the continuous voice and data services described above, let the bandwidth available be divided into $(N_r + N_m)$ channels as illustrated in Figure 4.10. Of these, N_r channels are kept aside for making reservations in a random access Slotted Aloha mode whereas the other N_m channels are used to actually carry information.

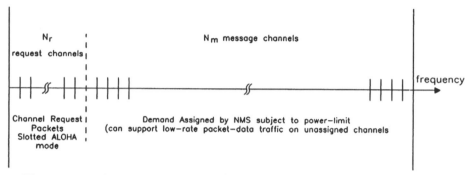

Figure 4.10: Bandwidth subdivision for Mobilesat continuous voice/data services.

The time axes at both the NMS and the mobile terminals are divided into frames. For the mobile terminals, each frame is further subdivided into request slots in order to operate the request channels in a Slotted Aloha mode. A possible organization of frames and request slots is shown in Figure 4.11 where the frame duration is T seconds and the request slots are of τ_r seconds.

Note that the NMS may send channel assignment information in a Bulletin Board which is transmitted at the beginning of each of its frames.

For the timing shown in Figure 4.11, if a mobile terminal channel request originates in frame i then channel assignments (if successful) for transmissions will start from frame $i+2$. Moreover, by monitoring the Bulletin Board, mobiles will know at the beginning of each of their frames the identity of the channels already assigned in that frame.

From Figure 4.11, the frame duration T is given by

$$T = 2R + \Delta \qquad (4.15)$$

where R is the one-hop propagation delay from the mobiles to the NMS (and any processing delay involved) and Δ is the duration of the bulletin board information on each frame from the NMS.

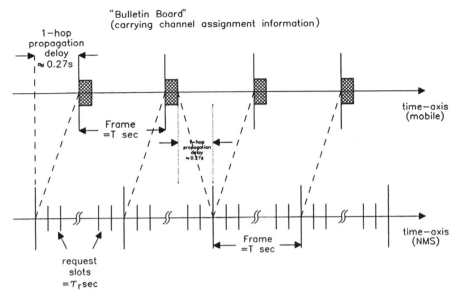

Figure 4.11: A possible frame structure and timing for Mobilesat continuous voice/data services.

Consider a request (data or voice call) arrival in Figure 4.11 where the corresponding mobile transmits this request in a randomly chosen request channel in the next slot. Assume that this slot is in frame i. If no other mobile transmits a request packet on the same channel and in the same slot, then this request should get through correctly to the NMS. (In this section, we analyze the access system assuming the channels are interference limited so that random noise effects may be ignored.)

The NMS tries to assign a message channel to this call. If a channel can be assigned then the requesting mobile is informed of this via the bulletin board transmitted from the NMS in its next frame. This bulletin board reaches the mobile just prior to its frame $i+2$ so that it can use the assigned channel from that frame onwards until the channel is relinquished at call termination. The NMS is assumed to send a negative acknowledgement (NAK) if the request cannot be granted a channel, that is the voice or data call is blocked.

The NMS implements a blocked calls lost system and it is therefore assumed that the blocked call leaves the system. (In reality, the blocked user may actually try to place the call again after some random delay. This is not taken into account in the following analysis but may be incorporated as an additional feedback on the request channels.)

If the request in frame i suffers a collision, the requesting mobile will neither get a channel assignment nor a NAK. In this case, the mobile repeats the request on a randomly chosen request channel following an appropriate random rescheduling delay.

It should be noted that the transponder power available on-board the satellite is very often a limiting constraint. For wide-band satellite transponders, the system may indeed be power-limited rather than being bandwidth-limited.

In such a situation, the transponder may have a wide bandwidth divided into a large number of channels but power limitations will require that only a fraction of them (at the desired signal to-noise ratio for acceptable quality) should be active at any given instant. If this threshold is exceeded, the output of the transponder will degrade to an unacceptable quality. For systems providing voice transmissions through such a power-limited transponder, a voice activation strategy may be used to support more calls than would otherwise be possible.

This strategy requires each user transmitting voice to have a voice activated switch at its transmitter. This is used to switch off the carrier whenever the voice source is idle. The carrier is switched on again when a speech burst arrives (that is the voice source becomes active once again). Some channels may be used to carry data in the system except that the data calls will keep their respective channels active for the entire call duration.

The channel assignment strategy assigns channels to voice calls or data calls such that the average total power for all assigned channels is below the system's power threshold. On occasion, when a number of voice calls have been assigned channels, this may lead to transient situations when the total power from the currently active voice and data calls exceeds the threshold. During that time, transmissions on all channels will be adversely affected. For voice, this implies that the voice signal will fall below acceptable quality. For data, there will be a loss due to errors which will have to be repeated using ARQ.

Channel assignments for voice calls and data calls are usually provided from the same set of identical channels. For analysis, let us assume that the two types of calls use the same power for their transmissions. Since the request channels are operated in a Slotted Aloha mode with requests of very small duration, the request traffic will be small enough so that its contribution to the total power may be neglected. The power threshold K may then be specified as the maximum number of simultaneously active channels (voice or data) such that the system's power-limit is not exceeded.

If n_d is the number of channels assigned to data calls and n_v the number of channels assigned to voice calls then the channel assignment strategy should ensure that

$$n_d + F n_v \le K \tag{4.16}$$

where F is the activity factor of a voice source. That is, F is the fraction of time a voice source is active.

A call request (data or voice) may be blocked if assigning it a channel would violate (4.16). Calls which are blocked are assumed to leave the system, that is a Blocked Customers Cleared (BCC) queueing discipline is followed.

At any instant, if the system has n_d channels assigned for data calls and n_v channels for voice calls, then the system may be considered to be in state (n_d, n_v). With the average power constraint of Equation (4.16), the region of allowable system states is shown in Figure 4.12. Note that n_d and n_v can only take on integer values (subject of course to the condition of (4.16)).

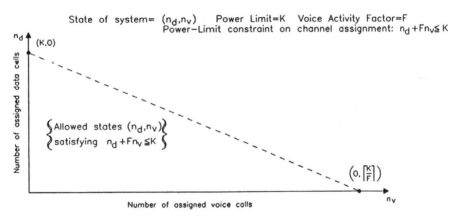

Figure 4.12: Region of allowable system states.

Consider a power-limited system as described above with N_r request channels and N_m message channels. The request channels are operated in the Slotted Aloha mode to carry requests for channel assignments (for voice or data calls). Following this, request packets are repeated until they get through correctly without collision. If the call can be assigned a channel, it is informed of the channel assignment by the central assigning station (the NMS). Otherwise, a NAK is sent to the mobile informing it that a channel could not be assigned.

If a channel is assigned, it is used for the duration of the call. On call termination, the channel is relinquished. Note that for a data call, repetitions required by the ARQ mechanism are considered to be done within the same call following the idealized assumptions noted earlier. (In an actual system, non-ideal behaviour of the ARQ protocol may require more repetitions and hence may cause greater inefficiencies.)

4.3.2   Analysis of system performance

In order to analyze the performance of the above system, we assume all arrival processes to be Poisson in nature and the call holding times to be exponentially distributed. Let

$$\lambda_v = \text{average arrival rate of voice calls}$$

and,

$$\lambda_d = \text{average arrival rate of data calls.}$$

The average arrival rate of new requests will then be $(\lambda_d + \lambda_v)$. For a request duration of τ_r, "safe" operation of the Slotted Aloha mode for request channels requires that

$$\frac{(\lambda_d + \lambda_v)\tau_r}{N_r} \ll \frac{1}{e} \qquad (4.17)$$

We will assume that this condition is satisfied in the system. Note that for typical values of N_r and τ_r the traffic contribution from requests will indeed be small. This justifies our earlier assumption in this regard.

Let,

$$1/\mu_v = \text{average holding time for a voice call}$$
$$1/\mu_{d0} = \text{average holding time for a data call without errors}$$

and,

$$\rho_v = \lambda_v/\mu_v = \text{offered voice traffic}$$
$$\rho_d' = \lambda_d/\mu_{d0} = \text{effective data traffic.}$$

Note that the presence of errors due to crossing of the power limit (assumed to be the only source of error) will generate repetitions because of ARQ requirements (to recover lost data). This will cause the actual holding time of a data call to be $1/\mu_d$ $(1/\mu_d > 1/\mu_{d0})$ and the corresponding "offered data traffic" to be ρ_d.

Let $P_L(\rho_d)$ be the fraction of a data stream which is lost due to errors and hence has to be repeated. Using this, we get

$$\rho_d = \rho_d'/[1 - P_L(\rho_d)] \tag{4.18}$$

which can be used to compute ρ_d given ρ_d'.

A continuous time bursty model for a single voice speaker indicates that the source will be alternatively active and idle. The active intervals and the idle intervals are both assumed to be exponentially distributed random variables with means $1/\sigma$ and $1/\gamma$ respectively. Let ν be the ratio of the mean durations of an active burst and an idle interval. In terms of the activity factor F, this may be expressed as

$$\nu = \frac{F}{1 - F} = \frac{\gamma}{\sigma} \tag{4.19}$$

Typical values used for F are 0.35-0.40.

It can also be shown that given N voice channels, the probability $P_v(N, i)$ that i of them are active at any given instant is given by

$$P_v(N, i) = \frac{1}{(1 + \nu)^N} \binom{N}{i} \nu^i \quad i = 0, 1, \ldots, N \tag{4.20}$$

Taking the queue of data and voice calls, the probability $P(n_d, n_v)$ that n_d data channels and n_v voice channels are currently assigned will be

$$P(n_d, n_v) = c \frac{\rho_d^{n_d} \rho_v^{n_v}}{n_d! \, n_v!} \tag{4.21}$$

where $c = \left[\sum_{i=0}^{K} \sum_{j=0}^{\lceil \frac{K-i}{F} \rceil} \frac{\rho_d^i \rho_v^j}{i! \, j!} \right]^{-1}$ and $\lceil x \rceil$ = integer part of x.

Using this, the blocking probabilities, P_{BD} and P_{BV}, respectively, of data and voice calls will be given by

$$P_{BD} = P(K, 0) + \sum_{i=0}^{K-1} \sum_{j=\lceil \frac{K-i-1}{F} + 1 \rceil}^{\lceil \frac{K-i}{F} \rceil} P(i, j) \tag{4.22}$$

$$P_{BV} = \sum_{j=0}^{K} P\left(j, \lceil \frac{K-j}{F} \rceil\right) \tag{4.23}$$

Note that for data and voice calls, the respective blocking probabilities indicate the fraction of successful requests which will still be denied a channel because of blocking. In the loss-system considered here, such blocked calls are assumed to leave the system without receiving service.

It is also interesting to note that the blocking probability for data calls will always be larger than that for voice calls. This arises because the "over-assignment" strategy and the bursty nature of voice give rise to system states in which the NMS might be willing to add another voice call but will not accept another data call.

The fraction of a data stream which is lost due to errors may be found as the probability that the total number of active channels exceeds K. For given n_d and n_v, this is given by

$$
P_{L|n_d,n_v} = \begin{cases} \displaystyle\sum_{i=K-n_d+1}^{n_v} P_v(n_v, i), & n_v > K - n_d \\ \\ 0 & \text{otherwise} \end{cases} \tag{4.24}
$$

Using this $P_L(\rho_d)$ is found to be

$$
P_L(\rho_d) = \sum_{n_d=0}^{K} \sum_{n_v=K-n_d+1}^{\lceil \frac{K-n_d}{F} \rceil} P(n_d, n_v) \sum_{i=K-n_d+1}^{n_v} P_v(n_v, i) \tag{4.25}
$$

which may be used in Equation (4.18) to find ρ_d for a given value of ρ_d'. Note that this also indicates the fraction of time a voice channel will degrade to an "unacceptable level" because the power limit is exceeded.

For a given number N_m of message channels, the threshold value K and the voice and data traffic values ρ_v and ρ_d', Equation (4.18) may be solved using Equation (4.25) to get the offered data traffic ρ_d. This also yields $P_L(\rho_d)$ which may be interpreted both as the fraction of time a voice call will degrade to a poor quality as well as the fraction of a data stream which will have to be repeated because of errors. This will be an important performance indicator of the system.

Other parameters of interest will be the blocking probabilities for data and voice calls which can be found using Equations (4.22) and (4.23), respectively. Knowing these probabilities, the data and voice traffic actually carried by the system may be found.

The data traffic carried by the system will be

$$\rho_{d0} \simeq \rho_d(1 - P_{BD})[1 - P_L(\rho_d)] \tag{4.26}$$

The voice traffic carried will be

$$\rho_{v0} = \rho_v(1 - P_{BV}) \tag{4.27}$$

Alternatively, the data and voice traffic carried by the system may be computed for given values of the offered traffic using

$$\rho_{d0} = \sum_{n_d=0}^{K} \sum_{n_v=0}^{\lceil \frac{K-n_d}{F} \rceil} n_d[1 - P_{L|n_d,n_v}]P(n_d,n_v) \tag{4.28}$$

$$\rho_{v0} = \sum_{n_d=0}^{K} \sum_{n_v=0}^{\lceil \frac{K-n_d}{F} \rceil} n_v P(n_d,n_v) \tag{4.29}$$

Note that ρ_{v0} as given by Equation (4.29) will be the same as that found from Equation (4.27). However, ρ_{d0} computed from Equation (4.28) will be more accurate than that given by the approximation of Equation (4.26). For large values, computations using Equation (4.26) will be considerably faster. We use this in subsequent calculations.

Finally, given the average duration of the holding times for data and voice calls, the average arrival rates for these calls may be found. For proper operation of the request channels, the sum of these arrival rates should be verified as satisfying the constraint given in Equation (4.17). In our subsequent computations of system performance, we will assume that this is indeed the case.

4.3.3 Numerical Results

Numerical computations using the above analysis were carried out for a typical system where the power limit constraint K was taken to be 200 and an activity factor of 0.4 was assumed for a single voice source. The total number of message channels, N_m, was assumed to be greater than 500 so that blocking due to lack of channels was not the limiting constraint. Instead, the power limit constraint of Equation (4.16) was assumed to be the reason for blocking, that is a successful request still not getting a channel assignment. In addition, the number of request channels were assumed to be such that the constraint of Equation (4.17) is satisfied. Moreover, it was assumed that the traffic on the request channels does not have a significant impact on the power limit of the system. Figure 4.13, shows plots of carried voice traffic ρ_{v0} versus the offered voice traffic ρ_v for different values of offered data traffic ρ_d for $K = 200$ and $F = 0.4$.

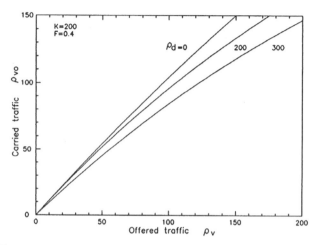

Figure 4.13: Load throughput characteristics for voice.

Similar curves are shown in Figure 4.14 for the data traffic.

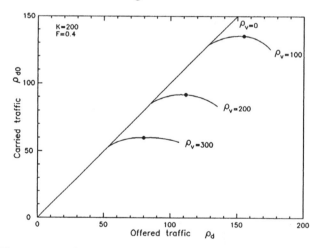

Figure 4.14: Load throughput characteristics for data.

It is interesting to note that, in the range of interest, for a given non-zero value of the offered voice traffic, the carried data traffic shows a maxima as a function of its offered traffic.

Apart from implying a limiting value of the carried data traffic as a function of the voice traffic, this also raises the possibility of potential instability in case the system is operated too close to this maximum value. (Voice traffic, on the other hand, tends to saturate to a limiting value without showing a similar maxima.)

This behaviour of the data traffic probably occurs because of the loss arising due to crossings of the power threshold. Note that for data, the "lost" portions of a data stream are repeated which, in effect, increases the data traffic further leading to more "loss".

For the same system as above, Figures 4.15 and 4.16 respectively show the blocking probabilities and the "loss" probability as a function of the offered traffic (voice and data).

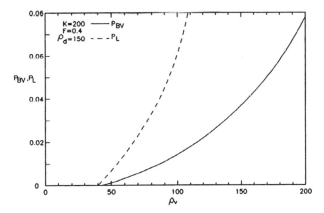

Figure 4.15: Probabilities of blocking and loss for voice.

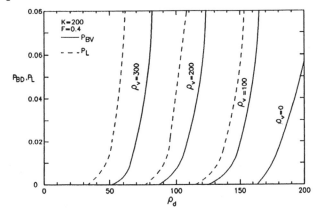

Figure 4.16: Probabilities of blocking and loss for data.

These probabilities are important performance indicators in terms of the users' perception of the system. The respective blocking probabilities indicate the grade of service provided by the system for data and voice calls. That is the fraction of time a request successful over the Slotted Aloha request channels will still have to leave the system without a channel assignment.

For voice calls, the loss probability is indicative of the quality of the voice calls carried by the system. That is the fraction of a call which is of poor quality because of the power-limit effects.

For data calls, this quantity indicates the fraction of a data stream which has errors because the power-limit is exceeded. As noted earlier, this lost data will need to be recovered through ARQ, thereby extending the effective duration of the data call.

Typically, for satisfactory operation, these probabilities should be below a specified limiting value, for example, 0.05. Applying such a limit, one can see that the range of satisfactory performance of the system is considerably more restricted than what has been shown earlier in Figures 4.12 and 4.13. (For example, note that at the maxima shown in Figure 4.14, these probabilities are already close to or greater than the typical limiting choice of 0.05.)

From the view-point of system operation, an important quantity will be the revenue that can be earned for various mixes of data and voice traffic. Let the revenue R earned by the system be given by

$$R = k(\rho_{v0} + \rho_{d0}/F) \qquad (4.30)$$

where k is a constant and where a data call is assumed to earn $1/F$ times more than a voice call. This is reasonable based on the fact that a voice source is only active with an activity factor F whereas a data source is always active.

For our subsequent discussion, we will assume k=1. For the given system, the maximum possible revenue is set by the total number of available channels K. In Figure 4.17 (and Figure 4.18), we have shown the revenue R earned as a function of the offered voice (data) traffic for different values of the offered data (voice) traffic.

For a given choice of the offered voice traffic ρ_v, one can find the operating point where the system will earn the maximum possible revenue. This maximum revenue and the "loss" probability at this optimum point are shown in Figure 4.19 as a function of the offered voice traffic.

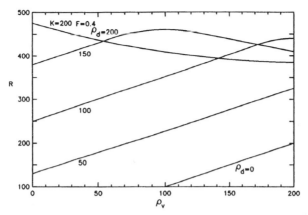

Figure 4.17: Revenue earned as a function of the offered voice traffic.

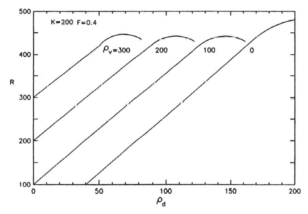

Figure 4.18: Revenue earned as a function of the offered data traffic.

Figure 4.19 also indicates the actual values of carried data and voice traffic if the system operates at this maximum revenue point. (Figure 4.19 does not show the blocking probabilities at this operating point. The reader might note that these are less than 0.05 for $\rho_v < 280$ but may be greater otherwise).

As expected, the revenue earned goes through a minimum value as more data traffic is mixed in with the voice traffic. The moral to this is that, with the revenue as defined in Equation (4.30), the system earns the most if it carries either pure voice or pure data. Mixing the two types of traffic leads to a loss of revenue. Obviously, the motivation to implement an integrated voice/data system has to be something other than pure revenue.

From Figure 4.19, it is also interesting to note that starting from the situation where the system is carrying mostly data and very little voice (the extreme left), the maximum revenue falls rapidly as more and more voice traffic is mixed into the system. This rapid decline is not observed when one considers mixing in data traffic to a system carrying voice with very little data.

A possible explanation for this difference may be that in the situation where the system has mostly data calls and a few voice calls, crossings of the power-limit threshold caused by activity on the few voice channels will simultaneously spell disaster for the large number of data channels.

This would severely affect the revenue earned by the system. The situation is not so disastrous when the roles of the data and voice calls are interchanged.

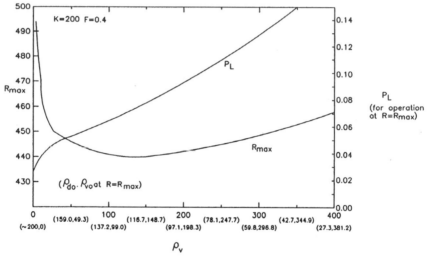

Figure 4.19: System performance at operating points maximizing revenue.

Finally, it should be noted that the maximum revenue points plotted in Figure 4.19 may not always be desirable points of operation if separate limits are specified for the loss probability and the blocking probabilities.

As can be seen from the Figure, if one specified a limit of 0.05 for the loss probability then it would be possible to operate at the maximum revenue point only when $\rho_v < 50$. For higher values of ρ_v, one would be forced to operate at a sub-optimum point to satisfy the constraint $P_L \leq 0.05$.

The revenue function of Equation (4.30) represents the system operator's point of view. The users of the system are more likely to judge the system's performance in terms of the probability indicators, namely the loss probability and the probabilities of blocking. If we choose 0.05 as the limiting values of the "loss" and the blocking probabilities, then operation at the maximum revenue point may not be possible except for fairly small values of P_v.

In Figure 4.20, we show the region of ρ_d and ρ_v, the offered data and voice traffic respectively, where these probability constraints are satisfied in the given system.

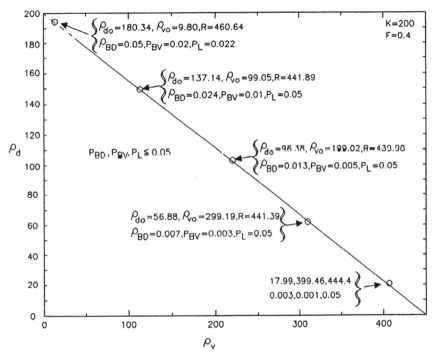

Figure 4.20: System operating region satisfying the constraints $P_{BD} \leq 0.05, P_{BV} \leq 0.05, P_L \leq 0.05$.

To earn the maximum revenue in this region, the system should be operated on points along the boundary. Figure 4.20 also shows the performance of the system at different operating points on this boundary. Note once again that the revenue goes through a minimum as the two types of traffic - data and voice - are mixed together. In Figure 4.20, the dotted boundary indicates the region where the active limiting constraint is $P_{BD} = 0.05$ and the other two probabilities are less than 0.05; the active constraint over the remaining boundary is $P_L = 0.05$.

This indicates that for earning the maximum revenue subject to these limiting probability constraints, the system is limited by the blocking probability of data when it is carrying mostly data with very little voice. Otherwise, it is constrained by the loss probability. (Note that since the voice blocking probability is always less than the data blocking probability, the former would never be the limiting constraint for the given specifications.)

4.4 SUMMARY

We have examined the performance of an integrated voice/data system carrying continuous voice and data calls over a power limited satellite channel. The system described corresponds to that where mobiles request channels from the NMS for their respective voice or data calls using a Slotted Aloha method and then use the assigned channel, if any, for the duration of the call. In order to maximize its utilization of the system's resources, voice sources use a voice activity switch to turn off their carriers when idle and the NMS uses an "over-assignment" strategy which relies on keeping the average transmitted power of all assigned (voice and data) channels below a system-specified power limit.

This "unconstrained" system was found to be liable to instability. This is a serious problem in so far as it indicates that if such a system is indeed implemented, one must also make provisions for additional signalling and control procedures to prevent such instabilities from arising (and have other procedures which would let the system recover even if it happens to stray into an unstable region of operation).

It is also interesting to note that compared to the situation where the system is either carrying pure voice or pure data, it earns less revenue when it is carrying a mixture of both. This is an unfortunate effect to happen in an integrated voice/data system but should be something that the system designer is aware of. It is also quite unfortunate that when the system is operated at its limits and is carrying mostly data, decreasing the data traffic slightly in order to add more voice traffic leads to a fairly sharp degradation in performance. Such a sharp decline is not present when one adds a little bit of data to a system carrying mostly voice.

In order to correct for some of the above problems, some changes can be made to the basic unconstrained system in Section 4.3.1. These changes essentially imply specifying additional constraints which further limit the state space of the system in such a way that some "undesirable" states are eliminated [5]. Access protocols incorporating such additional constraints will be more difficult to implement but can provide better performance.

4.5 PROBLEMS

1. Consider a Slotted Aloha system with n users. Let p be the probability that a user generates a packet for transmission in a slot. Assume that packets undergoing collisions in a slot are discarded, ie. retransmission attempts for these packets are not made.

   (a) For the offered load np, show that the throughput will be $np(1 - p)^{n-1}$. For a given value of n, what is the value of p at which the throughput will be maximum? What will be the value of this maximum throughput?

   (b) Let $n \to \infty$ and $p \to 0$ such that $np = G$ stays constant. Show that the throughput of (a) will then reduce in the limit to that given by Equation (4.9). Find the capacity, ie. the maximum throughput, for this case.

   (c) Comment on the similarity of the results in (b) with those of Equation (4.7) - (4.9) even though we have assumed the policy of discarding collided packets instead of retransmitting them.

2. Draw a block diagram of an RA-DAMA scheme by proceding as follows. Consider the representation of an Aloha system given in Figure 4.4. Now assume the output of the channel feeds into a Blocked Customers Cleared queue. The queue has finite capacity so some requests must be added to the number of retransmissions.

3. For a single voice source, a simple ON-OFF model may be used. Let "0" indicate that the voice source is silent (idle) and "1" indicates that it is active. In addition, it is assumed that the burst durations are independent, identically distributed (i.i.d) random variables with an exponential distribution where $\beta \Delta t$ is the probability of going from state "1" to "0" in a time interval of length Δt. Similarly, let the idle periods be exponentially distributed, i.i.d random variables with $\alpha \Delta t$ as the probability of going from "0" to "1" in a time interval Δt.

   Consider a system with M voice sources of the above type. In this system, we consider the system to be in state n if n voice sources are simultaneously active at that instant.

   (a) Draw the state transition diagram for this system for the states: $0, 1, 2, \ldots, M - 1, M$.

(b) Find the state probabilities at equilibrium, ie. the probability of their being n simultaneously active voice sources in the system at steady state for $n = 0, 1, \ldots, M$.

Relate this to Equation(4.20) in the text.

(c) Suppose we consider the system to be "overloaded" if more than K voice sources are simultaneously active, $K < M$.

What will be the fraction of time that the system will be overloaded?

4. Consider a generalization of the system of Problem 3 where the system has M voice sources and N data sources. We assume that the data sources are always active whereas the voice sources are bursty and are modelled in the same way as in Problem 3.

(a) Assume that the system gets overloaded whenever the total number of active sources (voice or data) exceeds K. What will be the fraction of time (ie. the probability) that the system will be overloaded?

(b) Using a voice activity factor F of 0.4, compute this overload probability as a function of M for $M = 0, 1, 2, 3, 4, 5$ where N=2 and K=5.

(c) For K=5, repeat the computation of (b) for $N = 0, 1, 2, 3, 4, 5$ and M ranging in values (integers) from 0 to atleast 25. If you can, with a suitable software package, draw a 3 dimensional plot of the overload probability (on the z-axis) as a function of M and N (drawn on the x and y axis).

(d) If a low overload probability is to be the design criterion for a system multiplexing voice and data calls as described here, can you use the results of (c) to determine a desirable operating region for ths system? One way to do this would be to keep only those states where the computed overload probability is low enough to be acceptable; the system is not allowed to operate with other values of M and N.

5. For each operating point (N,M) in the operating region of Problem 4(c), one can associate the revenue R(N,M) as given by Equation(4.32). Evaluate this for each point in this operating region. Use these values to determine a desirable set of operating points for this system.

Equation(4.32) provides one possible definition of revenue. Depending on what is interpreted as the "rewards" coming from the operation of this system, this revenue can be defined in different ways.

Suggest some alternative definition for this revenue function and repeat the computation of the first part for your new definitions of revenue.

6. Consider a Slotted Aloha system serving N users. Let n_i be the number of backlogged users at the end of slot i. Let q be the probability that a backlogged user will try sending a packet in slot $i + 1$.

Each one of the remaining $(N - n_i)$ thinking users can also generate a packet in slot $i + 1$ with probability p.

(a) With $n = n_i$ as the state of the system ($a \leq n \leq N$), draw its State Transition Diagram.

(b) With $\Delta n_{i+1} = n_{i+1} - n_i$, show that the various state transition probabilities will be given by

$$P\{\Delta n_{i+1} = 0\} = (N - n_i)p(1 - p)^{N-n_i-1}(1 - q)^{n_i}$$
$$+ (1 - p)^{N-n_i}(1 - q)^{n_i} \qquad (4.31)$$

$$P\{\Delta n_{i+1} = -1\} = (1 - p)^{N-n_i}n_iq(1 - q)^{n_i-1} \qquad (4.32)$$

$$P\{\Delta n_i = +1\} = (N - n_i)p(1 - p)^{N-n_i-1}[1 - (1 - q)^{n_i}] \qquad (4.33)$$

$$P\{\Delta n_i = +k\} = \binom{N - n_i}{k} p^k(1 - p)^{N-n_i-k}$$
$$\text{where } 2 \leq k \leq N - n_i \qquad (4.34)$$

(c) Show that the throughput in slot $i + 1$ will be given by

$$S_{out}^{i+1} = (N - n_i)p(1 - p)^{N-n_i-1}(1 - q)^{n_i} \qquad (4.35)$$
$$+ (1 - p)^{N-n_i}n_iq(1 - q)^{n_i-1}$$

The average number of new packets entering the system in slot (i+1) will be

$$S_m^{i-M} = (N - n_i)p \qquad (4.36)$$

and that the equilibrium operating points of the system may be obtained by solving $S_{out} = S_{in}$ for the number of backlogged users n. Note that this Equation can have multiple solutions each of which can be a possible operating point of the system.

REFERENCES

1. Abramson, N., "The Aloha system - another alternative for computer communications", *Proc. AFIPS Conf.*, Vol. 37, pp 282-285, 1970.

2. Aussat, *Mobilesat Draft Technical Specifications*, Sydney, August 1988.

3. Bertsekas, D. and Gallager, R.G., *Data Networks*, Prentice-Hall, 1987.

4. Bose, S.K., "Mobile Satellite Communications - A survey of literature in the area of mobile satellite communications", TRC Report No. 25/88, Teletraffic Research Centre, University of Adelaide, 1988.

5. Bose, S.K., "Multiple-Access Schemes for Mobile Satellite Communications", TRC Report No. 29/88, Teletraffic Research Centre, University of Adelaide, 1988.

6. Bose, S. and Rappaport, S.S., "High-capacity low-delay packet broadcast multiaccess', *Proc. Intl. Conf. on Comms.*, 1980.

7. Capetanakis, J.I., "Tree algorithms for packet broadcasting channels", *IEEE Trans. on Inf. Theory*, pp 505-515, Sept 1979.

8. Carleial, A.B. and Hellman, M.E., "Bistable behaviour of Aloha-type systems", *IEEE Trans. on Comms.*, Vol. COM-23, no. 4, pp 401-410, April 1975.

9. Dixon, R.C., *Spread Spectrum Systems*, Wiley Interscience, 1976.

10. Gallager, R.G., "Conflict resolution in random access broadcast networks", *Proc. AFOSR Workshop*, Provinceton, MA, pp 74-76, Sept. 1978.

11. Kleinrock, L. and Lam, S.S., "Packet switching in a multiaccess broadcast channel: Performance evaluation", *IEEE Trans. on Comms.*, Vol. COM.1-23, no. 4, pp 410-423, April 1975.

12. Li, V.O.K. and Yan, T-Y., "Adaptive mobile access protocol (AMAP for the message service of a land mobile satellite experiment (MSAT-X)", *IEEE J. Sel. Areas in Comms.*, Vol. SAC-2, pp 621-627, July 1984.

13. Li, V.O.K. and Yan, T-Y., "An integrated voice and data multiple access scheme for a land mobile satellite system", /it Proc. IEEE, Vol. 72, no. 11, pp 1611-1619, Nov. 1984.

14. Longo ed., G., "Multi-user Communication Systems", *CISM Courses and Lectures No. 265*, Springer-Verlag, 1981.

15. Massey, J.L., "Collision-resolution algorithms and random access communications", Multi-user Communication Systems, G. Longo ed., *CISM Courses and Lectures No. 265*, Springer-Verlag, 1981.

16. Moseley, J. and Humblet, P., "A class of efficient contention resolution algorithms for multiple access", *IEEE Trans. on Comms.*, Vol. COM-33, pp 145-151, Feb. 1985.

17. Parker, E.B., "Micro-earth stations as personal computer accessories", *Proc. IEEE*, Vol. 72, no. 11, pp 1526-1531, Nov. 1984.

18. Puente, J.G., et al, "Multiple-access techniques for commercial satellites", *Proc. IEEE*, pp218-229, Feb. 1979.

19. Rappaport, S.S., "Demand assigned multiple access systems using collision type request channels: traffic capacity considerations", *IEEE Trans. on Comms.*, Vol. COM-27, pp 1325-1331, 1979.

20. Rappaport, S.S. and Bose, S., "Demand assigned multiple access systems using collision type request channels: traffic capacity comparisons",*Proc. IEEE*, Part E, Vol. 128, no. 1, pp 37-43, Jan. 1981.

21. Raychaudhuri, D. and Joseph, K., "Channel access protocols for Ku-band VSAT networks: A comparative evaluation", *IEEE Comms. Magazine*, Vol. 26, no. 5, pp 34-44, May 1985.

22. Roberts, L.G., "Dynamic allocation of satellite capacity through packet reservations" *Proc. AFIPS Conf.*, Vol. 42, pp 711-716, 1973.

23. Roberts, L.G., "Aloha packet system with and without slots and capture", *Computer Comms. Review*, Vol. 5, April 1975.

24. Schwartz, J.M., et al, "Modulation techniques for multiple access to a hard limiting satellite repeater",*Proc. IEEE*, pp 763-777, May 1966.

25. Spilker, J.J., *Digital Communications by Satellite*, Prentice-Hall Information and System Sciences Series, Prentice-Hall Inc., 1977.

26. Wang, C.C., and Yan, T., "Performance analysis of an optimal file transfer protocol for integrated mobile satellite services", *Proc. Globecom 87*, Vol. 3, pp 1680-1686, 1987.

27. Yan, T-Y. and Li, V.O.K., "A reliable pipeline protocol for the message service of a land mobile satellite experiment", *IEEE J. Sel. Areas in Comms.*, Vol. SAC- 5, no. 4, pp 637-647, May 1987.

Chapter 5

DIGITAL MODEM DESIGN

by William G. Cowley
Australian Space Centre for Signal Processing
University of South Australia

This Chapter considers the modems used in digital communications systems. In particular, the design and implementation of modems for mobile satellite (mobilesat) systems will be examined. These modems provide a good illustration of the need for advanced approaches to modem design because of the propagation characteristics of the mobilesat channel. The emphasis will be on using discrete-time signal processing techniques. This provides greater flexiblity and allows performance closer to the theoretical optimum. In higher speed modems, approaches combining continuous-time and discrete-time processing will be examined.

Our attention will be focused on digital modems. Modern systems almost invariably encode analog signals (such as speech waveforms) into digital streams before modulation. In this way digital modulation and channel coding can be employed for analog as well as digital services as explained in Chapter 1. Since the most common family of digital modulation methods is M-PSK modulation, most of the discussion will be concentrated in this area. Many of the methods are also suitable, perhaps in a modified form, for systems which use amplitude as well as phase variations of the carrier (e.g. QAM).

The general layout of the chapter is as follows. In this section we will start with a descriptive review of modem functions and then complement this with a mathematical description of basic modem operation. We also look briefly at some particular problems of the mobile environment. In the next two sections we will examine filtering and frequency translation requirements in modems and how they can be implemented with digital signal processing techniques. This allows generic discrete-time modulator and demodulator structures, suitable for programmable implementation, to be suggested in Section 4. Synchronisation methods are then discussed in the context of maximum likelihood estimation. Sections 5 and 6 deal with carrier phase and symbol timing recovery respectively.

5.1 INTRODUCTION

To review the functions of a digital modem, consider the traditional block diagram of a PSK/QAM modulator and demodulator shown in Figure 5.1.

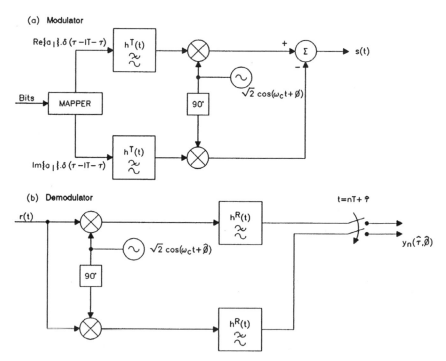

Figure 5.1: Conventional PSK/QAM modem.

A bit stream carrying information is presented to the modulator input. As explained in Section 1.3.2, the bits are mapped to complex-valued numbers (called symbols) from a finite set called a *signal constellation*. The real and imaginary parts represent Inphase and Quadrature (I and Q) components. Typical signal sets are shown in Figure 1.13.

Filters in the modulator have the task of limiting the bandwidth of the transmitted signal. Assume that regularly spaced impulses are input to the I and Q filters, and the impulse sizes are weighted according to the real and imaginary parts of the symbols being transmitted. The spacing between the impulses is called the symbol period, T_s secs. The modulator then translates the filtered signals to the carrier frequency by means of a quadrature mixer. The use of orthogonal signals in the mixers means that the I and Q message components are shifted to the same range of bandpass frequencies but can still be separated at the receiver.

Frequency translation and filtering operations at the receiver are very similar to those at the transmit end. As shown in Figure 5.1 (b), the received signal is converted into I and Q baseband signals in a quadrature mixer. Receive filtering then removes noise outside the message bandwidth. As explained in Section 1.3.2, satellite communication systems frequently use matched transmit and receive filtering with an overall response that satisfies a Nyquist criterion in order to eliminate inter-symbol interference (ISI) and give the optimum bit error rate. We will shortly verify that when Nyquist filtering is used and the sampling operation in the demodulator takes place at the symbol midpoint, the result doesn't depend on the value of adjacent symbols. The resulting samples, referred to as *soft-decision demodulator outputs* are usually passed to a decoder.

In Figure 5.1 an arbitrary time offset of τ seconds in the symbol timing and ϕ radians in the local oscillator phase are shown in the modulator. In general, the receiver has no knowledge of these quantities except what it can derive from the (noisy) received signal. The main synchronisation functions of the demodulator are therefore to estimate these time and phase offsets; the estimates will be denoted $\hat{\phi}$ and $\hat{\tau}$.

In most digital communications systems ϕ and τ are slowly varying compared to the symbol period. For example, if a slight frequency difference exits between the transmit and receive local oscillators (plus an initial phase offset), this can be interpreted as a linearly changing phase offset, the estimate of which must be continuously updated. Another type of synchronisation requirement is to identify unique symbol patterns which indicate the start of each frame.

Performance specifications of digital modems reflect the success with which filtering and synchronisation functions have been achieved. The modem *implementation loss*, IL, is the difference between the measured E_b/N_0 required to give a certain bit error rate and the theoretical E_b/N_0 to give the same error rate in an AWGN channel. The modem performance is often measured with modems connected back-to-back at the Intermediate Frequency. Modem imperfections are often caused by IF conversion problems, filtering imperfections (e.g. causing ISI) and noisy synchronisation estimates (causing timing and phase jitter). In modems which operate on well-behaved channels it is now possible to achieve an IL less than 1dB, [1] with careful design. The lowest value of E_b/N_0 for reliable operation is of interest and is called the *synchronisation threshold*. In systems involving burst modes of operation, such as TDMA satellite networks, the speed and reliability of synchronisation is obviously crucial.

5.1.1   Special Requirements of Mobilesat Modems

Modems operating in mobile satellite applications form a useful group to study since they present the designer with a difficult task. They often need to operate with the following disadvantages:

- Time varying channel responses due to multipath fading and/or obstruction loss.

- Varying frequency offsets due to the motion of the mobile terminal.

- Low signal to noise ratios.

- Burst mode operation due to packet transmission or voice activation.

- Low implementation loss and rapid recovery from fades

There seem to be only two advantages, both of which stem from the relatively low bandwidths of the signals involved. Firstly, low bandwidths imply that the fading is not frequency selective. This is called "flat fading" since all message frequencies are affected by the same amplitude change and phase shift at any instant of time (see Chapter 2). Frequency selective fading implies that adaptive equalisation, rather than fixed receive filters, is required to eliminate ISI.

The other advantage is the fact that low bandwidths mean that the received signal can be sampled at or above the Nyquist rate and processed with discrete-time techniques in programmable digital signal processors. This approach allows greater flexibility, more repeatability, and better performance. Conventional analog processing techniques in modems usually employ specialised phase locked loops (PLLs) to carry out synchronisation functions. These are poorly suited, for example, to the rapid phase changes of the mobile channel, or the need for rapid symbol timing recovery in burst-mode operation. It would be true to say that mobile satellite systems could not exist in their present form without the relatively recent advances in technology of digital signal processors.

5.1.2   Continuous-Time Formulation of Modem Functions

It is valuable to develop a continuous-time model of the functions shown in Figure 5.1. To write an expression for the transmitted signal, if the l−th symbol a_l, is input to the transmit filter whose impulse response is $h^T(t)$ at a time offset of τ seconds, the output will be

$$a_l \delta(t - lT_s - \tau) * h^T(t)$$

where $*$ represents the convolution operator. (When $h^T(t)$ is real-valued, independent filters can be used, as shown in the figure.) Simplifying the convolution shown above and summing over all symbols gives a filtered signal

$$c(t) = \sum_{l=-\infty}^{\infty} a_l h(t - lT_s - \tau) \tag{5.1}$$

The frequency translation operation can be written as $\mathrm{Re}\{c(t)\sqrt{2}e^{j\omega_c t}\}$, where ω_c is the carrier frequency, since this equals $\mathrm{Re}\{c(t)\}\sqrt{2}\cos\omega_c t - \mathrm{Im}\{c(t)\}\sqrt{2}\sin\omega_c t$ as shown. The reasons for the $\sqrt{2}$ scaling will be explained later. Our final expression for the transmitted signal is therefore

$$s(t) = \mathrm{Re}\left\{ \sum_l a_l h(t - lT_s - \tau)\sqrt{2}e^{j(\omega_c t + \phi)} \right\} \tag{5.2}$$

From Figure 5.1(b) it can be seen that the demodulator output for the n−th symbol can be written as

$$\left. \left(r(t)\sqrt{2}e^{-j(\omega_c t + \hat{\phi})} * h^R(t) \right) \right|_{t = nT_s + \hat{\tau}} \tag{5.3}$$

where

$$r(t) = s(t) + n(t) \tag{5.4}$$

and $n(t)$ is white Gaussian bandpass noise. Rather than substitute (5.2) directly into (5.3) and simplify, which becomes a little tedious, it is instructive to first show that the frequency translation operation can be replaced simply by a phase shift which equals the difference between the transmit and receive oscillators. Figure 5.2 shows the modem schematic redrawn in a more compact form.

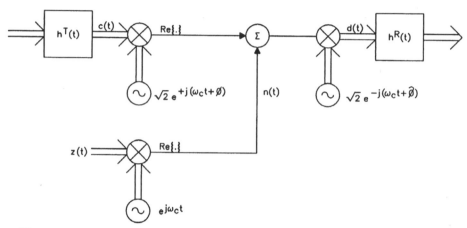

Figure 5.2: Frequency translation operations using complex-valued notation.

Considering the frequency translation path only, and neglecting the additive noise for the moment, the output can be written

$$
\begin{aligned}
d(t) &= \text{Re}\left\{ c(t)\sqrt{2}e^{j(\omega_c t + \phi)} \right\} \sqrt{2}e^{-j(\omega_c t + \hat{\phi})} \\
&= \frac{1}{2}\left(c(t)e^{j(\omega_c t + \phi)} + c^*(t)e^{-j(\omega_c t + \phi)} \right) 2e^{-j(\omega_c t + \hat{\phi})} \\
&= c(t)e^{j(\phi - \hat{\phi})} + \quad \text{double frequency term}
\end{aligned}
$$

The double frequency term can be neglected because of the receive filter. Thus the IF up and down conversion process can be replaced simply by a phase shift of the complex-valued baseband signal.

The effect of noise can also be taken into account by using a baseband representation with

$$
n(t) = \text{Re}\{z(t)e^{j\omega_c t}\} \tag{5.5}
$$

where $z(t) = w^I(t) + jw^Q(t)$, and w^I and w^Q are uncorrelated zero mean Gaussian processes whose power spectral densities are identical, each having twice the spectral density of $n(t)$. For details see Reference [2], Section 10.6.

It makes no difference if we include a phase offset (of say $\hat{\phi}$) in equation (5.5). Hence the received noise can be treated in the same manner as the transmitted signal of (5.2) except that the noise amplitude will be reduced by a factor of $\sqrt{2}$ (due to the absence of $\sqrt{2}$ in (5.5)).

The output of the demodulator for the l-th transmitted symbol during the n-th symbol period can consequently be expressed

$$\left(a_l h^T(t - lT_s - \tau) e^{-j\Delta\hat{\phi}} + z(t)/\sqrt{2} \right) * h^R(t) \Bigg|_{t = nT_s + \hat{\tau}}$$

where $\Delta\hat{\phi} = \hat{\phi} - \phi$. Letting $h^C(t)$ represent the combined filter impulse response i.e. $h^T(t) * h^R(t)$, the output for one symbol can be written

$$y_n(\hat{\tau}, \hat{\phi}) = a_l h^C \left((n - l)T_s + \Delta\hat{\tau} \right) e^{-j\Delta\hat{\phi}} + z(t) * h^R(t)/\sqrt{2} \Bigg|_{t = nT_s + \hat{\tau}}$$

where $\Delta\hat{\tau} = \hat{\tau} - \tau$. Now summing over all transmitted symbols gives an expression for the demodulator output

$$y_n(\hat{\tau}, \hat{\phi}) = \sum_{l=-\infty}^{\infty} a_l h^C \left((n - l)T_s + \Delta\hat{\tau} \right) e^{-j\Delta\hat{\phi}} + w_n \tag{5.6}$$

where w_n is a complex-valued noise sample of $z(t) * h^R(t)/\sqrt{2}$.

Notice that if there is no time offset ($\Delta\hat{\tau} = 0$) and if $h^C(kT_s) = 0$ when $k \neq 0$, the output sample simplifies to

$$y_n(\tau, \hat{\phi}) = e^{-j\Delta\hat{\phi}} h^C(0) a_n + w_n \tag{5.7}$$

This indicates that given perfect timing synchronisation and filtering the demodulator faithfully reproduces the transmitted symbols, scaled and rotated according to the phase offset, plus noise which can not be avoided. Furthermore if $\Delta\hat{\phi} = 0$ and the overall channel response is normalised so that $h^C(0) = 1$, the demodulator soft-decision output samples will be noisy version of the transmitted symbols.

A useful result concerning the size of the noise term in the above equation may be derived. If the real and imaginary parts of w_n each have variance σ_N^2, then if we assume ideal filtering, this is related to the energy per symbol over noise spectral density by

$$E_s/N_0 = \frac{|a|^2}{2\sigma_N^2} (h^C(0))^2 \tag{5.8}$$

where $|a|$ is the symbol magnitude.

5.2 FILTERING

Modem filters perform the crucial functions of limiting the transmit power spectrum and removing excess received noise. In this section we look at the properties of real filters and discuss design considerations for filter implementations.

5.2.1 Filter Requirements

We have just seen that the convolution of transmit and receive filter impulse responses must be zero at multiples of T_s to avoid ISI. It may also be shown (e.g. [3], sect. 4.2) that the optimum BER may be achieved with a matched filter implementation where

$$h^R(t) = (h^T(-t))^* \tag{5.9}$$

Again we will use a superscript to denote the particular filter in question; in this case the receive and transmit filter impulse responses. Since only real-valued responses are needed for PSK and QAM, this equation indicates that the receive filter response is simply a time-reversed version of the transmit pulse shape (possibly time shifted as well). Recall from Section 1.3 that to satisfy the ISI requirement, in the frequency domain the overall response should have vestigial symmetry about the Nyquist frequency $r_s/2$ and that raised cosine responses are frequently used. This may be written

$$H^N(f) = \begin{cases} T_s & |f| < (1-\alpha)/2T_s \\ \frac{T_s}{2}\left(1 - \sin\left(\frac{\pi T_s(f-1/2T_s)}{\alpha}\right)\right) & (1-\alpha)/2T_s < |f| < (1+\alpha)/2T_s \end{cases} \tag{5.10}$$

In many satellite communication systems $H^R(f)$ and $H^T(f)$, the Fourier Transforms of $h^R(t)$ and $h^T(t)$, are therefore approximations to $H^{RN}(f) = \sqrt{H^N(f)}$ (i.e. root-Nyquist filters).

Real filters can be realised with continuous-time or discrete-time implementations. In the former case, for high rates passive filters must be used, while active filter circuits can be employed up to several megasymbols per second. DSP hardware now exits which allows a wide range of data rates to be handled. Real filters can never have perfect root-Nyquist responses since their responses must be causal (i.e. $h(t) = 0$ for $t < 0$). Nevertheless the additional implementation loss due to the non-ideal filtering can be made quite small, as will be demonstrated shortly.

In conventional modulators, the input signals are often square pulses rather than the delta functions modelled in Section 5.1. To achieve the required transmit spectrum the continuous-time filter response must therefore be compensated with an inverse sinc function weighting. This complication doesn't arise when the filtering is implemented digitally. We should also mention that modems may perform filtering on the bandpass signals instead of the lowpass signals. Since linear operations are involved, the order may be swapped and the same effect is obtained (provided the appropriate bandpass filter response is used).

Consider a truncated and time shifted version of the root-Nyquist impulse response as shown in Figure 5.3 ($\alpha = 0.4$). The truncated version will be denoted by $\overline{h}^{RN}(t)$. Obviously if the truncation length is large enough the filtering imperfections when

$$h^T(t) = h^R(t) = \overline{h}^{RN}(t - t_D) \tag{5.11}$$

will be negligible. The total time delay in the system will be $2t_D$ which simply represents a linear phase shift in the overall channel response $H^C(f) = H^T(f)H^R(f)$.

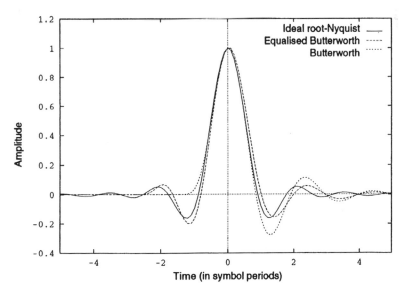

Figure 5.3: Impulse responses of ideal and actual filters.

Compare the Butterworth filter impulse responses in Figure 5.3 with h^{RN}.

This 6th order Butterworth filter has the required cutoff frequency of $r_s/2$. It will produce ISI since the shape of it's magnitude response doesn't exactly match the root-Nyquist shape and because it's phase response is non-linear (which produces group delay distortion). The latter may be corrected (or at least improved) by using a phase equaliser [4]. Then the equalised response resembles the ideal more closely than the unequalised. Simulations show that the Implementation Loss for these filters is small ($\simeq$ 0.2 dB). Equalised 6th order Butterworth filters therefore provide quite a reasonable approximation to the desired response for commonly used values of excess bandwidth parameter and are often employed.

Discrete-time realisations of modem filters use Finite Impulse Response (FIR) filters since their phase shift is linear provided that their coefficients are symmetrical. Recall that FIR filters have only feedforward signal paths and so the impulse response is simply given by the values of the multiplier coefficients. Further, the output signal y_n from a $L-$th order FIR filter with input x_n is given by the discrete-time convolution sum

$$y_n = \sum_{i=0}^{L} h_i x_{n-i}$$

In the present application, if the impulse response is chosen to be samples of $\overline{h}^{RN}(t - t_D)$ according to

$$h_i = \overline{h}^{RN}(t - t_D)\bigg|_{t = iT, i = 0, \cdots L}$$

then the FIR filter will have the desired response provided that the filter order L is large enough and the sampling interval $T\ (= 1/f_s)$ is small enough to avoid aliasing.

This leads us to consider how many samples per symbol period are necessary. Lets assume that an integer number of samples M_s is chosen per T_s, since this choice will have some advantages for filter implementation and synchronisation. Therefore the sampling rate of the FIR filter will be

$$f_s = M_s r_s \tag{5.12}$$

samples per second. Since discrete-time signals must be sampled at or above the Nyquist rate, and the bandwidth of filtered signals will be between $r_s/2$ and r_s, for values of α between 0 and 100%, we require $f_s \geq 2r_s$. The required value of M_s is therefore at least 2.

Exercise 5.1

Determine the frequency response of an FIR filter designed to approximate a raised cosine Nyquist response with $\alpha = .3$. Assume 21 taps and 4 samples per symbol period, where the coefficients $\tilde{h}_n^N$ are simply samples of a truncated version of $h^N(t)$.

Solution

The impulse response is plotted in Figure 5.4 (a). The filter response is given by the Discrete Time Fourier Transform (DTFT) pair [5] :

$$\tilde{H}(f) = \sum_{n=0}^{L} \tilde{h}_n e^{-j2\pi nf/f_s} \qquad (5.13)$$

where $L = 20$. The response is plotted in Figure 5.4 (b), together with the ideal response $H^N(f)$. Notice that the frequency response of the 21 tap filter approximately satisfies the vestigial symmetry condition. The Nyquist response was chosen in this example for clarity. Usually a root-Nyquist response would be employed if matched filtering was required. See Problem 2.

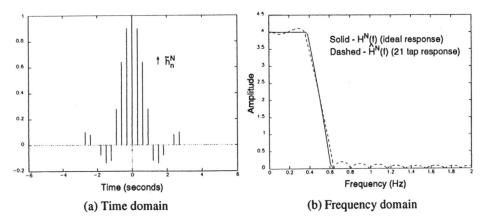

(a) Time domain   (b) Frequency domain

Figure 5.4: Discrete-time Nyquist filter using 4 samples per T_s.

Remember that $\tilde{H}(f)$ will be periodic in f_s (which is true for all digital filters due to the discretisation in time). As the number of samples per symbol increases, the separation between harmonics of the primary response lobe increases. It is difficult in practice to use the mimimum value of $M_s = 2$ due to the imperfect responses of analog anti-aliasing and reconstruction filters.

5.2.2   Design Techniques for FIR Filters

Space does not permit a detailed treatment of FIR filter design methods. These techniques are well described in standard DSP texts such as [5] or [6]. Quite satisfactory results can be obtained with the simple *window design* method. In this approach an inverse DTFT is applied to the desired response $H^{RN}(f)$, thus yielding an impulse response with infinite duration. The extent of this response, and consequently the filter order, is then reduced by applying a smooth time-domain window function, such as the Hanning Window. This method is therefore equivalent to that described in the previous section, except that a smooth window is used in the time domain rather than a rectangular one. Iterative CAD methods are also popular for filter design. The final optimisation of filter designs for modems is best carried out by simulations which include the effects of the overall channel response.

In a later section we will require FIR filters whose effect is to shift the received signal in time by up to $\pm T/2$ seconds. This is achieved with a frequency response which ideally has constant magnitude response and linear phase shift with respect to frequency. Since group delay is given by the derivative of phase shift, the result is a time-domain shift. Similar comments, regarding design techniques, apply to these *interpolating filters*.

The estimation of the computational load for FIR filter implementations is straightforward. The number of multipy/accumulates per second required is $L + 1$ times the sampling rate or $\approx LM_s r_s$. As pointed out, the order L must be large enough to give minimal distortion in the filter's frequency response. We have found, for example, that root-Nyquist FIR filters whose impulse response extends over about $7T_s$ give excellent responses (i.e. about 28 taps when $M_s = 4$) for $\alpha \simeq 40\%$. Therefore, if we let $L = N_D M_s$ (where N_D is about 7) we have

$$N_{MAC} \approx N_D M_s^2 r_s \tag{5.14}$$

This formulation highlights the importance of the number of samples per symbol on the computational load. In some cases it is possible to reduce the number of computations for filtering operations (see later).

5.3   LOW-IF SIGNAL INTERFACES

In a modem implemented with DSP techniques, some flexibility is possible in deciding where to convert between continuous-time and discrete-time signals. It appears from Figure 5.1 that following FIR transmit filters, D/A converters could be employed plus a conventional analog quadrature mixer. Similarly, in the receiver the signal could be sampled after the quadrature mixer and before (or after) the receive filters. These arrangements are well suited to higher data rates.

Alternatively, there are several reasons for realising the quadrature mixer with discrete-time processing in a low speed modem. On the transmit side this implies signal generation at a low IF which can then be translated to the normal IF by a synthesised up-converter. At the demodulator the bandpass low IF signal can be sampled using lowpass or bandpass sampling techniques and translated to baseband as required with a digital quadrature mixer.

Reasons for using a low-IF analog/digital interface include the following:

- elimination of imperfections in analog quadrature mixers, including amplitude and phase imbalances, feedthrough of the carrier component, non-linearities and DC offsets;

- greater flexibility in frequency translation;

- cost saving in use of only one A/D and D/A converter.

A variety of techniques exist for digitally synthesising and sampling bandpass signals. Only the basic principles will be provided in this section.

5.3.1 Synthesising the low-IF Signal

It has already been demonstrated that the transmit quadrature mixer may be viewed as multiplication between I and Q baseband signals and a complex exponential. This may be implemented in the discrete-time domain by multiplying the sampled baseband signal and a (complex) Digitally Controlled Oscillator (DCO). Suppose p_n denotes the discrete-time filtered baseband signal in the modulator. Let

$$q_n = p_n e^{j2\pi n f_1/f_s}$$

as shown in Figure 5.5. In the frequency domain:

$$
\begin{aligned}
Q(f) &= \sum_{n=-\infty}^{\infty} q_n e^{-j2\pi n f/f_s} \\
&= \sum_{n=-\infty}^{\infty} p_n e^{j2\pi n f_1/f_s} e^{-j2\pi n f/f_s} \\
&= P(f - f_1)
\end{aligned}
\tag{5.15}
$$

The effect is therefore a frequency translation by f_1 Hz as expected. The shifted signal q_n must be complex-valued since its magnitude spectrum is not an even function.

(a) Time domain

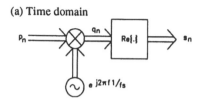

(b) Frequency domain

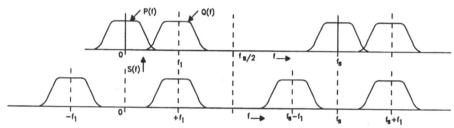

Figure 5.5: Frequency translation by f_1 Hz.

If we now take the real part of q_n, as occurs in the modulator quadrature mixer, the result is

$$
\begin{aligned}
S(f) &= \mathrm{DTFT}\left(\frac{1}{2}\left(q_n + q_n^*\right)\right) \\
&= \frac{1}{2}\sum_n q_n e^{-j2\pi n f/f_s} + \frac{1}{2}\sum_n (q_n e^{j2\pi n f/f_s})^* \\
&= \frac{1}{2}\left(Q(f) + Q^*(-f)\right) \\
&= \frac{1}{2}\left(P(f - f_1) + P^*(-f - f_1)\right)
\end{aligned}
$$

In words, a bandpass real-valued signal is generated, centered around f_1. Note that the size of the frequency shift should not result in $Q(f)$ stradling the foldover frequency $f_s/2$ else aliasing will result.

From an implementation point of view, we do not need to calculate the imaginary part of q_n since it is not used. The sin and cos terms needed for the frequency shifting would normally be stored in a table. We can also see that if f_1 and f_s have a large common divisor (e.g. f_s/f_1 is an integer), then only a small number of entries will be needed in the DCO table. Higher speed (non-programable) implementations can make good use of commercial Numerically Controlled Oscillator (or Direct Digital Synthesiser) chips which are now available.

So far we have not had to consider the operation of the D/A converter which converts s_n into a continuous-time signal. Ideally we could imagine this device producing weighted continuous-time impulses (delta functions) after which an ideal low pass or band pass filter would select the desired harmonic of $S(f)$. Real D/A converters employ a *zero-order hold (ZOH)* to retain the output voltage at constant amplitude until the next conversion. The resulting stepped waveforms are usually filtered by an *analog reconstruction filter* which, roughly speaking, removes the high frequency components in the stepped input signal.

(a) Time domain

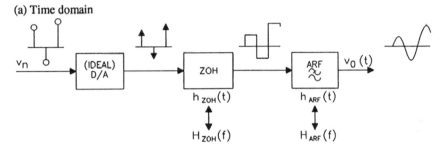

(b) Frequency domain

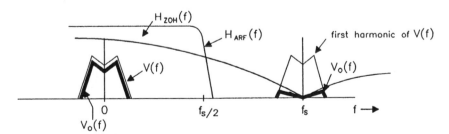

Fiure 5.6: Digital to analog conversion.

More precisely, the operation of real D/A converters may be dealt with by considering the combined filtering effect of the zero-order hold and the reconstruction filter (see Figure 5.6). The impulse response of the former is equivalent to a rectangular pulse of width T. Its magnitude response, $|H_{ZOH}(f)|$, is therefore a sinc function whose first null falls at f_s.

Consider first the case when the D/A is generating a lowpass signal, say from v_n. The effect of the zero-order hold will be to remove most, but not all, of the spectral components of the harmonic of $V(f)$ centered around $\pm f_s$, as shown in the figure. The zero-order hold can therefore be seen to carry out an imperfect filtering action and a reconstruction filter must be used to remove the remaining high frequency components. $H_{ARC}(f)$ should ideally have a cutoff at $f_s/2$ and include an inverse sinc weighting to compensate for the amplitude distortion introduced by $H_{ZOH}(f)$. In this case

$$\begin{aligned}|V_o(f)| &= |V(f)H_{ZOH}(f)H_{ARF}(f)| \text{ for all } f \\ &= |V(f)| \text{ for } |f| < f_s/2\end{aligned}$$

(There will also be a phase shift whose only practical effect will depend on the group delay of the reconstruction filter.) When the D/A converter is producing a bandpass signal such as $S(f)$ in Figure 5.5, rather than the lowpass case just considered, the filtering effect of the zero-order hold remains the same. A bandpass reconstruction filter will be required in most applications.

Schemes for discrete-time to continuous-time conversion can be designed using the principles outlined above. Points to watch include the effects of the sinc function rolloff across the message bandwidth, which must be kept within reasonable limits, the image rejection obtained by the reconstruction filter, and the possible group delay distortion caused by the reconstruction filter. In practice the "reconstruction" filter can be implemented after an analog frequency translation operation. Crystal filters are useful for this purpose provided very good shape factors can be achieved.

Exercise 5.2
Assume that a mobilesat system uses a symbol rate of 3300 symbols/sec and external constraints dictate that the transmit IF must be 28.8 kHz. Decide whether a synthesis scheme using $M_s = 12$ is possible and if so draw a sketch of the output spectrum and estimate the distortion present.

Solution
From (5.12) $f_s = 39.6$ kHz. The spectrum after low pass transmit filtering is represented in Figure 5.7 (a) with solid lines. Suppose that we shift the spectrum down by 10.8 kHz with a complex-valued multiplier, to give the dotted spectrum. After taking the real part of the shifted signal, we have the spectrum shown in part (b). The desired component centered around 28.8 kHz must be selected by a suitable reconstruction filter response.

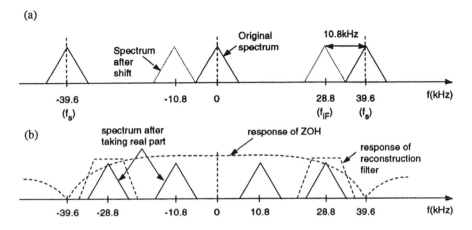

Figure 5.7: Spectra for Exercise 5.2.

The variation in ZOH response between, say, $f_{IF} - r_s/2$ and $f_{IF} + r_s/2$ is easily shown to be about 3dB. Could this be reduced? Suppose the D/A clock rate is doubled and a zero sample inserted after each original sample. This is equivalent to retaining the original sampling rate, and halving the effective D/A impulse response width. H_{ZOH} would therefore have its first null at 79.2 kHz. The variation in response from 27.15 to 30.45 kHz is now only 0.5 dB.

5.3.2 Sampling the low-IF Signal

The bandpass signal may be sampled by lowpass or bandpass sampling techniques. We rely on the fact that the information is not lost provided that a signal is sampled at a rate equal to at least twice its *bandwidth*. Consider the relation between the spectrum of a signal before and after the sampling device [5,sect 4.6]. If the continuous-time received signal $r(t)$ has Fourier Transform $R_c(f)$, and the sampled version r_n has DTFT $R(f)$, then

$$R(f) = \sum_{k=-\infty}^{\infty} R_c(f + kf_s) \tag{5.16}$$

Thus $R(f)$ is a superposition of shifted versions of $R_c(f)$, as expected. Figure 5.8 illustrates the situation when the input signal is bandpass but lies within 0 to $f_s/2$ Hz. After sampling, $R(f)$ can then be shifted down to DC (say by f_2 Hz) with a digital quadrature mixer in exactly the same fashion as described in the last section.

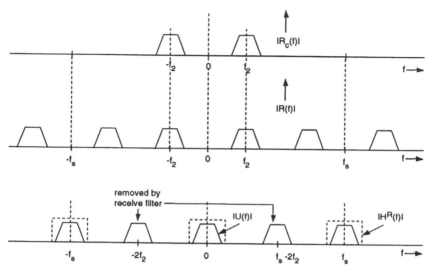

Figure 5.8: Sampling the low-IF signal.

It can be seen that after frequency translation, harmonics of the signal $U(f)$ will be present at $(0$ and $-2f_2) \pm m f_s$ where $m = 1, 2....$ The unwanted set of harmonics offset by twice the low-IF frequency, will be removed by the receive filter response $H^R(f)$ as shown.

Now imagine that the IF frequency is $f_2 + f_s$ Hz, instead f_2 Hz. Since the summation in (5.16) includes all shifted versions of $R_c(f)$, $R(f)$ is unchanged! The result is therefore to equivalent to a frequency shift of f_s followed by lowpass sampling. This effect may be readily interpreted; the frequency offset by f_s (or multiples thereof) has no effect as it simply gives a phase change of 360 degrees each sampling period.

As for the transmit side, it is convenient to arrange for the sampling rate to give an integer number of samples per symbol period. In addition, a suitable choise of intermediate frequency can produce significant simplifications in the implementation of frequency shifting and filtering operations. For example, if the effective low-IF frequency (f_2) is equal to $f_s/4$, the complex multipliers in the frequency shifter become $...1, -j, -1, j, ...$ This corresponds to selecting even samples to form the non-zero members of the I sample stream, odd samples for the Q stream, and alternating the signs of non-zero samples in each case. The number of non-zero multiplications per second in the shifting and filtering operations for both channels is the same as that represented by (5.14). Whilst other forms of bandpass sampling are possible, these simplifications make the scheme outlined above attractive.

Practical design considerations for the receive sampling include A/D specifications such as sampling jitter and slew rates, and the choice of a suitable
anti-aliasing filter before the sampler. Once again this filter can be located
before the final analog mixer stage. The design must ensure that noise and
adjacent channels are not aliased into the message spectrum.

Exercise 5.3

*Consider an IF sampling scheme with the following parameters: symbol rate
of 3300 symb/s, $\alpha = 0.4$, channel spacing = 7.5 kHz, 8 samples per symbol
period and $f_{IF} = f_s/4$. Assume that crystal filtering at 10.7 MHz has a
response which is 3 dB down at $\pm$ 3.75 kHz, and 65 dB down at $\pm$ 8.75 kHz.
Determine the spectrum of the received signal after lowpass filtering.*

Solution

(a) Spectrum before A/D

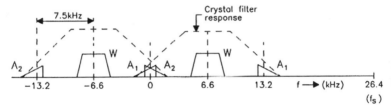

(b) Spectrum after frequency shift

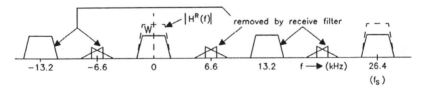

Figure 5.9: Spectra for Exercise 5.3.

*From the information supplied $f_s = 26.4$ kHz and $f_{IF} = 6.6$ kHz. The IF
signal bandwidth will be approximately $1.4 \times 3300 = 4620$ Hz. The gap between adjacent channels will be approximately 2880 Hz, assuming no Doppler
shifts. The received signal spectrum is shown in the first part of Figure 5.9.
Remnants of the adjacent channels, left after the crystal filter, are marked
as A1 and A2. The wanted signal is labelled W. After shifting and filtering,
shown in the second part of the figure, only the wanted signal is retained.
In practice, further savings in the number of computations might be possible
since the output signal could be significantly decimated.*

5.4   LOW SPEED MODEM ARCHITECTURES

We will now make use of the frequency shifting and filtering concepts discussed so far to outline a possible block diagram of a digital modem which is suitable for programmable implementation. During this discussion we will have reason to introduce interpolation and decimation operations in the modem and explain why they are required. The section is mainly directed towards modems implemented with digital signal processors, however some of the techniques can also be applied to high speed DSP implementations.

Although synchronisation algorithms have not been discussed yet, we can treat carrier-phase and symbol-timing offset estimators as "black boxes" whose estimates will be used in the demodulator outlined below. There are many approaches to carrying out the estimation functions and we will return to the examine the internal working of these "black boxes" later in the chapter.

5.4.1   Discrete-Time Modulator Processing

Most of the transmit processing has already been outlined. We can now combine the filtering, frequency translation and digital to analog conversion operations to give the block diagram shown in Figure 5.10.

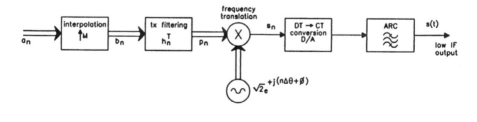

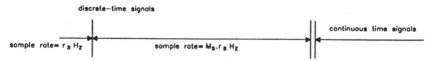

Figure 5.10: Discrete-time modulator structure.

The operation of the first block can be explained as follows. The modulator requires one complex-valued symbol per symbol period and yet Section 5.2 established that at least 2 samples per T_s are required to satisfy the Nyquist sampling criterion. Suppose that $M_s - 1$ zeros are inserted after

each input symbol and that the resulting sequence is filtered as shown in
Figure 5.10. The relation between the a_n and b_n is therefore

$$b_n = a_{n/M_s} \text{ if } n = 0, \pm M_s, \pm 2M_s, \cdots$$
$$b_n = 0 \text{ otherwise} \qquad (5.17)$$

The process of inserting zeros and then lowpass filtering is called *interpolation*. Note that the sampling rate has been increased by a factor of M_s.
Provided that the discrete-time transmit filter has the required response,
and that ideal reconstruction is assumed, this processing will give the same
result as the continuous-time modulator in Section 5.1.

We can point out that significant implementation savings may be made
in the transmit filter computation. Obviously most of the samples entering
the FIR filter are zero and so a large number of the multiply-accumulate
operations can be omitted. The interpolation is equivalent to a symbol-
spaced filter which is evaluated M_s times per T_s, each evaluation using a
different set of filter coefficients. The number of operations (for each of I
and Q streams) is

$$N_{TX-MAC} = N_D M_s r_s \qquad (5.18)$$

where N_d is again the filter impulse response duration in symbol periods.
Comparing this to (5.14) we see a factor of M_s saving.

In fact a further simplification can be obtained for simple signal constellations such as BPSK and QPSK, by remembering that the non-zero members
of b_n will be simply ± 1. This means that multiplications can be avoided
altogether, if desired. This is not as important for digital signal processor
implementations since multiplications are as quick as additions. However, for
high speed modulators the small number of values for the input signal, plus
the limited order of the FIR filter, implies that only a finite number of output samples from the filter are possible. These outputs may be precalculated
and stored in a look up table (LUT) so allowing very simple implementations
of transmit digital filtering up to tens of megasymbols rates [7][8].

5.4.2 Discrete-Time Demodulator Processing

The discrete-time and continuous-time modulator structures have been
similar due to the direct mapping of filtering and frequency translation between domains. At the receiver, extra synchronisation functions are required
and greater variety in demodulator structures is possible.

In the high data-rate case, for example, frequency translation and filtering are best carried out in the continuous-time domain. With a block diagram similar to Figure 5.1 (b), the demodulator could still employ discrete-time synchronisation algorithms which require only 1 sample per symbol period. Rather than attempting to phase lock the local oscillator, as shown, it could be free running with the phase offset removed later from the soft-decision samples (e.g. [7]).

When more elaborate processing is required in the receiver, and the data rates are lower, more discrete-time processing can be employed. This might allow, for example, more optimum filtering, better burst mode synchronisation, or superior performance in time-varying channels. Reasons for using an IF sampling scheme have already been outlined in Section 5.3. Figure 5.11 shows one possible discrete-time architecture for low rate PSK signals.

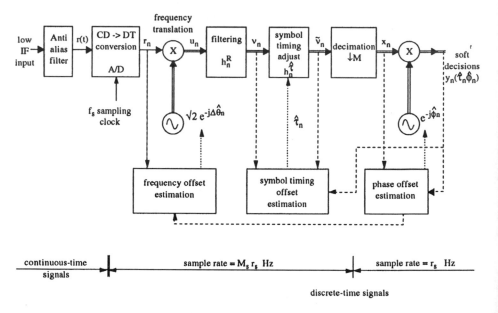

Figure 5.11: Discrete-time demodulator structure.

The combined effects of Doppler shifts plus long term drift in the various up and down converters, means that frequency offset in a mobile communications system, for example, must be continuously estimated and removed. This may be incorporated into the down conversion, as shown in the figure. Notice that it might be necessary to remove the frequency offset in a low rate system prior to receive filtering since it could be a significant fraction of the signal bandwidth. If the estimated frequency offset is $\hat{f}_0$ Hz, the phase step per T in Figure 5.11 will be

$$\Delta\hat{\theta} = 2\pi\frac{\hat{f}_0}{f_s} \tag{5.19}$$

For correct symbol timing, the sampling instants must be translated according to the estimated timing offset. The sampling adjustment (sometimes called "resampling") can be achieved by an FIR filter as described in Section 5.2.2. The time shift will be a fraction of the sampling interval and will cause one complex-valued sample to occur at the symbol midpoint. Denote its impulse response by $h_n^{\hat{\tau}}$. Algorithms for estimating the timing offset are examined in Section 5.6.

The process of selecting the desired sample and discarding the other $M_s - 1$ samples after low pass filtering is called *decimation*. It is represented in Figure 5.11 by the $M \downarrow$ box. This processing can be summarised as

$$
\begin{aligned}
\tilde{v}_n(\hat{\tau}) &= u_n * h_n^R * h_n^{\hat{\tau}} \\
x_n(\hat{\tau}) &= \tilde{v}_{nM}(\hat{\tau})
\end{aligned}
\tag{5.20}
$$

Finally the phase offset remaining after IF sampling and frequency translation can be corrected so that the soft decision outputs (with appropriate scaling) cluster around the nominal signal set constellation. This is achieved by a simple phase rotation by ϕ_n for the $n-$th output as shown. Estimation of the phase offset is discussed in the next section.

$$y_n(\hat{\tau}, \hat{\phi}) = x_n(\hat{\tau})e^{-j\hat{\phi}_n} \tag{5.21}$$

Note that the sampling rate changed by a factor of M_s in (5.20). A number of interpolation and decimation stages could be employed, instead of the single stage methods outlined so far. This may improve the computational efficiency [1]. The result in (5.14) highlights the importance of the oversampling factor in the overall processing rate and indicates that in general it should be kept to the minimum practical value for that stage of processing. On the other hand, the interpolation process may be quite simple if M_s is large enough (see Problem 4). The most efficient arrangement will therefore depend on the particular synchronisation and filtering algorithms selected.

Dashed lines in the demodulator block diagram indicate signals that the estimation algorithms might use in forming estimates of frequency offset, carrier phase offset and timing offset. Notice that in all cases it is possible to use signals before the estimator in question, or after it. This gives rise to *feedforward* or *feedback* processing methods respectively. In the latter case algorithms try to track the current estimate and therefore have a feedback loop. The feedforward methods are not subject to loop "hangup" problems but generally require more computations.

5.5   COHERENT DEMODULATION METHODS

Our subject in this section is the estimation of phase offset on the received signal constellation. Ideally we would like to have a perfect estimate at all times i.e. $\hat{\phi} = \phi$ but this is not possible because of noise in the received signal. A great many techniques for carrier phase synchronisation have been developed; we will give an outline of some methods and will have to omit full derivations in many cases. We will try to provide a unified view of these techniques via maximum likelihood estimation theory.

The model we developed in Section 5.1 assumed coherent demodulation since the incoming symbol's phase was compared to a coherent reference. Almost all modern demodulators work in this manner.

Under very poor channel conditions where it is difficult to establish a phase reference, some schemes have used differentially coherent demodulation. In this case the present symbol's phase is simply compared to the previous symbol's phase. We will not consider this approach any further, although there is nothing to prevent it being implemented in the discrete-time domain. Most of our discusion will be concerned with the AWGN channel since this is the most tractable channel model.

5.5.1   Decision-Directed Maximum Likelihood Estimation

Maximum likelihood estimation techniques form a useful framework for considering how the receiver can make estimates of the carrier phase and symbol timing offsets (ϕ and τ) present in the received signal. In the maximum likelihood approach ϕ and τ are treated as deterministic but unknown parameters. We assume that the received signal is corrupted by additive Gaussian noise as before.

The basis of the maximum likelihood (ML) approach is that the conditional probability of receiving a signal $r(t)$ over some time interval T_0, given an assumed set of transmit signal parameters such as values of ϕ, τ and the transmitted symbols a_l, is maximised with respect to the set of transmit parameters. The conditional probability

$$\text{Prob}\{r(t), t \in T_0 \mid \phi, \tau, a_l \text{ for } lT \in T_0\}$$

is called the *likelihood function*; hence the name maximum likelihood estimation.

It can be shown that in many cases, including this one, the ML criterion is equivalent to the maximum a posteriori probability (MAP) criterion which maximises the conditional probability distribution of the unknown parameters, such as ϕ and τ, given the observed signal [3]. If the transmit symbols are assumed to be unknown at the receiver and the likelihood function is maximised with respect to a_l as well, we can see that the whole process of demodulation and detection is carried out as one processing step!

Unfortunately this joint estimation is not usually computationally feasible. In one approach, which we will follow at the moment, the transmitted symbols are assumed to be known at the receiver and that knowledge is then used to achieve receiver synchronisation.

In practice, estimates of the transmitted symbols are made and used instead of the actual transmitted symbols. This "decision directed" (DD) method (also called "data-aided") works well as long as most of the symbol decisions are correct (i.e. as long as the SNR is not too low). In addition it is possible to separately estimate the timing and phase offsets, or to carry out joint estimation.

If we let the assumed values of time and phase offset at the modulator be $\hat{\tau}$ and $\hat{\phi}$, with a resulting transmitted signal of $s(t, \hat{\tau}, \hat{\phi})$, then it can be shown (e.g. [9], [3]) that the significant components of the likelihood function can be expressed as

$$\Lambda(\hat{\tau}, \hat{\phi}) = e^{2/N_0 \int_{t \in T_0} r(t) s(t, \hat{\tau}, \hat{\phi}) dt} \tag{5.22}$$

The likelihood is therefore maximised by choosing $\hat{\tau}$ and $\hat{\phi}$ to give the largest cross correlation between the received signal and the hypothesised transmit signal. In line with the comments above, we will assume that the transmit symbols are known at the receiver. The alternative case will be considered in the next section.

Obviously the logarithm of the likelihood function, Λ_L, can be maximised instead of Λ. Substituting (5.2) to (5.4) in the equation above, gives a criterion relating the actual transmit parameters to the hypothesized parameters.

Then using (5.6), which indicates how the actual receiver makes use of $\hat{\phi}$ and $\hat{\tau}$, after some manipulation leads to

$$\text{maximise } \Lambda_{Ls}(\hat{\tau}, \hat{\phi}) = \sum_k \text{Re}\{y_k(\hat{\tau}, \hat{\phi})a_k^*\} \qquad (5.23)$$

where the summation is carried out over the observation interval of interest, which is assumed to be a large number of symbol periods, and Λ_{Ls} is simply a scaled version of Λ_L. Note that the summation requires only one sample per symbol period.

This expression may be used to derive estimators for timing and phase offsets. The decision-directed approach means that we will use symbol estimates in place of the actual a_k in (5.23). In particular, suppose we differentiate Λ_{Ls} with respect to the parameter of interest and set the result equal to zero. Considering the phase recovery case, using (5.21) and assuming ideal symbol timing for the moment, plus constant phase offset over the observation interval so that we can drop the subscript from $\hat{\phi}_k$, then

$$\begin{aligned}
\Lambda_{Ls}(\tau, \hat{\phi}) &= \sum_k \frac{1}{2}(x_k(\tau)e^{-j\hat{\phi}}a_k^* + x_k^* e^{j\hat{\phi}}a_k) \\
\frac{\partial \Lambda(\tau, \hat{\phi})}{\partial \hat{\phi}} &= \sum_k \frac{1}{2j}(x_k(\tau)e^{-j\hat{\phi}}a_k^* - x_k^* e^{j\hat{\phi}}a_k) \\
&= \sum_k \text{Im}\{x_k(\tau)e^{-j\hat{\phi}}a_k^*\} \qquad (5.24)
\end{aligned}$$

This result can give rise to either feedforward or feedback decision-directed phase recovery schemes. In the former case, setting the partial derivative equal to zero gives

$$\sum_k \text{Re}\{x_k(\tau)a_k^*\}(-\sin\hat{\phi}) + \sum_k \text{Im}\{x_k(\tau)a_k^*\}(\cos\hat{\phi}) = 0$$

so

$$\hat{\phi} = \arctan \frac{\sum_k \text{Im}\{x_k(\tau)a_k^*\}}{\sum_k \text{Re}\{x_k(\tau)a_k^*\}} \qquad (5.25)$$

A practical implementation of this feedforward scheme could be as follows [10]. The summations are evaluated over a sliding window of symbols with a new phase estimate being generated each symbol period. As each new "unrotated" complex sample x_n becomes available, the previous value of $\hat{\phi}$ would be used to rotate x_n and make a new hard decision for a_n. (5.25) could then be evaluated to give an updated estimate of phase offset. Other variations would also be possible. For example, if the channel contains blocks of known symbols (unique words etc), or if known reference symbols are inserted into the transmit stream to assist synchronisation in poor channels, then the receiver can evaluate (5.25) without having to guess the transmitted symbols.

Now consider using (5.24) in a decision-directed feedback phase recovery approach. In higher speed modems, evaluation of the summations in the preceding method could become computationally difficult. In addition when the channel is "well-behaved" the phase offset will only change very slowly and a simple gradient algorithm can be used to track the phase offset. A classic approach in this situation would be to make an estimate of the partial derivative in (5.24) and use this to adjust the current parameter estimate. The new estimate of $\hat{\phi}$ is therefore

$$\hat{\phi} + \gamma \frac{\widehat{\partial \Lambda_{Ls}}}{\partial \hat{\phi}}$$

where γ is a small positive number called the adaption constant. (Since Λ_{Ls} is convex with respect to $\hat{\phi}$, then if $\hat{\phi}$ is too small, $\partial \Lambda_{Ls}/\partial \hat{\phi}$ will be positive; conversely if $\hat{\phi}$ is too large, then $\partial \Lambda_{Ls}/\partial \hat{\phi}$ will be negative.) The simplest method of estimating the derivative is to make the summation over one symbol only. Consequently the algorithm becomes

$$\hat{\phi}_{n+1} = \hat{\phi}_n + \gamma \text{Im}\{y_n a_n^*\} \qquad (5.26)$$

Notice that the algorithm uses the demodulator output symbol y_n, and its associated hard-decision a_n, to update the phase offset estimate and is therefore a feedback method. We can interpret the phase detector function of the right hand side of this algorithm in a simple way. If we write the current a_n as $e^{j\theta_a}$, then we could express y_n as $re^{j(\theta_a+\phi)}$ where ϕ is the phase offset, and r is a scaling factor. Then $\text{Im}\{y_n a_n^*\}$ equals $\text{Im}\{re^{j(\theta_a+\phi)}e^{-j\theta_a}\}$ or approximately $r\phi$ for small ϕ. Consequently the detector output is proportional to the phase offset, provided the phase offset is not too large. The size of γ is a tradeoff between the desired rate of convergence and the required variance of $\hat{\phi}$. This algorithm effectively uses an exponentially weighted window on the received symbols to estimate the phase offset. This phase recovery method is very suitable for higher speed, continuous mode modems, and can be implemented with relatively little dedicated hardware at rates of many tens of Megasymbols/s [7].

5.5.2 Non-Decision-Directed Phase Recovery

As stated above, the decision-directed method works well until the symbol estimates at the receiver become error prone. Under these conditions it is appropriate to treat the a_l as unknown random variables. In this case the likelihood function of (5.22) must be averaged according to the probability distribution of the transmit symbols. This gives an optimum but not readily implementable result (e.g. [11]). Nevertheless it is worth giving the relevant expression in a simple case such as BPSK as it allows a number of simpler schemes to be realised. Assuming independent, unit amplitude BPSK symbols, the log likelihood function can be shown to be

$$\Lambda_{Ls}(\hat{\phi}) = \sum_k \ln \cosh(\frac{2}{N_0}\text{Re}\{y_k(\hat{\phi})\}) \tag{5.27}$$

The problem with this result is that differentiating with repect to $\hat{\phi}$, as we did before, leads to a non-linear set of equations. A commonly used approximation in this case is that $\ln \cosh x \simeq x^2/2$ for small $|x|$. Hence, at low signal to noise ratios the replacement of the $\ln \cosh$ function by a squarer can still give a close to optimum result for BPSK. Another common approach is to assume that the symbols have a Gaussian distribution, rather than their true discrete distribution function. With this assumption the likelihood function naturally has a squared term instead of the $\ln \cosh$ function [3].

Therefore using the approximation

$$\Lambda_{Ls}(\hat{\phi}) \simeq \sum_k (\text{Re}\{y_k(\hat{\phi})\})^2 \tag{5.28}$$

where the constant terms such as N_0 have been omitted, and defining the real and imaginary terms of x_k as x_k^I and x_k^Q respectively, so that $y_k(\hat{\phi}) = (x_k^I + jx_k^Q)e^{-j\hat{\phi}}$, we obtain

$$\Lambda_{Ls}(\hat{\phi}) \simeq \sum_k (x_k^I \cos \hat{\phi} + x_k^Q \sin \hat{\phi})^2 \tag{5.29}$$

This can be differentiated with respect to the phase estimate to obtain

$$\frac{\partial \Lambda_{Ls}(\hat{\phi})}{\partial \hat{\phi}} = \sin 2\hat{\phi} \sum_k (x_k^Q)^2 - (x_k^I)^2 + \cos 2\hat{\phi} \sum_k 2x_k^I x_k^Q \tag{5.30}$$

and if this is equated to zero we obtain

$$2\hat{\phi} = \arctan \frac{\sum_k 2x_k^I x_k^Q}{\sum_k (x_k^I)^2 - (x_k^Q)^2} \tag{5.31}$$

Once again this presents the possibility of both feedforward and feedback implementations for non-decision-directed phase estimators. The last equation represents a feedforward phase estimator making estimates over a block of symbols containing an arbitary phase offset. Note that our derivation above only covered the case of BPSK and in fact this estimator fails for more complex constellations (see below). In the case of a tracking feedback estimator, another gradient algorithm could be used. It can be easily verified that the right side of (5.30), when the summation is dropped, is equal to $\mathrm{Im}\{(y_k(\hat{\phi}))^2\}$. This phase detector function can be used in a feedback loop in the same way as shown in the last section.

The operation in (5.31) is equivalent to estimating the phase offset after a BPSK signal (with symbols nominally at 0 or π radians) is squared. Squaring the signal removes the effect of the modulation since each symbol is now mapped to 0 radians and only the doubled phase offset is left. The operation is analogous to the traditional analog method of phase recovery for BPSK by means of a squarer followed by a phase locked loop operating at twice the carrier frequency. This "modulation stripping" has a natural extension to higher order PSK signal constellations with M phases - the signal is simply raised to the M-th power and the resulting phase offset (multiplied M fold) is tracked. This is illustrated in Figure 5.12 for QPSK with a phase offset of 20 degrees. Notice that all symbols, after modulation stripping, are mapped to point Z.

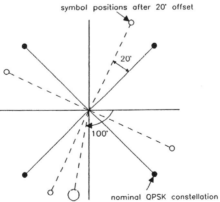

Figure 5.12: Illustration of "modulation stripping".

The phase estimator is

$$M\hat{\phi} = \arctan \frac{\sum_k \mathrm{Im}\{(x_k)^M\}}{\sum_k \mathrm{Re}\{(x_k)^M\}} \qquad (5.32)$$

Whilst (5.32) implies that the magnitude of each symbol is raised to the M power, this need not be the case. For example, as long as the symbol phase is multiplied by M, the magnitude could be left the same or normalised to unity (by taking $|x_k|^1$ or $|x_k|^0$). Different magnitude mappings have slightly different performance and the best alternative depends on the signal to noise ratio and the value of M (see [12] for details).

A practical aspect of the phase estimators in (5.31) and (5.32) is worth mentioning. The multiplied phase offset changes M times as fast as the real phase offset. For each 2π change in real phase offset, $|x_n|^M$ makes M orbits around the unit circle; these must be counted and a correction factor of $2\pi/M$ applied to $\hat{\phi}$ for each orbit [13].

5.5.3 Performance of Phase Estimators

We will now take a brief look at the performance of some of the phase recovery schemes just discussed. By examining a decision-directed and a non-decision-directed method for BPSK, some general characteristics of synchroniser performance can be observed.

Let us analyse the decision-directed feedforward approach in (5.25) with the aim of determining the variance of the estimates $\hat{\phi}$. Since we are considering BPSK, the numerator becomes $\sum_k x_k^Q a_k$ and the demoninator is $\sum_k x_k^R a_k$. We will assume ideal symbol timing and channel filtering (i.e. overall Nyquist response and $h^C(0) = 1$). The x_k samples are uncorrelated and for the present purposes we can let the K transmitted symbols within the observation interval be 1. For the moment also imagine that there is no static phase offset on the received symbols. Thus (using (5.7)

$$x_k = 1 + w_k^I + j w_k^Q \qquad (5.33)$$

and the phase estimate becomes

$$\hat{\phi} = \arctan \frac{w_K^I / K}{1 + w_K^R / K}$$

where the summed noise terms w_K represent $\sum_k w_k$ for each of the I and Q components. It can be shown that the noise components are uncorrelated from symbol to symbol given the ideal filtering assumed above. We can see from the equation above that as the observation interval (K) becomes large, the denominator noise terms can be neglected. In addition the numerator terms become very small and so we can drop the arctan function. The estimator will be unbiased as the noise terms are zero mean. Finally therefore, since the variance of w_K^I is K times that of w_n^I (i.e. $K\sigma_N^2$), the result is

$$\sigma_{\hat{\phi}}^2 = \frac{\sigma_N^2}{K}$$

or, using (5.8),

$$\sigma_{\hat{\phi}}^2 = \left(2K\frac{E_s}{N_0}\right)^{-1} \tag{5.34}$$

Remember that we assumed there was no phase offset in deriving this result. It is fairly easy to show that the phase estimate has the same variance for any phase offset, as we would expect. The result in (5.33) is interesting for a couple of reasons. It can be shown that the *Cramer-Rao Bound* (CRB), which is a lower bound for efficient estimators [14], gives this same result when applied to estimating the phase of an unmodulated carrier over K symbol periods (where K is large). In other words, decision-directed schemes can sometimes avoid the performance penalty associated with modulating the transmitted signal with information bearing symbols. (We could expect that the perturbation of carrier phase by multiples of $2\pi/M$ radians would upset the estimation of phase offset!) However, this attractive performance of a decision-directed phase estimator only applies in practice when no errors are made at the receiver in estimating the transmitted symbols, i.e. when the signal to noise ratio is sufficiently high. This threshold naturally depends on the signal constellation in use.

Consider now the non-decision-directed scheme of section 5.5.2. For simplicity we again consider the case of no static phase offset, and make the same assumptions as above. From (5.31) and (5.33) we have

$$2\hat{\phi} = \arctan \frac{\sum_k 2(1 + w_k^I)w_k^Q}{\sum_k (1 + w_k^I)^2 - (w_k^Q)^2}$$

Using similar arguments to those given above, if K is large the noise terms from the denominator, and the arctan nonlinearity, can be neglected, giving

$$\hat{\phi} = \sum_k (w_k^Q + w_k^Q w_k^I)/K$$

Again, the mean value of the estimator is zero. In taking the expected value of $\hat{\phi}^2$, almost all the products between different noise terms will drop out as the noise components are independent between symbols and I and Q paths. The only terms left are

$$E\{\hat{\phi}^2\} = E\{\sum_k \left((w_k^Q)^2 + (w_k^Q w_k^I)^2\right)/K^2\}$$

so the result becomes

$$\sigma_{\hat{\phi}}^2 = \frac{\sigma_N^2}{K}(1 + \sigma_N^2) \tag{5.35}$$

Observe that the result is the same as the decision-directed case except for an extra term which represents the effect due to squaring the signal and has been called the *squaring loss* [31]. As pointed out before, squaring is the means whereby the modulation is removed so that the phase can be estimated. This analysis becomes difficult for more complex signal constellations, or when shorter observation intervals do not allow the simplifying assumptions used above. Simulation may then used to assess the performance of synchronisers.

Exercise 5.4

Simulate the performance of the feedforward phase estimators in (5.25) and (5.32) for QPSK modulation when 20 symbols are used in the observation interval.

Solution

Figure 5.13 shows a set of simulation results for the feedforward phase estimators, plus the Cramer-Rao Bound.

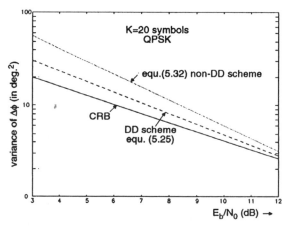

Figure 5.13: Comparison of feedforward phase estimators.

It can be seen that at high signal to noise ratios the squaring loss is very small and might as well be neglected. In an intermediate range of signal to noise ratios, the decision-directed method shows improved performance. At the lowest E_b/N_0 in Figure 5.13 the non-decision-directed estimator is starting to phase slip significantly (see next section). The performance difference between the two estimators is considerably larger for QPSK than BPSK.

5.5.4 Ambiguity Removal

We can see, for example from (5.32), that estimates of phase offset have a range of $2\pi/M$ radians. Actual phase offsets outside this range will be mapped back into $|\hat{\phi}| < \pi/M$. All phase estimators for M-PSK have an M fold ambiguity in their operation. The result of this effect is that the soft-decision symbols from the demodulator could be rotated by any multiple of $2\pi/M$ radians. A change in the symbol mapping by an extra $2\pi/M$ radians is called a *phase slip*.

There are a number of methods of resolving phase ambiguity, and all practical systems must ensure that the output symbols are not permuted with respect to the input symbols. The most common method is *differential encoding and decoding* of symbols. In effect, the differences between phases of input symbols are transmitted, instead of the absolute phases. This method is easy to implement [15] but causes an increase in demodulated bit error rate with respect to the uncoded case. Systems that use unique words, such as TDMA systems, may resolve the phase ambiguity by searching for any of the permuted versions of the unique word. Finally, FEC can be used in some applications to resolve phase ambiguity.

5.5.5 Advanced Methods of Phase Recovery

We have concentrated on the AWGN channel as this situation is relatively easy to analyse compared to any other channel model. In mobilesat modem applications, some of the preceeding feedforward estimators can give satisfactory performance provided that the channel statistics are relatively stationary during the observation window [1].

Ideally however, the phase estimation approach should take into account the channel model. Promising recent developments for fading channels include Kalman filter approaches to phase estimation in the mobile channel [16], [17], and the use of reference symbols for channel state estimation (e.g. [18].

5.6   SYMBOL TIMING RECOVERY

In most digital modulation schemes there is no discrete spectral component at the symbol rate. The demodulator therefore has no direct (or prior) knowledge of exactly when the received symbols start and end, and it must determine this information from the noisy received signal. The extent to which the process of symbol timing recovery is successful will determine how much additional degradation in system performance is introduced due to timing errors. In almost all situations the receiver knows the incoming symbol rate and so it must only deduce the phase of the received symbols relative to its own local symbol clock reference.

In this section we will examine feedback and feedforward approaches to the timing recovery problem. If we assume that the signal is present for relatively long periods and that a timing estimate has been made so that the symbol midpoints are approximately known, then the task simplifies to that of making small corrections to the symbol sampling point. A function of the received signal which indicates how the timing should be adjusted is called a *timing detector*. The output from such a detector can be employed to control the sampling device. The resulting timing loop can be modelled as a feedback system and analysed or simulated to determine the sampling jitter and/or bit error rate degradation.

Alternatively, suppose that the duration of the received signal is relatively short (e.g. tens or hundreds of symbol periods) such as in a TDMA or packet radio system. In these applications, rapid timing acquisition is of greatest importance. Until recently, a preamble signal component was always used in burst mode systems to allow timing estimates to be made before the data started; in modern systems this need not be the case (e.g. Inmarsat C) since with discrete-time methods the received signal burst can be sampled and processed after the burst. We will look at a class of timing estimators that use a feedforward processing approach suitable for burst mode operation.

5.6.1   Timing Detectors for Tracking Approaches

We will first consider decision-directed methods that follow from the maximum likelihood equations of the previous section. To derive a timing detector, take the partial derivative of $\Lambda_{Ls}(\hat{\tau})$ with respect to $\hat{\tau}$, giving

$$\frac{\partial \Lambda(\hat{\tau})}{\partial \hat{\tau}} \;=\; \sum_k \frac{\partial}{\partial \hat{\tau}} \mathrm{Re}\{y_k(\hat{\tau}) a_k^*\}$$

$$= \sum_k \mathrm{Re}\{\dot{y}_k(\hat{\tau})a_k^*\} \qquad (5.36)$$

where

$$\dot{y}_k(\hat{\tau}) = \frac{\partial}{\partial\hat{\tau}} y_k(\hat{\tau})$$

Alternatively, using the low SNR approximation for the BPSK likelihood expression in (5.28), gives

$$\frac{\partial \Lambda(\hat{\tau})}{\partial \hat{\tau}} \approx \sum_k \frac{\partial}{\partial \hat{\tau}} (\mathrm{Re}\{y_k(\hat{\tau})\})^2$$

$$= \sum_k 2\mathrm{Re}\{y_k(\hat{\tau})\}\mathrm{Re}\{\dot{y}_k(\hat{\tau})\} \qquad (5.37)$$

The continuous-time versions of these equations have provided one approach to timing recovery for many years. For example, using the PAM case for simplicity, (5.37) could be realised by forming the product between the filtered received signal and its derivative, and using this to control a VCO which provides the sampling clock [19]. To make use of this method in modems that use discrete-time processing, approximations can be made to the differentiation operations, as follows.

A rough estimate of the current signal derivative, using only one sample per symbol period, would be

$$\dot{y}_n \approx \frac{y_{n+1} - y_{n-1}}{2T}$$

(The notation has been simplified by dropping the explicit dependence of y on $\hat{\tau}$.) This approximation can be used in (5.36) to give a decision-directed timing detector. Removing the summation to give an "instantaneous" estimate of $\partial\Lambda(\hat{\tau})/\partial\hat{\tau}$ gives

$$\frac{\partial\Lambda(\hat{\tau})}{\partial\hat{\tau}} \propto \mathrm{Re}\{y_{n+1}a_n^* - y_{n-1}a_n^*\}$$

resulting in the decision-directed tracking algorithm

$$\hat{\tau}_{n+1} = \hat{\tau}_n + \beta\mathrm{Re}\{y_n a_{n-1}^* - y_{n-1}a_n^*\} \qquad (5.38)$$

where β is the adaption constant. (Notice that the indices in the first signal component have been decreased by one in order to make a practical algorithm; this can be justified from $E\{y_{n+1}a_n^*\} = E\{y_n a_{n-1}^*\}$.) Again, the symbols values in (5.38) are not actually known at the receiver, but estimates of a_n and a_{n-1} are used instead.

Alternative derivations of this class of algorithm can be found in [20] or [21]. Better performance can be obtained by using, for example, two samples per T to improve the derivative estimate, however a significant advantage of (5.38) for high speed modems is that only one sample per symbol is needed. It works for all PSK signal sets. A larger size will produce a faster but noisier response.

Notice that the algorithm uses the symbol-sampled signal y *after* phase recovery. It will not work reliably if a phase offset is present. This means, for example, that if this method is used in the demodulator shown in Figure 5.11 to adjust the timing interpolator, carrier phase and symbol timing recovery are dependent on each other. For optimum acquisition time this is not desirable. However if the symbol timing algorithm is simply tracking small timing variations in a continuous signal, little degradation in overall performance would be expected.

In the non-decision-directed case we can also illustrate how another timing detector is derived. Suppose that two samples per T are available and we use the approximation

$$\dot{y}_n \approx \frac{y_n - y_{n-1}}{T}$$

This is an estimate of the derivative of the signal halfway between the $(n-1)$-th and the n-th symbols; i.e. at the $y_{n-1/2}$ sample location. Using (5.37) our new timing detector is

$$y^I_{n-1/2}(y^I_n - y^I_{n-1})$$

A more useful version of this, suitable for BPSK or QPSK in tracking mode, is

$$\hat{\tau}_{n+1} = \hat{\tau}_n + \beta(y^I_{n-1/2}(y^I_n - y^I_{n-1}) + y^Q_{n-1/2}(y^Q_n - y^Q_{n-1})) \qquad (5.39)$$

This algorithm [22] has been widely used in digital modems. It has an advantage over the one sample decision-directed method in that phase offset on y_n has no effect (see Problem 8). It can therefore be employed before carrier phase recovery.

The algorithms just mentioned can be readily implemented, even at quite high rates. Dedicated hardware implementations can use loop-up-table approaches so that only one memory access time is required per symbol period (e.g. [8]). In this case the timing detector output can be used to drive a numerically controlled oscillator, or an analog VCO, which controls the symbol sampling clock.

A statistical analysis of these two timing detectors is presented in [23] so that their performance as a function of signal to noise ratio and excess bandwidth may be assessed. In addition a modification of the single sample per T_s decision-directed algorithm is discussed. This version is insensitive to phase offset but has reduced performance. A one sample per T_s non-decision-directed algorithm has recently been reported [24].

Exercise 5.5
A high speed demodulator with analog receive filtering has a simplified block diagram shown in Figure 5.14. Show that this conforms to the timing algorithms described this section, and identify the loop constant β. Assume that the voltage controlled crystal oscillator (VCXO) has a sensitivity of k_f Hz per Volt, and the D/A converter gain is k_0.

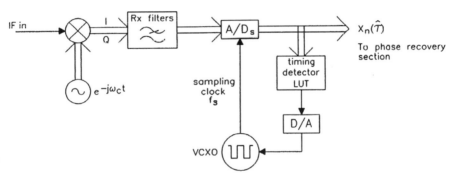

Figure 5.14: Timing loop for Exercise 5.5.

Solution
Either one or two sample per symbol timing detectors could be employed in this design. Assume that a one sample per T_s algorithm is used, and so the nominal VCXO frequency will be r_s Hz. Let the timing detector output for the n-th symbol be denoted ψ_n. The VCXO frequency will then become

$$f_s(\psi_n) = r_s + k_f k_0 \psi_n$$

If the sampling instant for the n-th symbol is $\tilde{t}_n$, then

$$
\begin{aligned}
\tilde{t}_{n+1} &= \tilde{t}_n + \frac{1}{r_s + k_f k_0 \psi_n} \\
&= \tilde{t}_n + T_s \left(1 - \frac{k_f k_0 \psi_n}{r_s} \right)
\end{aligned}
$$

where we have used the approximation that $|k_f k_0 \psi_n \ll r_s|$. *If we write this in terms of a normalised timing offset*

$$\tilde{\tau}_n = \frac{\tilde{t}_n - nT}{T_s}$$

the timing update becomes

$$\tilde{\tau}_{n+1} = \tilde{\tau}_n + \beta \psi_n$$

in agreement with (5.38) or (5.39), where $\beta = -k_f k_0 / r_s$. *Lack of space precludes a full analysis of synchronisation loops in this chapter. Linearised analysis of timing and phase loops in discrete-time demodulators, along the lines given in [25] or [26] allows timing and phase jitter to be estimated, provided that the detector characteristics are known. Frequently a combination of linearised analysis and simulation is required in order to design practical modems.*

5.6.2   FeedForward Symbol Timing Approaches

Returning to our expression for the likelihood function in (5.23), a "brute force" decision-directed approach to estimate the optimum timing would be to evaluate (5.23) for a large number of timing offsets, and pick the one that gave the maximum value of $\Lambda_{Ls}(\hat{\tau})$. This would require either a very large number of samples per T_s, or an iterative scheme in which interpolation is used to expand the number of samples in the region of timing offset giving the largest $\Lambda_{Ls}(\hat{\tau})$. Since the symbol timing quantisation might typically be expected to be not greater than about 1% of T_s, the latter approach is preferable. Remember that some form of phase recovery is required to make the hard decisions needed in (5.23).

Alternatively, a similar iterative scheme could be used with the low SNR approximation of $\Lambda_{Ls}(\hat{\tau})$ in (5.28), or the more exact log cosh form of (5.27) could be used. This gives an iterative non-decision-directed feedforward timing method which effectively finds the sampling instant that maximises the average eye opening over the observation interval. In addition, prior phase recover is not required.

These schemes could be compared to the "conventional" feedforward technique of subjecting the received signal to a non-linear operation which produces a discrete spectral component at the symbol rate. This component may then be extracted with a narrow bandpass filter, the output of which gives the timing phase. Typically, squarer or magnitude non-linearities have been used (e.g. [27]). The Q of the bandpass filter determines the filter's impulse response duration and so the effective observation interval.

This approach can also be used, of course, with discrete-time processing. For example, the received signal, sampled at 4 samples per T, could be squared and then bandpass filtered by standard digital filters. Alternatively, since a Discrete Fourier Transform can be interpreted as a bank of bandpass filters, and we only require the filtered output at r_s Hz, one "bin" of a DFT can be evaluated and the phase determined from the real and imaginary components. This turns out to be quite easy to implement (e.g. [16], also see Problem 9).

A similar method of efficiently evaluating a DFT at one bin to extract symbol timing has been used in a 2 Msym/s TDMA burst demodulator [28]. In this case the timing preamble at the beginning of the burst produced a discrete spectral component and so no squaring operation was required. The DFT output also gives the magnitude of the preamble component, which can be used for burst detection.

Before giving some results to illustrate the performance of these feedforward timing estimators, it is useful to state the lower bound concerning the variance of symbol timing estimation. Consider a situation where a single known symbol is transmitted, e.g. $a = 1$. The demodulator therefore receives $h^T(t - \tau)$. The Cramer-Rao Bound in this case can be shown to be [29]

$$\sigma_{\hat{\tau}}^2 \geq \left(\frac{2}{N_0} \int_{-\infty}^{\infty} (2\pi f)^2 |H^T(f)|^2 df \right)^{-1} \tag{5.40}$$

The integral term depends on the pulse energy and shape; it can also be written as $\int_{-\infty}^{\infty} (\dot{h}^T(t))^2 dt$. With raised cosine pulse shaping, the bound is only mildly effected by the rolloff factor. (Practical timing schemes may however be strongly effected by the rolloff factor.) When K pulses are used in a timing estimator, where K is large, the lower bound on timing variance is reduced by a factor of K.

Figure 5.15 (from [30]) shows a comparison of timing estimate variance for 64 symbol and 256 symbol packets. Results are shown for the two recursive maximum likelihood schemes described above, after 6 iterations. Performance of the squarer plus DFT method is also plotted, plus the appropriate CRB. The results show that the decision-directed method gives a significant improvement in performance unless the signal to noise ratio is very low. Although this is gained at the expense of requiring phase recovery, the iterative likelihood maximisation algorithm could be extended to include block decision-directed phase estimation as well.

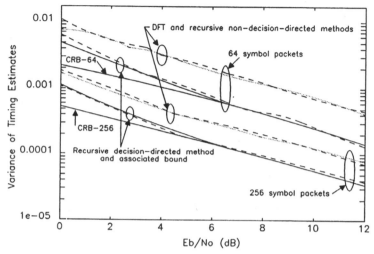

Figure 5.15   Performance of feedforward timing estimators for 64 and 256 symbol packets.

5.7   PROBLEMS

1. (a) By considering (5.7) and (5.8) for BPSK, with $|a| = 1, \Delta\hat{\phi} = 0$, and $h^C(0) = 1$, derive an expression for the BER as a function of E_b/N_0. Check that the result is equivalent to (1.34).

   (b) Repeat part (a) for a static phase error, $\Delta\hat{\phi} \neq 0$. Show that this situation causes a translation of the standard BPSK error curve (Figure 1.22) to the right by an amount dependent on $\Delta\hat{\phi}$. (Obviously the effect of phase errors is more pronounced in MPSK when $M > 2$.)

2. Repeat the FIR design given in Exercise 5.1 (for example using MATLAB), but this time use a Hanning window in the time domain instead of the rectangular window. An expression for $h^N(t)$ may be found in [3, section 6.2.1.]. Compare the frequency domain responses, with and without the Hanning window, to the ideal raised cosine Nyquist response.

3. Show that, given suitable assumptions, the spectrum of $S(f)$ derived in Section 5.3.1, is the same as the continuous-time processing scheme shown in Figure 5.1.

4. Assuming $M_s = 8$ samples per symbol period, consider a very simple interpolation scheme where a linear combination of two adjacent samples is used to carry out the interpolation. Assuming maximum shift (by $T/16$) is required, calculate the response of this interpolating "filter" and compare to the optimum. How would this scheme perform when M_s is smaller?

5. Verify that Equations (5.31) and (5.32) are equivalent for BPSK.

6. Could the feedforward phase estimator in (5.32) be used with Aviation-BPSK [29]? What size constellation should be used? Could you devise a modified algorithm where the signal phase is still only doubled? Can you extend these results to $\pi/4$-QPSK?

7. The expected value of a timing detector, as a function of time offset τ, is called its *S Curve*. Derive an expression for the S curve of the timing detector in (5.38), assuming no decision errors or phase offset, and BPSK signals. How does the S curve vary as α changes?

8. Prove that the 2 sample per symbol-period algorithm in (5.39) is insensitive to carrier phase offset.

9. Assume that a symbol timing scheme uses 4 samples per symbol period and squares the samples in order to generate a discrete component at the symbol rate. Using the DFT, determine an expression for the phase of this component as a function of the squared samples, $s_0, s_1, \ldots s_N$.

10. Real modulators (and demodulators) may have imperfect amplitude balance between the I and Q channels, and likewise a phase shift not exactly equal to $90°$. Imagine a test arrangement where quadrature sinusoids of equal amplitude are input to the I and Q mixers in Figure 5.1(a). This produces SSB tone modulation. For small amplitude and phase errors, derive an expression for the ratio of wanted sideband level to unwanted sideband level.

11. Using the approach outlined in section 5.4.1, estimate the total number of computations per second for Exercise 5.2.

12. As for question 7, the expected value of a phase detector as a function of phase offset is called its S curve. Derive and sketch the S curves for phase detectors $I_m\{y_n(\hat{\phi})a_n^*\}$ and $I_m\{(y_n(\hat{\phi}))^2\}$ for BPSK signals. (Assume no timing offset and no decision errors in the former case.)

REFERENCES

1. Rice M.A., Miller M.J., Cowley W.G., and Rowe D., "Modem Design for a MOBILESAT Terminal," *Proc. IMSC-90*, Ottawa, 1990.

2. Papoulis A., *Probability, Random Variables and Stochastic Processes*, McGraw-Hill, 1965.

3. Proakis J.G., *Digital Communications*, McGraw-Hill, 2nd Ed, 1989.

4. Blinchikoff H.J., *Filtering in the time and frequency domains*, Wiley, New York, 1976.

5. Roberts R.A. and Mullis C.T., *Digital Signal Processing*, Addison–Wesley, 1987.

6. Oppenheim A.V. and Schafer R.W., *Discrete-Time Signal Processing*, Prentice-Hall, 1989.

7. Cowley W.G., Morrison I.S. and Lynes D.C., "Digital Signal Processing Algorithms for a Phase Shift Keyed Modem," *Proc. ISSPA-87*, Brisbane, August, 1897.

8. Bolding G., Cowley W.G. and Morrison I.S., "A Flexible 85 MSym/s Demodulator for Earth Resources Applications," *Proc. IREECON*, Sydney, Sept. 1991.

9. Stiffler J.J., *Theory of Synchronous Communications*, Prentice-Hall, 1971.

10. Guren H.C., "Burst Channel Receiver for Aeronautical Communications," *Proc. ASSPA 89*, Adelaide, April, 1989.

11. Kam P.Y., "Maximum Likelihood Carrier Phase Recovery for Linear Suppressed-Carrier Digital Data Modulations," *IEEE Trans. Comm.*, Vol. COM-34, No. 6, June 1986.

12. Viterbi A.J. and Viterbi A.M., "Nonlinear Estimation of PSK-Modulated Carrier Phase with Application to Burst Digital Transmission," *IEEE Trans. Info. Theory*, Vol. IT-29, No. 4, July 1983.

13. McVerry F., "Performance of a Fast Carrier Recovery Scheme for Burst-Format DDQPSK Transmission over a Satellite Channels," *Proc. Int. Conf. on Digital Proc. of Sigs in Comm.*, Loughborough, April 1985.

14. Helstrom C.W., *Statistical Theory of Signal Detection*, Pergamon Press, 1968.

15. Feher, K., *Digital Communications; Satellite/Earth Station Engineering*, Prentice–Hall, 1983.

16. Oerder M. and Meyr H., "Digital Filter and Square Timing Recovery," *IEEE Trans. Comm.*, Vol. COM-36, No. 5, May 1988.

17. Aghamohammadi A., Meyr H. and Ascheid G., "Adaptive Synchronisation and Channel Parameter Estimation Using an Extended Kalman Filter," *IEEE Trans. Comm*, Vol. 37, No. 11, Nov. 1989.

18. Young R.J., Lodge J.H. and Pacola L.C., "An Implementation of a Reference Symbol Approach to Generic Modulation in Fading Channels," *Proc. IMSC-90*, Ottawa, 1990.

19. Franks L.E., "Carrier and Bit Synchronization in Data Communications—A Tutorial Review," *IEEE Trans. Comm.*, Vol. COM-28, No. 8, August 1980.

20. Kobayashi H., "Simultaneous Adaptive Estimation and Decision Algorithm for Carrier Modulated Data Transmission Systems," *IEEE Trans. Comm.*, Vol. COM-19, No. 3, June 1971.

21. Mueller K.H., and Muller M., "Timing Recovery in Digital Synchronous Receivers," *IEEE Trans. Comm.*, Vol. COM-24, No. 5, May 1976.

22. Gardner F.M., "A BPSK/QPSK Timing-Error Detector for Sampled Receivers," *IEEE Trans. Comm.*, Vol. COM-34, No. 5, May 1986.

23. Cowley W.G. and Sabel L.S., "The Performance of Two Symbol Timing Recovery Algorithms for PSK Demodulators," *submitted to IEEE Trans. Comm.*, 1992.

24. Moeneclaey M. and Batsele T., "Carrier Independent NDA Symbol Synchronisation for MPSK Operating at Only One Sample per Symbol," *Proc. Globecom 90*, paper 407.2, San Diego, Dec, 1990.

25. Lindsey W.C. and Chie C.M., "Digital Phase Locked Loops," *Proc. IEEE*, Vol. 69, No. 4, April 1981.

26. Moeneclaey M., "A Simple Bound on the Linearized Performance of Practical Symbol Synchronizers," *IEEE Trans. Comm.*, Vol. COM-31, No. 9, Sept. 1983.

27. Wintz P.A. and Luecke E.J., "Performance of Optimum and Suboptimum Synchronizers," *IEEE Trans. Comm.*, Vol. COM-17, No. 3, June 1969.

28. Bolding, Asenstorfer P.M. and Cowley W.G., "A Time Division Multiple Access Modem for use in the M-SAT Network," *Proc. IREECON*, Melbourne, 1989.

29. Blahut R.E., *Principles and Practice of Information Theory*, Addison-Wesley, 1987.

30. Sabel L.P. and Cowley W.G., "A Recursive Algorithm for the Estimation of Symbol Timing in PSK Burst Modems," *Proc. Globecom*, 1992.

31. Lindsey W.C., and Simon M.K., *Telecommunication Systems Engineering*, Prentice-Hall, 1973.

32. Crochiere R.E. and Rabiner L.R., *Multirate Digital Signal Processing*, Prentice-Hall, 1983.

Chapter 6

SPEECH CODEC SYSTEMS

by Andrew Perkis
The Norwegian Institute of Technology
Division of Communications
Trondheim, Norway

This Chapter examines the development of efficient speech compression algorithms, with particular focus placed on new techniques developed over the past 10 years. The main emphasis is placed on techniques that are likely to have applications in narrowband satellite communications, developing the theory as well as describing practical aspects of design and performance evaluations.

Speech coding research in recent years has focused on the development of speech coder/decoder (codec) systems which represent high quality speech while lowering the bit rates, utilising the explosion of computing power that has recently been made available. It is thus possible today to obtain near toll quality speech at bit rates as low as 4.8 kbit/s. This represents a compression ratio of 13.3 compared to the standard 64 kbit/s logarithmic Pulse Code Modulation (PCM) scheme used in public switched telephone networks.

Out of this research several applications have emerged, with mobile communications being one of the most important of these. All new proposed mobile communication schemes will be digital and operate in narrow bands, requiring some form of speech compression.

For example, of the cellular services, the PAN European scheme [1] employs a 13.1 kbit/s Residual Pulse Excited Linear predictive (RPE-LTP) speech coder [2], while the digital cellular mobile communication scheme proposed for the US will use a 7.95 kbit/s Vector Sum Excited Linear Predictive (VSELP) speech coder [3].

For mobile satellite communications, speech coder robustness is even more of a concern, due to the occurrence of comparatively poorer channel conditions than for the land mobile systems.

The codec design methodology will, however, in principle be the same for the two scenarios. A particularly robust scheme known as Improved Multiband Excited (IMBE) [4] speech coding has been chosen for the Inmarsat Standard M service and the Australian Mobilesat system.

There are a number of excellent books available on speech coding in general. However, none of them cover speech codec systems considering the specific needs apparent in satellite communications that will be emphasised in this Chapter.

A major part of this Chapter is devoted to special design considerations and methodology for low bit rate speech codecs intended for satellite services.

Section 6.1 gives a brief overview of speech coding in general, presenting the two major classes of speech coders, waveform coders and parametric coders.

A more in depth coverage of waveform coding can be found in [6], with the main emphasis on PCM, Adaptive Transform Coding (ATC) and Sub Band Coding (SBC) at bit rates in the range 16-64 kbit/s.

Vocoders and Linear Predictive Coding (LPC) techniques are covered in depth in [7] with the main emphasis on LPC analysis techniques with applications to speech coders in the range 1-2.4 kbit/s.

Section 6.2 describes new coding schemes such as Code Excited Linear Predictive (MPE) coding, Code Excited Linear Predictive (CELP) coding and Multiband Excited (MBE) coding.

Section 6.3 is devoted to evaluating the perceived techniques for quality of speech codec systems both by objective and subjective means, where much of the approach concentrates on methodology. A detailed description of quality evaluation of speech codecs can be found in [9].

Section 6.4 covers the design methodologies of low bit rate speech coding schemes. Finally, Section 6.5 examines the special design considerations one should take into account when considering a speech codec system for use in satellite communications, with special emphasis on mobile services.

6.1 SPEECH CODING PRINCIPLES

All people rely on some form of communication, that is, messages carried through a channel in some way. A large part of this communication is carried out using speech. The fundamental study of speech communication and the characteristics of it, including the study of the speech waveform itself, traditionally belongs to the fields of linguistics and acoustics, and to some extent psychology and sociology. The engineering or technical aspect of speech communications therefore has to extract knowledge from all these fields, making speech communications a highly interdisciplinary science.

The speech communications signal can be defined as the time varying electrical waveform, resulting from the output of a transducer. The transducer transforms the acoustic pressure wave resulting from the human speech production mechanism, into a time varying voltage or current representing our speech signal.

Speech processing can be defined as the mathematical manipulation of this signal. Speech processing can be divided into three main areas: *speech coding, speech synthesis* and *speech recognition*. This Chapter aims to look at the speech coding part, also referred to as *speech compression*.

Through Nyquist's Sampling Theorem and our knowledge of the speech signal, we know that a speech waveform can be sampled using a finite sampling frequency. That is, a time continuous waveform s(t) can be represented by a time series s(nT), where T is the reciprocal of the sampling frequency f_s. Once this is achieved, the signal can be represented infinitely well by means of discretisation of the continuous amplitude, through the process known as *waveform quantization* (referred to as *quantization*), producing the digital signal s(n).

Speech coding can then be defined as the process of representing analog waveforms by a sequence of binary digits [6]. The simplest of the speech codec systems consists of a sampler and an instantaneous quantizer, taking each sample of the speech waveform and representing every amplitude value by a digital code. An example of such a scheme is Pulse Code Modulation (PCM) coding.

On the other hand there exist speech codec systems that try to fit a mathematical model to segments of the speech signal by varying the model parameters and then only transmitting an optimum set of these. In such schemes the receiver reproduces a synthetic segment of speech based on each received parameter set. An example of such coder is the Linear Predictive vocoder.

The fundamental concept leading to such a variety of schemes is the fact that the precise form of the digital representation of the speech waveform is perceptually highly irrelevant and statistically redundant. The *irrelevancy* of the signal lies in the fact that the human ear cannot differentiate between changes in magnitude below a certain level. This can be exploited by designing optimum quantization schemes, in that only a finite number of levels are necessary.

The *redundancy* is due to the statistical characteristics of the source. The signal is slowly varying, resulting in a highly correlated signal. This correlation can be exploited by introducing prediction schemes. For example if every second sample is compared to the first, only the difference need be quantized. If the difference is small, the quantizer for this signal can have fewer levels than that for the actual values themselves, giving a coding gain.

Figure 6.1 illustrates the main goal of speech coding, namely to be able to transmit a perceptually relevant and statistically non redundant signal over the channel to the receiver.

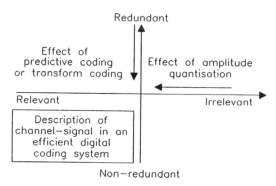

Figure 6.1: Illustration of fundamental coding goal.

Irrelevancy is minimised by quantization and redundancy removed by prediction. Note that quantization introduces a loss of information, while prediction is reversible and will in general preserve all the information in the signal.

Speech codec systems can be described as belonging to one of two different coding classes:

1. *Waveform coders* - Coding schemes that aim to mimic the speech waveform in the best possible way, by transmitting parts of the actual time domain magnitudes.

2. *Parametric coders* - Coding schemes that analyse the speech waveform to extract perceptually important parameters, based on a model of the source, and transmitting these model parameters only. The speech waveform is synthetically reproduced in the receiver based on these parameters.

Waveform coders represent the most general coding scheme and also the most commonly used, as they provide high quality speech at reasonable bit rates in the range 16-64 kbit/s. They also have the ability to code music and other audio signals.

There is, however, an emerging need to lower the transmission rate of communication systems, especially for mobile services. This has produced a boom for research into low bit rate coding schemes for speech. It has led to the increasing use of parametric coding schemes for new services being offered in telecommunications networks.

When choosing a speech codec system for a given application there are generally five factors that play an important role in the decision process: speech quality, bit rate, complexity, delay and finally, coder robustness.

Figure 6.2 illustrates the tradeoff in quality versus bit rate for three classes of coding schemes.

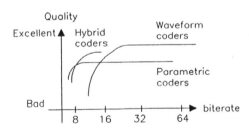

Figure 6.2: Quality versus bit rate for speech codec systems.

Note that Hybrid schemes are a combination of parametric coding and waveform coding schemes.

Waveform coders give the best quality at high bit rates, while their quality deteriorates rapidly below 16 kbit/s. Parametric coders outperform waveform coders at very low bit rates, but due to model limitations, reach a saturation in quality as the bit rate increases. Hybrid coding schemes aim to bridge the gap in quality at low bit rates, by combining the two methods.

6.1.1   Waveform Coding

All waveform coders have one major advantage over parametric coders in that they are relatively source independent. This is one of the most significant differences from parametric coders which are totally dependent on a model of the source, and will only work well for signals originating from that source.

The simplest of the waveform coders use PCM. In such waveform coders the output sample $\hat{s}(n)$ is reproduced based on transmitted information related to the magnitude of the encoder input signal s(n). The reconstruction error, e(n), is then defined as

$$e(n) = s(n) - \hat{s}(n) \tag{6.1}$$

The performance of waveform coders can be evaluated as the ratio of the signal variance, σ_s^2, to the reconstructed error variance, σ_e^2. The performance factor, denoted the signal-to-quantization-noise ratio (SNR), in dB, is defined as

$$SNR = 10 \log_{10} \frac{\sigma_s^2}{\sigma_e^2} \qquad [dB] \tag{6.2}$$

The reconstruction error, e(n), reflects the coder distortion, and σ_e^2 is therefore often referred to as a measure of the coder noise level. The design of waveform coders rely among other criteria, on the minimisation of σ_e^2 as a design factor. For these coders there is an almost linear correspondence between transmission rate and quality. A standard 64 kbit/s telephone circuit gives an SNR of 38 dB and this will decrease by 6 dB per 8 kbit/s. Subjectively this gives standard telephone quality speech, commonly referred to as *toll quality*.

More advanced waveform coding schemes are Differential PCM (DPCM) and Adaptive DPCM (ADPCM). These are time domain waveform coding schemes which utilise the correlation in the speech signal by only quantizing the difference between consecutive samples, either using fixed or adaptive predictor coefficients. These coders are capable of producing toll quality speech down to 32 kbit/s. They all have been standardised by the CCITT at 64, 56, 48 and 32 kbit/s. They all represent low complexity coders with no delay.

Frequency domain waveform coders offer higher compression ratios at the cost of a higher complexity and some extra delay. For example, the Sub Band Coder (SBC) and the Adaptive Transform Coder (ATC) both operate on a vector of speech samples *s(n)*, where each vector *s(n)* consists of N input samples $[s(n - N), \ldots, s(n - 1), s(n)]$. Each vector is split into frequency bands, where the energy in each band determines the quantization scheme to be used for each sample in that band.

These coders introduce side-information, since the sub band energies have to be transmitted in addition to the waveform information. SBC and ATC schemes give acceptable quality speech, referred to as *communications quality*, down to 16 kbit/s.

For further compression, more statistical information has to be taken into consideration as in the Adaptive Predictive Coder (APC) [10, 11], by using highly adaptive algorithms and longer delays.

6.1.2 Parametric Coding

Unlike waveform coding schemes, parametric coding schemes depend on a model of the source. Parametric speech codec systems are thus based on a mathematical model of the human speech production mechanism. The encoder will in general analyse the incoming speech signal on a frame by frame basis to extract perceptually important information relating to the model. The frames consist of a speech vector *s(n)* of length N. The model parameters for each frame, are then quantized and transmitted. The receiver will reproduce the speech signal synthetically according to the model and the received model parameters.

The analysis in these coders is often performed by synthesising the speech signal in the encoder and comparing the synthetic speech signal $\hat{s}(n)$ with s(n). The optimum model parameters are then obtained by minimising the error signal e(n). Again a design criteria based on minimising σ_e^2 is the most common. This type of coders will be referred to as analysis-by-synthesis coders.

The parametric coding schemes are naturally divided into two classes,

- *vocoders:* codecs transmitting model parameters only, and

- *hybrid codecs:* transmitting information on the error signal in addition to the model parameters.

In evaluating parametric coding schemes, the SNR defined as in Equation 6.2, proves to be an inadequate measure. Since the speech waveform is synthetically reproduced using parameters representing perceptually important information, with no regard to the actual shape of the waveform, it is only meaningful to judge the quality using listening tests.

The parametric coding schemes can again be divided into the following types depending on the model used:

- *Time domain* coding schemes use a combination of short term and long term prediction of the signal, based on a mathematical model of the speech production mechanism. These techniques are generally referred to as Linear Predictive Coding (LPC). The LPC10 [12] and LPC++ are low bit rate vocoders standardised for military and commercial purposes, respectively, operating at 2.4 kbit/s. The coders give a *synthetic quality*, lacking naturalness, but maintaining a high degree of intelligibility. Both coders are more complex than most waveform coders, but still simple enough for low cost systems.

- *Frequency domain* coders rely on a harmonic modelling of a spectral representation, S(z), of the signal s(n). There are a number of encoders in this class. One type known as sinusoidal coders [16] produce $\hat{s}(n)$ by tuning a bank of sine wave generators, based on a set of received amplitude, frequency and phase values.

  These parameters are extracted in the encoder and updated on a frame by frame basis. Another type known as harmonic coders [17] differ from sinusoidal coders in the form of spectral representation, by fitting a set of "generalised harmonics" to the speech spectrum and transmitting these.

  Multiband Excitation (MBE) coders divide the speech spectrum up into bands, each band being the width of the estimated fundamental angular frequency, ω_0, of the speech segment. The speech spectrum is then modelled by a spectral amplitude per band and an excitation spectrum. MBE coders provide communication quality speech at bit rates around 3.2-9.6 kbit/s.

Hybrid coding schemes lie in between vocoders and waveform coders. The Multipulse Excited (MPE) linear predictive coder [13] and the Code Excited Linear Predictive (CELP) [14] coder both transmit some information representing the model error, e(n). This information is used in the decoder, in addition to the model parameters, to produce the synthetic speech signal $\hat{s}(n)$.

These coders are capable of giving communication quality at bit rates as low as 4.8 kbit/s, and toll quality at rates above 8 kbit/s. At the lower rates, there is a degradation in quality in comparison to toll quality, but still a degree of naturalness. The coders all have a high degree of intelligibility and are highly complex. There is a codec delay corresponding to the frame rate, usually in the order of 12-30 msec. (The Low Delay CELP chosen as the CCITT standard at 16 kbit/s represents an exception from this with a delay of merely 0.625 msec [15].).

6.1.3 Speech Coding for Satellite Communications

Many fixed satellite communication services access the Public Switched Telephone Network (PSTN). They are therefore confined to using CCITT standard toll quality speech codec systems. So far, the approved standards for toll quality communications are 64 kbit/s PCM, 32 kbit/s ADPCM and 16 kbit/s LD-CELP.

For mobile satellite communications, it is necessary to operate within narrow frequency bands and at low carrier to noise ratios. This is in order to make efficient use of the available spectrum and to keep costs down. In addition, the effects of shadowing and multipath propagation in mobile systems influence the design criteria for codec systems. The effects of these influences can be summarized as follows:

- *Narrow band communication* - places strict bandwidth limitations on mobile satellite communications, necessitating the use of speech codecs operating at low bit rates.

- *Low carrier to noise ratios* - results in a high probability of random errors, that is, a high bit error rate (BER) at the receiver end.

- *Shadowing and multipath propagation* - the resultant fading gives long error bursts , making it possible to lose whole blocks of information at the receiver end.

These three limitations necessitate the design of robust, low rate speech codec systems [18]. This means trading off quality versus bit rate. It also means placing a lot of importance on efficient error protection algorithms in addition to optimising the algorithms for parameter robustness as well as quality.

As mentioned in previous Sections, parametric coders have demonstrated considerable promise for giving high quality speech at bit rates as low as 4.8 kbit/s. Possible speech coder candidates for such systems could be the Multiband Excited (MBE) coder or the Code Excited Linear Predictive (CELP) coder.

These two coding schemes are both members of the parametric coding class, representing time domain and frequency domain approaches, respectively. Both coders are described in more detail in Section 6.2.

6.2   PARAMETRIC SPEECH CODING ALGORITHMS

In order to develop sophisticated speech coding algorithms with high compression ratios yielding good quality, it is crucial to have a fundamental understanding of the speech signal itself. Generally, a speech signal can be classified as a continually time varying nonlinear production.

To be able to analyse this process it is therefore important to find an equivalent mathematical model that describes the physiological process of speech and has

- minimum complexity and

- maximum accuracy.

The Linear Speech Production model presented by Fant [19] satisfies these criteria. Further, the time domain interpretation of the model equations by Atal [20] links the process of speech production to the well known field of Linear Prediction. This modelling forms the basis for the class of time domain parametric coding schemes.

By a frame-wise analysis of the incoming speech signal, the model parameters can be computed, quantized and transmitted. At the receiver, the de-quantized model parameters are used to produce a synthetic speech signal. The frequency domain coding schemes, in contrast, rely on some form of spectral estimation to find the model parameters.

6.2.1   Linear Predictive Coding Schemes

The speech signal, regarded as a time varying nonlinear process, is modelled by a linear combination of an excitation source and a spectral shaping vocal tract system [19]. In this model, the voiced sounds, such as vowels, are generated by passing quasi-periodic pulses of air through the vocal tract. The vowels differ in the location of spectral peaks in the short term spectrum, the formants, and in the fundamental frequency or period. The unvoiced sounds are generated by passing turbulent air through the vocal tract. The turbulent air is created by a temporary closure and sudden release of the vocal tract, creating a flat, short term spectrum.

The vocal tract is modelled by an all pole filter, denoted using z-domain notation, as $\frac{1}{A(z)}$. Then the speech signal spectrum is synthesised as

$$S(z) = \frac{U(z)}{A(z)} \qquad (6.3)$$

where U(z) represents the spectrum of the excitation source. The synthesis filter is specified by

$$\frac{1}{A(z)} = \frac{1}{1 - \sum_{k=1}^{p} a_k z^{-k}} \qquad (6.4)$$

where $\{a_k\}$ are the all pole filter coefficients and p is the filter order. From Equation 6.3 we can identify the excitation source U(z) as

$$U(z) = S(z)A(z) \qquad (6.5)$$

The all zero filter, A(z), is usually referred to as the analysis filter or inverse filter, and the process represented by Equation 6.5 as speech analysis. Conversely, the process of Equation 6.3 is referred to as speech synthesis and $\{a_k\}$ as the predictor coefficients.

6.2.2 Application of LPC in Low Bit Rate Speech Coding

Equation (6.3) gives a means of synthetically producing a speech signal, provided we can model the excitation source and vocal tract appropriately. Alternatively, Equation (6.5) shows us how to determine the excitation source given the speech signal and the vocal tract parameters. Both these aspects are used in speech coding algorithms. In LPC vocoders a set of parameters describing U(z) and A(z) are quantized and transmitted on a frame by frame basis, achieving a compression ratio of up to 25 as compared to PCM. The MPE and CELP coders differ from the LPC vocoder in that they use a more accurate modelling of the excitation source, reducing the compression ratio to approximately 13-14.

By using the all pole synthesis model for speech production, it is possible to achieve an efficient representation of the speech signal in terms of a set of unique, slowly time varying parameters, namely the predictor coefficients $\{a_k\}$, k= 1, ..., p. The speech coding problem is reduced to the estimation of these parameters, based on the available incoming speech signal. This problem is linked to the field of linear prediction. By setting $a_0 = 1$ in Equation 6.3 and performing the inverse z transform, the speech signal is given as a linear combination of past values and an input signal, u(n), as

$$\hat{s}(n) = -\sum_{k=1}^{p} a_k s(n-k) + Gu(n) \qquad (6.6)$$

where G is a gain factor ensuring the synthesised speech has the same energy level as the original speech signal.

The predictor coefficients, $\{a_k\}$, can be determined by defining the prediction error as the difference between s(n) and its predicted value $\hat{s}(n)$. For a given speech signal this becomes a minimisation problem, which can be solved by the method of least squares [21].

This process will be referred to as the *LPC analysis* in the remainder of this Chapter. In the frequency domain, the predictor coefficients give a short term spectral estimate of the speech signal, as $\frac{1}{|A(z)|^2}$ (For a theoretic treatment of its correctness, refer to standard parametric spectral estimation techniques).

As we have discussed earlier, the LPC synthesis model relies on voiced sounds being generated by exciting the vocal tract filter by a sequence of periodic pulses. The period of the pulse generator is related to the fundamental frequency of the speech signal, commonly denoted as the pitch frequency or period of the speech signal. In practical systems pitch analysis can be viewed as the process of extracting some form of long term correlation in the speech signal (long term in comparison to the LPC analysis extracting short term correlation in the form of $\{a_k\}$). The pitch predictor most commonly used is the one tap pitch predictor given as [22]

$$P(z) = 1 - \rho z^{-M} \qquad (6.7)$$

where ρ is the pitch coefficient and M the pitch lag. The pitch synthesis filter is given as

$$\frac{1}{P(z)} = \frac{1}{1 - \rho z^{-M}} \qquad (6.8)$$

There are several techniques for finding the pitch parameters . An autocorrelation method seems to be the most popular [38].

6.2.3 Analysis-by-synthesis Speech Coding

Due to the difficulty of getting an accurate estimate of the pitch parameters, several coding schemes have emerged in order to try to overcome this shortcoming. Two low rate coding algorithms based on the LPC model are the MPE and CELP coding schemes.

They both have a common structure in that the analysis function in the encoder is performed by comparing a completely synthesised speech segment with the original signal in order to determine the coder parameters. This gives rise to their family name, *analysis-by-synthesis* coders.

Another commonality is that both coders rely on the LPC analysis to estimate the short term spectral envelope, and differ only in their representation of the prediction error. A common synthesis stage for hybrid analysis-by-synthesis coders is shown in Figure 6.3.

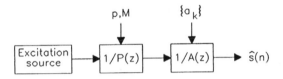

Figure 6.3: The synthesis stage of an analysis-by-synthesis coder

The quality improvement of analysis-by-synthesis coders over vocoders lies in the transmission of the residual signal in terms of the excitation source, making them hybrid schemes. In the MPE coder the residual signal, $e(n)$ is quantized by a sequence of pulses, while in the CELP this residual is quantized by a code vector chosen from a fixed codebook.

The Multipulse coder was first introduced by Atal and Remde [13]. The multipulse improved version of the LPC vocoder. In this model, a major problem is the rigid classification of sounds into voiced or unvoiced types. This is not only difficult, but often impossible, as these two modes are often mixed.

The Multipulse coder represents this excitation signal on a frame by frame basis, by a given sequence of pulses. The amplitude and location of the pulses are determined using an iterative analysis-by-synthesis procedure.

This procedure does not rely on any a priori information of the speech signal and is the same for all types of sounds. The analysis determines the excitation sequence by a closed loop matching technique which minimises a perceptual distortion metric representing the subjectively important differences between the original and synthetic speech signal.

The quality of the synthesised speech is determined by the number of pulses per frame. As an upper limit, one pulse per sampling instant will make it possible to duplicate the original waveform. In general, quality is traded against number of pulses per frame and thus the complexity and bit rate. The MPE coder typically uses a frame size of 16 msec, at a sampling frequency of 8 kHz, requiring approximately 1 pulse/msec to produce high quality speech at bit rates in the range 9.6-16 kbit/s.

6.2.4   Stochastic Coding Schemes

To reduce the bit rate further the Code Excited Linear Predictive (CELP) coder, often referred to as the stochastic coder, was first introduced by Atal and Schroeder [22]. The CELP coder represents the excitation signal by a random sequence of numbers, where the best sequence is chosen from a codebook of stored sequences using an iterative analysis-by-synthesis procedure. The analysis determines the excitation sequence by a closed loop matching technique which minimises the same perceptual distortion metric as in the MPE coder. The quality of the synthesised speech is defined by the size and structure of the codebook. In general an increase in the codebook size will raise the quality of the synthetic speech quality at the cost of an increased complexity and bit rate.

A block diagram of the CELP coder is shown in Figure 6.4.

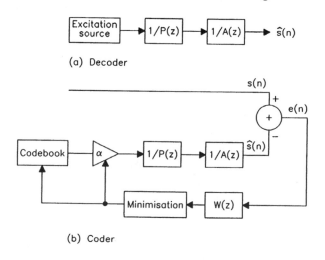

Figure 6.4: Block diagram of a CELP coder/decoder

The incoming speech signal is buffered to create N sample frames of the speech signal for LPC analysis. Through this analysis the predictor coefficients, $\{a_k\}$, are computed for each frame. Pitch analysis and determination of the excitation signal is performed at subframe intervals, splitting the signal into K subframes of length $N_s = N/K$ samples prior to the synthesis stage. For each subframe a code j, j=1, ..., K, a code vector, $c_j(n)$ of length N_s, from the codebook is used as the input to the synthesis filters 1/A(z) and 1/P(z), producing the synthetic speech signal $\hat{s}_j(n)$ at the output.

The resulting error signal between the original and the synthetically produced speech signal, $e_j(n)$, is perceptually weighted by W(z) and fed back to the excitation generator. The error minimisation procedure determines the code vector that minimises this weighted error. The index number i specifying this code vector, is transmitted together with the quantized pitch information, predictor coefficients and gain factor to enable the decoder to regenerate a synthetic speech signal at the receiving end.

The error signal, e(n), is not a good representative measure of the perceptual difference between the synthetic and original signal. In order to be able to incorporate our knowledge about the ear's auditory masking properties and limited frequency resolution, a frequency weighted *Mean Square Error (MSE)* criterion is usually used as the distortion measure.

Due to the fact that we can tolerate more noise in frequency regions where the speech has high energy, and less in the low energy areas, the weighting filter, W(z), should be chosen to de-emphasize the error in the high energy areas and emphasize the low energy areas. Thus the error signal should be weighted according to the inverse of the speech spectrum itself. This can be achieved by using a weighting filter of the form:

$$W(z) = \frac{A(z)}{A(z/r)} \tag{6.9}$$

where r is in the range [0,1] and controls the degree to which the error is de-emphasised in any frequency region. A typical value for r is 0.9.

An optimum codebook vector is computed by minimising the weighted mean square error between the original signal $s_j(n)$ and $\hat{s}_j(n)$ at subframe intervals through an exhaustive search through the codebook. For each vector $c_j(n)$ a stochastic gain α_j is computed to minimise the energy of the weighted difference signal $E_j(z)$, where

$$E_j(z) = W(z)[S_j(z) - \alpha_j \frac{C_j(z)}{A(z)P(z)}] \qquad j = 1, ..., J \tag{6.10}$$

where $C_j(z)$ is the z-transform of $c_j(n)$ and J is the total number of vectors in the codebook. The code vector which minimises $||E_j||^2$ is then selected, and the parametric representation of the excitation signal is given as the codebook index to this vector.

The CELP coder typically uses a frame size of 20 msec, splitting this into 4 subframes of 5 msec each (K=4), requiring a codebook of approximately 1024 code vectors (10 bits) to produce good communication quality speech at bit rates in the range 4.8 - 8 kbit/s. In the original CELP coder, the codebook vectors where populated with white Gaussian random numbers with unit variance, giving rise to its being called stochastic codebook. This coder is extremely complex, due to the filtering operation of 1024 40-sample (5 msec) vectors, through a 10th order filter, every second (approximately 82 million multiply/add instructions per second).

Since 1984 a variety of alternative strategies have been proposed in order to increase the performance and reduce the complexity. The two major quality improvements are the substitution of the pitch analysis (long term correlation) by an adaptive codebook search and the design of an alternative codebook structure.

The long-term correlation can also be computed using an analysis-by-synthesis procedure [23], in which case the coder performance is increased at the cost of added complexity. This is done by searching an adaptive codebook consisting of the most recent excitation vectors, similar to the stochastic codebook search. The name *self-excitation* is used for this process. An optimum lag, L, is found, and for this the optimum self excitation gain factor, B, is calculated. The long-term correlation lag is restricted to be at or above the subframe length and the correlation is calculated on original and synthesised speech rather than the short-term residual as for the pitch analysis.

A major complexity reduction is achieved by utilising the fact that the codebook vectors are identically and independently distributed (iid) Gaussian numbers. Then the codebook vectors can be constructed as overlapping vectors, only making it necessary to filter one new sample for each new vector in the synthesis process. The searches for the optimum code vector and self excitation vector will be similar if the exhaustively-searched stochastic codebook is replaced by a long codebook, i.e., a codebook consisting of overlapping code vectors [24]. By this we will gain the combined advantages of both memory savings and a reduction in coder complexity. It is shown in [25] that a vector overlap of all but 2 samples of the stochastic code vectors will not affect the speech quality.

The optimum code vector and gain factor are traditionally found by an analysis-by-synthesis procedure following the long-term correlation analysis. The performance can, however, be increased by performing a joint optimisation of the self excitation and stochastic gain factors during the codebook search.

For each candidate codebook index i, and with a given lag L, the optimum combination of the two gain factors α and β can be found that minimises the energy of the weighted difference between the original and synthetic speech. This can be done by solving a set of linear equations, given for example, in [26]. By this, the coder complexity is nearly doubled. The increase in speech quality does, however, make the method interesting.

Motivated by the low bit rate of the CELP coder and the excellent speech synthesis performed by the higher bit rate MPE coders, a novel approach combining these techniques was suggested in 1987 [27]. It is called a Self Excited Stochastic Multipulse (SESTMP) coder. The name "Stochastic Multipulse" was chosen for what is more commonly known as "sparse vectors", namely stochastic code vectors consisting of only a small number of non-zero samples .

By constructing such spare vectors we gain both a reduction in coder complexity and an increase in speech quality. The code vectors are generated by distributing a fixed number of Gaussian, zero-mean samples uniformly over the vector.

As opposed to the sub-optimum sequential analysis-by-synthesis procedure in the MPE coder, the pulse positions and amplitudes are simultaneously optimised under the constraints of a fixed number of pulses in each code vector and a common gain factor.

6.2.5 Frequency Domain Coding Schemes

In the Linear Predictive coding schemes the speech signal is synthetically produced as the convolution of an excitation source and a vocal tract system. This means that the spectrum of the output of such a model is the product of the excitation spectrum and the frequency response of the vocal tract system. In this way it is to be expected that the output spectrum reflects the properties of both these systems. We would therefore expect to be able to parametise the speech signal by analysing it in the frequency domain.

A parametric frequency domain speech codec system can be viewed as a coding scheme where, in general, parts of the waveform are regenerated in the receiver as sums of sinusoids with time varying amplitudes and instantaneous phase values. The synthetic speech waveform will be given by

$$\hat{s}(n) = \sum_{n=1}^{L} A_n(n) cos[q_j(n)] \tag{6.11}$$

where $A_n(n)$ represents the magnitude characteristics of the waveform, $q_j(n)$ represents the instantaneous phase and L represents the model order of the system.

Multiband Excited (MBE) Coding

The MBE speech coding algorithm [4] belongs to this class of coders, and is capable of producing communications quality speech at medium (8 kbit/s) to low (2.4 kbit/s) bit rates. It produces speech of comparable quality to CELP at around 5 kbit/s, but through formal listening tests, has been shown to be significantly more robust to random and burst bit errors [28]. It is thus an excellent speech coding algorithm for speech transmission over mobile satellite channels, and has been chosen for the Aussat Mobilesat and Inmarsat Standard M systems.

The spectrum of a short (20 msec) speech segment obtained through an FFT, consists of impulses regularly spaced across the frequency axis, that are slowly varying in amplitude. In addition, there are parts of the spectrum that seem to consist of random, noise like energy. The spectrum can thus be considered to be the convolution of a slowly varying spectral envelope, and an excitation spectrum that consists of a combination of harmonically related impulses and noise.

The impulses correspond to voiced energy. That is, for a totally voiced segment of speech, the excitation spectrum consists entirely of harmonically related impulses. The areas of the excitation spectrum containing noise correspond to unvoiced energy in the speech.

The MBE model divides the spectrum into bands, the width of each band being made equal to the fundamental angular frequency, ω_0. There is one spectral amplitude sample for every band across the spectrum. The excitation spectrum is also divided into bands, and each band is classified as either voiced or unvoiced, using a binary decision. Thus the MBE model parameters are the fundamental frequency, spectral amplitude samples, and voiced/unvoiced decisions.

One significant feature of this algorithm is that the number of parameters in the spectral amplitude sample and voiced/unvoiced decision groups is defined by the pitch. That is, the number of parameters that need to be transmitted is variable. Also, the MBE model is entirely parametric, and thus is a form of vocoder. The model parameters are typically updated every 20 msec.

MBE Encoder

The most important parameter in the MBE model is the pitch. Hence great importance is attached to its estimation. Pitch estimation for the MBE algorithm is done using an analysis by synthesis approach. For a given pitch value, an all voiced (entirely periodic) spectrum is constructed and compared to the original speech spectrum using a mean squared error criterion such as:

$$E = \frac{1}{2\pi} \int_{-\pi}^{\pi} [|S_n(z)| - |\hat{S}_n(z)|]^2 dz \qquad (6.12)$$

where $S_n(z)$ is the Fourier transform of the windowed segment of input speech being analysed, and $\hat{S}_n(z)$ is the all voiced synthetic spectrum. The all voiced synthetic spectrum is constructed by dividing the input spectrum into bands of width ω_0.

For each band, the spectral amplitude sample is estimated, and used to synthesis the corresponding band of the synthetic spectrum. A range of possible pitch values are tried, and for each one the error between the original and synthetic spectrum is measured on a mean squared error basis. The pitch value that minimises this mean squared error is chosen.

Pitch tracking is usually incorporated in MBE coders to improve the reliability of the pitch estimate determined above. This can introduce extra delay into the coder as the pitch of several past and future frames is examined to determine the most likely pitch track.

Voiced/unvoiced decisions are determined by comparing the mean square error between the original speech spectrum and the all voiced synthetic spectrum discussed above. If the difference is below a certain threshold, the band is declared voiced, otherwise the band is declared unvoiced.

Voiced and unvoiced decisions tend to be clustered. Some variations of the MBE coder take advantage of this by making only one voiced/unvoiced decision for every three bands, saving on the number of bits required for transmission of the voiced/unvoiced information.

More recent MBE coders such as the algorithm adopted by Inmarsat and Aussat Improved MBE (IMBE) do not transmit any phase information, reducing speech quality slightly but lowering the bit rate.

MBE Decoder

The MBE decoder uses a combination of time and frequency domain techniques. Voiced harmonics in the spectrum are synthesised using techniques similar to that of Equation (6.11). That is, it uses a bank of harmonic sinusoidal oscillators. The amplitude of each oscillator is determined by the corresponding spectral amplitude sample. The phase, frequency and amplitude of each sinusoid is interpolated across each frame such that the sinusoids are continuous across frame boundaries, where the model parameters are updated. For unvoiced harmonics, the amplitudes of the sinusoidal oscillators are set to 0.

The unvoiced components of the synthesised signal are produced by modifying the short time Fourier transform of a windowed time domain noise sequence. The sections of the Fourier transform declared voiced are set to zero, and the unvoiced sections are modified to represent the corresponding unvoiced spectral amplitude samples. An inverse Fourier transform is then used to produce the time domain unvoiced signal using the overlap-add method. The unvoiced time domain signal is then summed with the voiced speech to construct the final synthesised speech signal.

6.3   SPEECH CODEC EVALUATION

Speech codec evaluation measures serve two purposes, firstly to be able to evaluate and compare coder performances, and secondly as a tool for the designer in order to be able to optimise a speech codec system for a given bit rate. For both purposes, it would be useful to have a simple objective measure, where an absolute value would show how well the coder performs for any given condition.

Unfortunately such a measure does not exist due to the subjective nature inherent in judging speech quality. For practical applications it is therefore common that the designer makes do with a waveform distortion measure, such as the signal-to-quantization-noise ratio (SNR) or a frequency domain distortion measure like the logarithmic root mean square spectral distortion (SD).

However for more precise coder evaluation, it is generally accepted that some form of subjective criterion must be used. Unlike objective measures, subjective measures take into account the users of the speech codec systems, and rely on their hearing and perception of the speech quality to determine what is good or bad. The most formalised of these measures are the Mean

Opinion Score (MOS) tests, which have been standardised by CCITT [9] and the Paired Comparison Tests (PCT) [30].

The most common speech evaluation measures will be defined in this Section.

6.3.1 Signal to Noise Ratio (SNR)

The simplest of the objective measures is the long term signal to quantization noise ratio (SNR). The SNR (in dB) is defined as

$$SNR = 10log\frac{\sum_{n=0}^{N-1} s^2(n)}{\sum_{n=0}^{N-1}[s(n) - \hat{s}(n)]^2} \tag{6.13}$$

where N is the number of samples, s(n) is the input sampled speech signal and $\hat{s}(n)$ is the reconstructed speech signal. That is, the SNR is the ratio between the long term signal energy and the long term noise energy, the noise being defined as the difference between the input waveform and the output waveform samples, s(n) and $\hat{s}(n)$, respectively. This measure is strongly influenced by the high energy components of the speech signal. That is, the SNR measure weights the large energy parts of the signal more heavily than the low energy parts. A waveform coder design, thus, tends to be unfair to low energy segments such as fricatives.

The Segmental SNR (SEGSNR) is an improved version of the SNR. The SEGSNR is based on dynamic time weighting. Specifically, a logarithmic weighting converts component SNR values to dB values prior to averaging, so that high SNR values corresponding to well coded large signal segments do not camouflage coder performance for the lower amplitude segments. That is, the SEGSNR compensates for the de-emphasis of low energy signal performance in the long term SNR. The SEGSNR is defined as

$$SEGSNR = \overline{SNR(m)} \qquad (dB) \tag{6.14}$$

where $\overline{SNR(m)}$ is the average of SNR(m) and SNR(m) is the long term SNR for segment m.

Despite refinements, performance measures based solely on waveform distortion tend to be inadequate, especially for low bit rate parametric speech coders. The SNR measures can in some cases, be used for design purposes as a relative number, but are unsuitable for cross-comparisons between parametric coding schemes and as absolute measures of quality. Part of this inadequacy is due to the fact that a low value of SNR does not necessarily mean a perceptually poorer quality.

6.3.2   Spectral Distortion Measures

The logarithmic rms spectral distortion (SD) measures the difference between the original signal and the reconstructed signal in the frequency domain. The SD can be computed as

$$SD = 10\sqrt{\sum_{i=0}^{N-1}[log10\frac{S(\omega_i)}{\hat{S}(\omega_i)}]^2 d\omega} \qquad (6.15)$$

where $S(\omega)$ is the original short term Fourier spectrum and $\hat{S}(\omega)$ is the reconstructed short term Fourier spectrum.

This measure is often used for performance evaluation of LPC based coders and as a general design criterion in quantizer design. That is, it is used to evaluate the performance dependancy on the LPC coefficients $\{a_k\}$. The SD is accepted to be the objective measure most comparable to the subjective measures [31].

For LPC based coders the spectra are defined as $\frac{1}{|A(\omega)|^2}$ and $\frac{1}{|\hat{A}(\omega)|^2}$, respectively, where $A(\omega)$ is the frequency response of the LPC analysis filter. In speech coding applications, N is usually chosen to be 256. It is however, important to note that keeping the average $SD < 1$ dB does not always guarantee non-noticeable quantization noise. The variance of the SD is also important, as is the need to check for occasional peak values. Equation (6.15) can also be used as a general distortion measure, where $S(\omega)$ and $\hat{S}(\omega)$ can be estimated using any suitable spectral estimation method.

6.3.3   Subjective measures

Since speech codecs are usually used in real time telecommunications systems, the quality of these systems can best be evaluated by some subjective criteria, taking into account the quality as perceived by human beings. Of course when we are talking about evaluating quality, every listener will have his or her own opinion of what is good or bad.

Due to this fact, some formal test procedures have been devised to attempt to average out these preferences. These aim to give absolute measures of quality utilizing an untrained group of people to evaluate the quality. Two of these procedures are the Paired Comparison Test (PCT) and the Mean Opinion Score (MOS). The PCT is more often used in the design process, while the MOS test is used to formally evaluate codec quality, and compare codec performance [29].

6.4   QUANTIZER DESIGN

This Section deals with the design and performance evaluation of scalar quantizers based on the Max optimisation scheme.

Designing a low bit rate voice codec for any given application requires two distinctly different steps:

- deriving a suitable coder model and

- designing quantizers for the model parameters.

Most narrowband speech coders operating at bit rates in the range 4-8 kbit/s rely on a combination of waveform coding and analysis-by-synthesis techniques. These hybrid coding schemes generally transmit information describing the waveform in addition to model parameters, commonly consisting of spectral parameters and gain factors. Quantization of this information is crucial for the coder performance, and should be optimised for each individual coding scheme. In quantizing these parameters there are two distinctly different approaches to be considered:

- instantaneous quantization of the parameter (scalar quantization), or

- delayed decision quantization.

For the purpose of low bit rate speech coding, scalar quantization is usually preferred due to its low complexity and relatively low distortion, and the fact that no delay is introduced. This Section will therefore only deal with scalar quantization. For a good treatment of delayed decision quantization refer to [6].

6.4.1   Model Parameters

Once a model for a parametric coding scheme is established, one still has to determine the model order. In the LPC based coders for instance, the order of the LPC analysis has to be chosen. Another example of model order is the number of voiced/unvoiced decisions in the MBE coder.

The model orders will of course affect the speech quality significantly, and in some cases the model might fail to function if not chosen properly. Probably the most important fact to be aware of is the fundamental trade-off between quality, complexity and bit rate in setting these parameters. Higher order models require more information to be transmitted and thus an increase in bit rate. Therefore in the CELP coder, a larger codebook size will give better speech quality at the cost of increased complexity and bit rate. Therefore, in designing a model based speech coder, exhaustive simulations must be run to determine the optimum order of the model.

This must be chosen to meet the required voice quality, maximum allowable complexity and desired bit rate for a given application.

6.4.2   Quantizer Design

Amplitude quantization is the procedure of transforming a given signal sample x(n) at time n into an amplitude y(n) taken from a finite set of possible amplitudes. In a speech codec this procedure is split into two fundamental processes: quantization and dequantization. Quantization describes the mapping of the amplitude continuous value into an L-ary number k, the quantizer index.

Unlike the sampling process, quantization inherently introduces a loss of information, which we previously referred to as the quantization noise.

There are two fundamentally different quantization schemes, uniform and non uniform quantization. Uniform quantizers have decision intervals of equal length Δ and the reconstruction levels are the midpoints of the decision interval. Although being the most common means of quantization, and also the conceptually simplest scheme for implementation, uniform quantizers are generally not optimum in the sense of minimising the quantization error.

A smaller error variance can be achieved by choosing smaller decision intervals where the probability of occurrence of the amplitude x(n) is high, that is, where the probability density function (pdf), $p_x(x)$, is high. Larger decision intervals are used for regions where $p_x(x)$ is lower. This type of quantizers are often referred to as pdf-optimised quantizers or Max quantizers [32].

It is important to note that in these quantisers, the input x and quantised value q will be statistically dependant.

The design methodology for the pdf-optimised quantizer can be found in [6]. The design is based on an iterative solution of two equations, using a database of the model parameter in question for estimating its pdf.

The first equation states that the decision levels are half way between neighbouring reconstruction values, and the second equation states that the reconstruction value should be the centroid of the pdf in the appropriate interval.

6.4.3 Training and Test Databases

A crucial design parameter in the design of Max-quantizers, is the pdf of the parameter in question. In order to be able to estimate this pdf we need to design a database of actual values of our parameter. To get an accurate estimate it is important that the database, which is called the *training database*, be large enough, and also representative of the real data to be encoded.

The size of the training database must be chosen such as to ensure that the number of values results in statistically valid estimates of the pdf parameters. To give a rough idea of how many "large enough" would be, consider the total number of data values, denoted M, and the number of decision intervals, denoted N. As a rule of thumb it is considered necessary to have at least 1000 entries in each interval, requiring a total of 1000N values. However, considering that N could be around 200, in practical designs this is not always feasible and smaller subsets will have to be sufficient.

Choosing a representative training database will reflect in the performance of our quantizers. In other words the training database should reflect the variety of speakers one would expect to use the system. It is therefore important that examples of both sexes and of a wide range of different speakers and utterances are represented in the database.

A test database is also needed for evaluation purposes. Ideally, the test database should consist of different speakers and different utterances from the training database. As this is often difficult to achieve, a good compromise is to use a test database with a subset of the same speakers but with different utterances. The size of the test database can be significantly smaller than the training database.

6.4.4 Quantizer Evaluation

Two ways of evaluating designed quantizers are used in practice. One is to use the spectral distortion measure given in Equation (6.15). This is often used for spectral quantizers. Alternatively, the time domain quantization error could be used. This is often used for assessing gain quantizers.

By ensuring that the SD is below 1 dB, a just noticeable limit, the degradation caused by quantization will be largely non-detectable. Thus for quantization of the LPC parameters it is only necessary to allocate a sufficient number of bits to obtain a SD below 1 dB.

6.5   CODECS FOR MOBILE SATELLITE COMMUNICATIONS

6.5.1   Introduction

In speech codec design for a given application, there are two major considerations. Efficient transmission over a communication channel requires compression of the signal by removing redundant and irrelevant information. However, to ensure a high quality transmission of the compressed information over noisy channels, redundancy must be added for bit error correction or detection. The channel coding algorithms used for error protection are designed to optimise the tradeoff between the transmission bit rate and the resultant speech quality for a given channel characteristic.

This Section is concerned with considering the major impacts these tradeoffs have on the voice coding algorithm. It also discusses strategies and systems for minimising the degradation caused by channel errors.

It is important to stress that the algorithms and techniques discussed are specific to satellite communications. However, some schemes can readily be transferred to other applications with only minor modifications.

6.5.2   The Scenario

For transmission over a satellite link, it is generally necessary to operate within narrowband channels at low carrier to noise ratios. Therefore the main concern for these applications is robustness [18]. Robustness is achieved by the combined optimization of source codec, channel codec and modem subsystems.

In mobile satellite communications, the signal envelope suffers from rapid fading fluctuations which result in error bursts . The subjective impact of burst errors on decoded speech quantity is far more serious than that of random errors. Burst errors can result in a total loss of intelligibility.

Design for reliable transmission of the voice coder parameters over a noisy channel, requires that it is possible to characterise the channel, as discussed in Chapter 2. Then forward error correction (FEC) schemes can be designed based on these models, and tailored for a given speech coder. System robustness can be enhanced by determining which speech encoder output bits are most critical so far as the effect of channel errors on reconstructed speech. Then an FEC scheme can be chosen to selectively protect only the most critical bits [32].

During severe fading conditions, characterised by deep fades and shadowing, speech codecs will eventually break down due to long bursts of information being received containing errors. In such a system, graceful degradation is usually obtained by muting the system, ultimately leading to disconnection.

An alternative method which can be incorporated in the decoder to enhance system performance, is to utilize an error masking technique [33]. During error masking the quality will be slightly degraded, but overall the system throughput is improved.

Some of the most recent advances in voice codec design consider the impact of a realistic channel. One approach [34] suggests a method of joint optimisation of the source and channel coding stages. The method has been applied to the quantization of voice codec parameters, increasing the coder performance by two orders of magnitude over a conventional convolutional coder, in terms of logarithmic RMS spectral distortion [35, 36]. Other reported schemes include Pseudo Gray coding [37] and matched convolutional channel coding [38].

6.5.3 Selective FEC

In digital transmission of audio signals, there will always exist two types of signal degradation: quantization errors (including modelling errors for model based source coders) and transmission errors. In the design process it might therefore seem as a reasonable criterion that the quantization noise power is chosen to be equal to the noise power due to transmission errors. However, this would often result in a required bit rate which is far too high.

The more realistic solution to this problem is to implement an unequal error protection scheme, called *selective FEC*, in which only a limited number of bits are protected by FEC. The design process will then consist of the following steps:

1. Design a source coder yielding acceptable quality under error free conditions, giving a specified long term average quantization noise power level.

2. Determine the error sensitivity of the source coder output bits.

3. Design a channel codec to yield low quality degradation up to a given BER, striving to keep the error level constant from frame to frame independent of varying channel conditions.

A practical example of selective FEC is the scheme used in the Inmarsat Standard M system [39]. Here the source coder has an information rate of 4150 bit/s, utilizing 20 msec voice frames each containing 83 bits. Of these, the 21 most sensitive bits are coded using the powerful Golay code, and 55 bits are coded by a Hamming code, leaving only 7 bits unprotected. The FEC coding has a redundancy rate of 45 bit/frame or 2250 bit/s, which together with the signalling channel, synchronisation, and timing bits. (1600 bits/s) results in a channel rate of 8000 bit/s.

Before selective FEC schemes can be utilised, it is necessary to determine the parameter error sensitivity of the source coder parameters. Several methods for error sensitivity measurements have been proposed [1, 40, 38]. As in all audio and speech processing, subjective quality is hard to classify by an objective measure. Therefore most of the tests rely on subjective testing, usually by Paired Comparison listening tests. This means that the actual determination of error sensitivity is time consuming. Also, the source coder design must be fixed before testing can commence. The testing methodology can be illustrated by the GSM procedure [1].

The GSM procedure is entirely based on subjective evaluations by a Paired Comparison listening test (PCT). The procedure consists of the following steps:

1. Choose a representative 2 second speech segment, which consists of k=50 frames per second with 260 bits per frame , giving a total of 26000 bits in the test sequence.

2. Invert bit numbered i in all k frames for i = 0, 1 ..., 259.

3. Let PCTi denote a Paired Comparison listening test for the segment with an error in bit i. Run PCTi: i = 0, 1, ..., 259 for a range of different listening conditions involving different noise levels.

By running 260 PCTs, each of the bits will be classified into one of 6 relevance classes, each class then being defined as having a given error sensitivity, given by the noise reference levels.

Successive test signals are the same speech segment corrupted by a deterministic bit error sequence. The test uses three listeners. When all the speech frame bits are classified into one of 6 categories, a selective FEC design can proceed.

6.5.4 Error Masking

In a land mobile satellite channel, burst errors will occur due to shadowing. These burst errors are known to be far more serious than random errors as often whole blocks of data are destroyed. Particularly detrimental for speech codec systems, is the fact that conventional FEC schemes may be unable to correct long bursts of errors effectively. This leads to a technique designed to overcome the effect of residual errors. It is called *error masking*.

This technique relies on the fact that the source signal is a highly redundant signal with a large amount of irrelevancy in addition to having a high fram to fram correlation. Let us assume that at the receiving end, an error detection technique is used to detect the presence of an error burst. This may be done by use of an error detection code or by utilizing the demodulator to estimate the channel conditions. Thus the speech decoder will be informed if the current bit frame is invalid due to an error burst. Then two possible error masking methods can be used.

- *Frame Repetitions*: When a burst is detected within a frame, the frame is considered invalid. Its contents are discarded, and the data is copied from the previous frame for reuse. The method implies storage of previous parameters at the decoder, but is very simple in use and requires almost no overhead. Such a scheme has been used in the Inmarsat Standard M system.

- *Sub Packet Repetitions*: The repetition of complete frames may seem a bit wasteful. The frames are constructed from distinct sub-packets, each containing certain user defined parameter values. This scheme will therefore treat each sub-packet as a complete entity, and only repeat those sub-packets tagged as erroneous.

The above repetition schemes are non-redundant. That is, no extra information needs be transmitted over the channel. There are, however, reasons for adding extra protection to the most sensitive information. For example, by far the most sensitive set of parameters in LPC coders, is the spectral information.

This has a major impact on both the intelligibility of the signal and the pleasantness. A retransmission scheme can therefore be designed which provides for retransmission of errored spectral information bits in each speech frame, introducing a form of time diversity in the system. While increasing the overall bit rate, it can greatly improve the quality of the decoded signal transmitted over a bursty channel [18].

The method is as follows. Prior to transmitting a given sub packet, the spectral information from the previous frame is re-transmitted for possible use by the decoder, thereby reducing the probability for a single burst destroying both. A penalty is a slight increase in the decoder delay. For the best results, a combination of side information re-transmission and sub packet repetition could be used.

6.6 PROBLEMS

The following computer problems are intended to give "hands on" experience with some of the topics introduced in the Chapter, by solving various problems using a computer. For the best tutorial results, it would be advantageous to use the $MATLAB^{TM}$ script language for the simulations. In this program, plotting routines and several of the necessary functions, like FFTs, AR modelling and filtering functions, already exist. However, it is possible to solve the problems using any programming language, provided these functions exist, or the user is willing to write them.

There is a certain necessity to do the problems in the correct order as many of the functions required in the later problems are introduced earlier. This will hopefully motivate the reader to write well documented structured programs. As a bonus, it will provide the reader with a small but very useful software library to build on in further developments of speech codec system simulations.

1. AR-modelling

1. Write a function which generates and plots an autoregressive AR(1) process, x(n), using a = 0.9, and n = 1,128, where:

   x(n) = u(n) + ax(n-1)

   and u(n) is a random variable with unit variance and zero mean, N(0,1).

2. Test the function by calculating and plotting the variance normalized auto-correlation function (acf), $\rho_{xx}(k) = \frac{R_{xx}(k)}{R_{xx}(0)}$, of the process for k = 1,128.

3. Calculate and plot the energy normalized power spectral density function (psd), $|S_{xx}(z)|^2$, of the process.

2. Spectral Estimation

1. Write a function to estimate the prediction (LPC) coefficients of a Pth order AR model, using any least square method, and plot its magnitude spectrum $|1/A(z)|^2$.

2. Test the function by plotting the estimated and actual magnitude spectrum of a cosine with given frequency f_0, sampled at 8 kHz.

3. Compare the estimate for varying model orders using the logarithmic RMS spectral distortion (in dB) defined in Equation 6.15.

3. Linear Prediction

In linear predictive analysis the speech samples, s(n), are related to the excitation source, u(n), by the simple difference equation, given as,

$$\hat{s}(n) = -\sum_{k-1}^{p} a_k s(n-k) + Gu(n)$$

The synthesis filter transfer function, H(z), for this equation will be $H(z) = \frac{G}{A(z)}$. By matching the energy in the original signal s(n), with the energy in the linearly predicted samples, $\hat{s}(n)$, we find the gain, G, as

$$G^2 = R_{ss}(0) - \sum_{k=1}^{p} a_k R_{ss}(k)$$

where $R_s s(k)$ is the acf of s(n).

1. Write a function to calculate

   (a) the short term spectrum of a speech segment, S(z), using an M point FFT on a speech frame of approximately 20 msec.

   (b) the LPC coefficients $\{a_k\}$

   (c) the gain, G, of the synthesis filter.

   The function should have the speech segment and LPC order as inputs.

2. Test the function by plotting the short term spectrum, $|S(z)|^2$, together with the estimated spectrum $|G/A(z)|^2$ for various model orders.

4. Speech Processing

1. Write a function that simulates the block diagram given in Figure 6.5. Verify your program by:

   (a) Comparing the energy level in s(n) and e(n)
   (b) Plotting $|S(z)|^2$ and $|E(z)|^2$ in the same diagram
   (c) Comparing s(n) and $\hat{s}(n)$

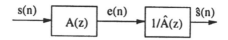

Figure 6.5

2. As a first indication of the distortion introduced by quantization, modify the function to include the simulated fixed wordlength effects as indicated in Figure 6.6. Calculate the logarithmic RMS spectral distortion between the original spectrum, A(z), and the distorted spectrum, $\hat{A}(z)$, using 0, 1, 2 and 3 decimal resolution in the LPC coefficients.

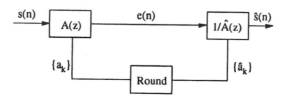

Figure 6.6

3. To simulate the effect of coding the residual signal, by an excitation source, try substituting the error signal, e(n), in Figure 6 with uniformly distributed random numbers in the range [-A,A]. To evaluate the results, calculate the SNR and SEGSNR between the original and synthesised speech signal for various values of A.

5. Channel Distortion

Consider the system shown in Figure 6.7, which is a simple model of an LPC based coding scheme, where only the LPC coefficients are subjected to errors.

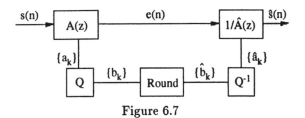

Figure 6.7

1. Write a channel simulator based on a discrete memory less AWGN channel that has a binary input and output, and symmetric transition probabilities.

2. Test the channel simulator by inserting errors in the LPC coefficients before the speech synthesis stage (as shown in the figure) and calculating the Log RMS spectral distortion between $A(z)$ and $\hat{A}(z)$ (Q - indicates any suitable quantizer).

3. Establish the error sensitivity curve of the LPC coefficients by plotting the logarithmic RMS spectral distortion as a function of the BER.

REFERENCES

1. GSM. *Recommendation: 06.10*, Version 3.1.2, Sept. 1988.

2. Kroon, P., Deprettere, E.F. and Sluyter, R., "Regular-Pulse Excitation—A Novel Approach to Effective and Efficient Multipulse Coding of Speech", *IEEE Trans. Acoustics, Speech, and Signal Processing*, Vol. ASSP-34, 1986.

3. Electronics Industries Associates (EIA) Project number 2215, IS-54, "Cellular system", 1989.

4. Griffin, D.W. and Lim, J.S., "Multiband Excitation vocoder", *IEEE Trans. Acoustics, Speech, and Signal Processing*, Vol ASSP-36, August, 1988, pp. 1223.

5. Rhamachandrand, R.P. and Kabal, P., "Pitch prediction filters in speech coding", *IEEE Trans. Acoustics, Speech, and Signal Processing*, Vol 37, no 4, April, 1989.

6. Jayant, N.S and Noll, P., *Digital Coding of Waveforms: Principles and Applications to Speech and Video*, Prentice-Hall, 1984.

7. Max, I., "Quantizing for Minimum Distortion", *IEEE Trans. Information Systems*, Vol IT-6, pp7-12, 1960.

8. Atal, B.S., Cuperman, V. and Gersho, A. (Editors) *Advances in Speech Coding* , Kluwer Academic Publishers, 1991.

9. CCITT *Methods used for assessing telephony transmission performance*, Red book, Supplement No. 2, Vol. 5, 1985.

10. Atal, B.S. and Schroeder, M.R., "Predictive Coding of Speech Signals and Subjective Error Criteria", *IEEE Trans. Acoustics, Speech, and Signal Processing*, Vol. ASSP-27, no 3, 1979.

11. Atal, B.S., "Predictive Coding of Speech at Low Bit Rates", *IEEE Trans. Commun.*, Vol. COM-30, 1982.

12. Tribolet, J.M. and Crochiere, R.E., "Frequency domain coding of speech", *IEEE Trans. Acoustics, Speech, and Signal Processing*, Vol. ASSP-29, Oct. 1979.

13. Atal, B.S. and Remde, J.R., "A New Model of LPC Excitation for Producing Natural-Sounding Speech at Low Bit Rates", *Proc. IEEE Conf. Acoustics, Speech, and Signal Processing*, Paris, 1982.

14. Secker, P. and Perkis, A., "A robust speech coder incorporating source and channel coding techniques", *Proc. 3rd Australian Int. Conf. on Speech Science and Technology*, Melbourne, Nov, 1990.

15. CCITT question U/N (16 kbit/s speech coding) "Description of a Low Delay Code Excited Linear Predictive Coding (LD-CELP) Algorithm", Source AT&T, 1989.

16. Makhoul, J., "Linear Prediction: A tutorial review", *Proc. of the IEEE*, Vol 63, no 4, April, 1975.

17. Almeida, L.B and Tribolet, J.M, "Non stationary spectral modelling of voiced speech", *IEEE Trans. Acoustics, Speech, and Signal Processing*, Vol ASSP-31, June, 1983.

18. Perkis, A., "A system description of speech coding for mobile satellite communications", *Proc. of Norwegian Symp. Signal Processing (NORSIG)*, Bergen, 1992, Norway.

19. Fant, G., "The acoustics of speech", *Proc. Third Int. Congress on Acoustics*, 1959.

20. Atal, B.S and Hanauer, S.L., "Speech analysis and synthesis by linear prediction", *Journ. Acoust Soc. Amer.* , Vol 50, August, 1970.

21. Perkis, A., "Speech Coding for Mobile Satellite Communications: A Novel Scenario", *Proc. Int, Symp. Signal Processing and its Applications (ISSPA)*, Brisbane, 1990.

22. Schroeder, M.R. and Atal, B.S., "Code-Excited Linear Prediction (CELP): High-Quality Speech at Very Low Bit Rates", *Proc. IEEE Conf. Acoustic, Speech, Signal Processing*, 1985.

23. Tremain, T.E., "The government standard LPC-10", *Speech Technology*, April, 1982.

24. Lin, D., "New Approaches to Stochastic Coding of Speech Sources at Very Low Bit Rates", *Signal Processing III: Theories and Applications*, 1986.

25. Kleijn, W.B., Krasinsky D.J. and Ketchum, R.H, "Improved Speech Quality and Efficient Vector Quantization in SELP", *Proc. IEEE. Conf. Acoustics, Speech, and Signal Processing*, 1988.

26. Davidson, G. and Gersho, A., "Complexity Reduction Methods for Vector Excitation Coding", *Proc. IEEE. Conf. Acoustics, Speech, and Signal Processing*, 1986.

27. Ribbum, B., Perkis, A., Paliwal, K.K and Ramstad, T., "Performance studies on stochastic speech coders", *Speech Communications*, Vol. 10, No. 3, August, 1991.

28. Bundrock, A. and Wilkinson, M., "Evaluation of voice codecs for the Australian mobile satellite system", *Proc. Inmarsat Working Party Meeting*, Ottawa, June, 1990.

29. Kitawaki, N., Honda, M. and Itoh, K., "Speech Quality Assessment Methods for Speech Coding Systems", *IEEE Communications Magazine*, Vol. 22, No. 10, 1984.

30. IEEE Recommended Practice for Speech Quality Measurements, *IEEE Trans. on Audio and Electroacoustics*, pp. 227-246, Sept. 1969.

31. Rahman, M. and Bulmer, M.,"Error models for land mobile satellite channels", *Mobile satellite conference*, Adelaide, 1990.

32. McAulay, R.J. and Quatieri, T.F., "Speech analysis/synthesis based on a sinusoid representation", *IEEE Trans. Acoustics, and Speech, Signal Processing*, Vol. ASSP-34, No. 4, August, 1986.

33. Perkis, A. and Ng, T.S., "A Robust Low Complexity 5.0 kbps Stochastic Coder for a Noisy Satellite Channel",*Proc. IEEE Region 10 Conf. on Communications (TENCON)*, Hong Kong, 1990.

34. Ribbum, B., Perkis, A. and Svendsen, T., "Self excited stochastic multipulse coders", *Journal of Electrical and Electronics Engineering Australia*, Vol 11, No 3, Sept., 1991.

35. Ayangolu, E. and Gray, R.M., "The design of joint source and channel trellis waveform coders", *IEEE Trans. Inform. Theory*, Vol. IT-33, No. 6 Nov., 1987.

36. Secker, P. and Perkis A., "Joint Source and Channel Coding of Line Spectrum Pairs", *Speech Communication*, Vol. 11, No. 2-3, June, 1992.

37. Singhal, S. and Atal, B.S., "Improving Performance of Multipulse LPC Coders at Low Bit Rates", *Proc. IEEE Conf. Acoustics, Speech, and Signal Processing*, 1984.

38. Zeger, K. and Gersho A.,"Pseudo-Gray coding", *IEEE trans. on Commun.*. Vol 38, No. 132, Dec., 1990.

39. Cox, R.V., Hagenauer, J., Seshadri, N. and Sundberg, C.-E., "Sub Band speech coding and matched convolutional coding for mobile radio channels", *IEEE Trans Acoustics, Speech, and Signal Processing*, Vol 39, No. 8, Aug., 1991.

40. Inmarsat (1991) "Inmarsat-M voice coding system description", Version 2, 5 February, 1991.

41. Quackenbush, S.R., Barnwell III, T.P. and Clements, M.A., *Objective measures of speech quality*, Prentice Hall, 1988.

42. Perkis, A. and Ribbum, B., "Applications of Stochastic Coding Schemes in Satellite Communications", *Advances in Speech Coding*, Kluwer Academic Publishers, 1991.

Chapter 7

ERROR CONTROL CODING

by Branka Vucetic
School of Electrical Engineering
University of Sydney NSW 2006

Chapter 5 has dealt with modulation techniques that provide efficient transmission and low nonlinear distortion in satellite communication systems. The error performance of these modulation techniques in the presence of noise could be improved only by increasing signal power. As the signal power is at a premium in satellite systems, this approach is not practical. Error control coding is a branch of communications which deals with reliable transmission of digital signals. The primary goal of error control techniques is to maximize the reliability of transmission within the constraints of signal power, system bandwidth and complexity of the circuitry.

In satellite communication systems, error control coding is used to reduce the transmitter power and the size of antennas at a spacecraft or mobile vehicles to obtain a specified error probability. It is achieved by introducing structured redundancy into transmitted signals. This usually results in a lowered data transmission rate or increased channel bandwidth, relative to an uncoded system.

An alternative for bandlimited channels is the use of trellis codes which do not require the sacrifice of either data rate or bandwidth. The cost is the complexity of the circuitry. In this Chapter we will examine some of the codes capable of achieving low error probabilities. The practical implementation of their encoding and decoding will also be discussed.

7.1 INTRODUCTION

The block diagram of a communication system with error control coding is
shown in Figure 7.1.

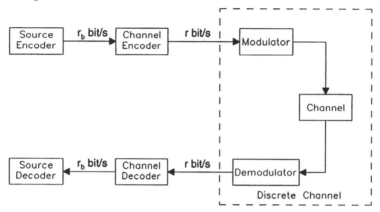

Figure 7.1: Block diagram of a digital communication system with error control coding.

A binary message sequence from the output of the source encoder is generated at the data rate of r_b bit/s. The error control encoder assigns to each
message of k digits a longer sequence of n digits called a *code word*. A good
error control code generates code words which are as different as possible
from one another. This makes the communication system less vulnerable to
channel errors. Each code is characterized by the ratio $R = k/n < 1$ called
the *code rate* . The data rate at the output of the error control encoder is
$r_c = r_b/R$ bit/s.

The modulator, as described in Chapter 5, maps the encoded digital sequences into a train of analog waveforms suitable for propagation. The
signal is generally corrupted by noise in the channel. At the demodulator
the modulated waveforms are detected individually. The demodulator typically generates a binary sequence as its best estimate of the transmitted
code words. Then the error control decoder makes estimates of the actually
transmitted messages.

The decoding process is based on the encoding rule and the characteristics
of the channel. The goal of the decoder is to minimize the effect of channel
noise. If the demodulator is designed to produce hard decisions, then its
output is a binary sequence. The subsequent channel decoding process is
called *hard decision decoding*.

The three blocks consisting of the modulator, the waveform channel and the demodulator can be represented by a *discrete channel*. The input and the output of the discrete channel are binary sequences at r_c bit/s. If the demodulator output in a given symbol interval depends only on the signal transmitted in that interval and not on any previous transmission, it is said that this channel is *memoryless*. Such a channel can be described by the set of transition probabilities $p(i \mid j)$ where i denotes a binary input symbol and j is a binary output symbol.

The simplest channel model is obtained when the probabilities of error for binary symbols 0 and 1 are the same and the channel is memoryless. This channel model is known as the *binary symmetric channel (BSC)*. The BSC is completely described by the transition probability p as shown in Figure 7.2.

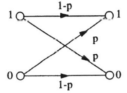

Figure 7.2: A BSC channel model.

Hard decisions in the demodulator result in some irreversible information loss. An alternative is to quantize the demodulator output to more than two levels and pass it to the channel decoder. The subsequent decoding process is called *soft decision decoding.*

The discrete memoryless channel model with multilevel demodulator output is shown in Figure 7.3.

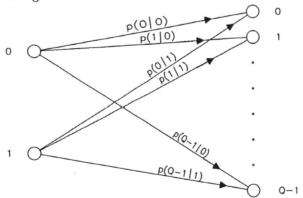

Figure 7.3: Discrete memoryless channel model.

It has binary inputs while the output consists of Q symbols. A Q-level quantization scheme, with $Q = 8$, 16 and 32, is most frequently used in soft decoding systems. It is important to note that in both of the above systems, coding is performed separately from modulation/demodulation and a performance gain is obtained at the expense of increased signal bandwidth. If bandwidth efficiency is essential, a more effective signal design is obtained by combining coding and modulation into a single entity. In this approach, binary messages are encoded into signal sequences over a certain modulation signal set. This results in increased noise immunity of the signal without increasing the channel bandwidth. This combined coding and modulation is called *coded modulation.*

The two most frequently used types of codes are block and convolutional codes. The main difference between the two of them is memory of the encoder. For block codes each encoding operation depends only on the current input message and is independent of previous encodings. That is, the encoder is a memoryless device. In contrast, for a convolutional code, each encoder output sequence depends not only on the current input message, but also on a number of past message blocks.

Error control can be exercised in several ways. *Forward error correction (FEC)* relies on error correction at the receiving end only. After error correction the decoded message is delivered to the user.

In *automatic repeat request (ARQ)* systems, error detecting codes are used. If errors are detected in a received message, retransmission is requested to ensure error free data is delivered to the user, the request signal is transmitted over a feedback channel. *Hybrid ARQ* systems apply a combination of FEC procedures with error detection and retransmission to achieve high reliability and data throughput.

7.2   BLOCK CODES

The encoder of an (n, k) block code accepts a *message* of k symbols and transforms it into a longer sequence of n symbols called a *code word.* In general, both messages and code words can consist of binary or nonbinary symbols. Block codes are almost invariably used with binary symbols due to implementation complexity constraints. In an (n, k) block code there are 2^k distinct messages. Since there is a one-to-one correspondence between a message and a code word, there are 2^k distinct code words. The *code rate* $R = k/n$ determines the amount of redundancy.

An (n, k) linear block code is *linear* if

1. the component-wise modulo-2 sum of two code words is another code word, and

2. the code contains the all-zero code word.

In the language of linear algebra, an (n, k) linear block code is a k-dimensional subspace of the vector space V_n of all the binary n-tuples. An (n, k) linear block code can be generated by a set of k linearly independent binary n-tuples $g_0, g_1, \ldots, g_{k-1}$. The code words are obtained as linear combinations of these k n-tuples. The k vectors generating the code $g_0, g_1, \cdots, g_{k-1}$ can be arranged as rows of a $k \times n$ matrix as follows

$$\mathbf{G} = \begin{bmatrix} \mathbf{g}_0 \\ \mathbf{g}_1 \\ \vdots \\ \mathbf{g}_{k-1} \end{bmatrix} = \begin{bmatrix} g_{00} & g_{01} & \cdots & g_{0,n-1} \\ g_{10} & g_{11} & \cdots & g_{1,n-1} \\ \vdots & \vdots & \vdots & \vdots \\ g_{k-1,0} & g_{k-1,1} & \cdots & g_{k-1,n-1} \end{bmatrix} \tag{7.1}$$

The array $\mathbf{G}$ is called the *generator matrix* of the code. Then, the code word $\mathbf{v}$ for message $\mathbf{c}$ can be written as

$$\mathbf{v} = \mathbf{c} \cdot \mathbf{G} = c_0 \cdot \mathbf{g}_0 + c_1 \cdot \mathbf{g}_1 + \cdots + c_{k-1} \cdot \mathbf{g}_{k-1} \tag{7.2}$$

Exercise 7.1

An $(6, 3)$ linear block code can be generated by the generator matrix

$$\mathbf{G} = \begin{bmatrix} \mathbf{g}_0 \\ \mathbf{g}_1 \\ \mathbf{g}_2 \end{bmatrix} = \begin{bmatrix} 0 & 1 & 1 & 1 & 0 & 0 \\ 1 & 0 & 1 & 0 & 1 & 0 \\ 1 & 1 & 0 & 0 & 0 & 1 \end{bmatrix}$$

If the message is $\mathbf{c} = (0\ 1\ 1)$ find the code word. Find all other code words.

Solution

The message $\mathbf{c} = (0\ 1\ 1)$ is encoded as follows:

$$\begin{aligned} \mathbf{v} &= \mathbf{c} \cdot \mathbf{G} \\ &= 0 \cdot (011100) + 1 \cdot (101010) + 1 \cdot (110001) \\ &= (000000) + (101010) + (110001) \\ &= (011011) \end{aligned}$$

Table 7.1 presents the list of messages and code words for the (6, 3) linear block code.

Messages	Code Words
(c_0, c_1, c_2)	$(v_0, v_1, v_2, v_3, v_4, v_5)$
(0 0 0)	(0 0 0 0 0 0)
(1 0 0)	(0 1 1 1 0 0)
(0 1 0)	(1 0 1 0 1 0)
(1 1 0)	(1 1 0 1 1 0)
(0 0 1)	(1 1 0 0 0 1)
(1 0 1)	(1 0 1 1 0 1)
(0 1 1)	(0 1 1 0 1 1)
(1 1 1)	(0 0 0 1 1 1)

Table 7.1: A (6, 3) linear block code.

7.2.1 Linear Systematic Block Codes

A linear *systematic* block code has the additional structure that the message sequence is itself part of the code word. In addition to the k-digit message sequence the code word contains the $(n - k)$-digit parity check sequence. This format allows direct extraction of the message from the code word. The systematic format is shown below

$$\overbrace{(c_0, c_1, \ldots, c_{k-1})}^{\text{message}} \longrightarrow \underbrace{v_0, v_1, \ldots, v_{n-k-1}}_{\text{parity--check}}, \overbrace{c_0, c_1, \ldots c_{k-1}}^{\text{code word}}_{\text{message}}$$

Systematic format

The generator matrix for a systematic block code has the following form

$$\mathbf{G} = [\mathbf{PI_k}] \qquad (7.3)$$

$\mathbf{I_k}$ is the kxk identity matrix and $\mathbf{P}$ is a kx$(n - k)$ matrix of the form

$$\mathbf{P} = \begin{bmatrix} p_{00} & p_{01} & \cdots & p_{0,n-k-1} \\ p_{10} & p_{11} & \cdots & p_{1,n-k-1} \\ p_{20} & p_{21} & \cdots & p_{2,n-k-1} \\ \vdots & \vdots & \cdots & \vdots \\ p_{k-1,0} & p_{k-1,1} & \cdots & p_{k-1,n-k-1} \end{bmatrix} \qquad (7.4)$$

where $p_{ij} = 0$ or 1. The (6,3) code given in Exercise 7.1 is a systematic code.

Let $\mathbf{c} = (c_0, c_1, c_2)$ be the message to be encoded. The corresponding code word is then,

$$\begin{aligned} \mathbf{v} &= (v_0, v_1, v_2, v_3, v_4, v_5) \\ &= \mathbf{c} \cdot \mathbf{G} \end{aligned}$$

The last three digits of the code word are identical to the original message. That is,

$$v_5 = c_2; \quad v_4 = c_1; \quad v_3 = c_0$$

while the first three digits are linear combinations of the message bits. That is,

$$v_2 = c_0 + c_1; \quad v_1 = c_0 + c_2; \quad v_0 = c_1 + c_2$$

The above equations specify the parity digits and are called the *parity check equations*. The encoder implementation based on the parity check equations is shown in Figure 7.4.

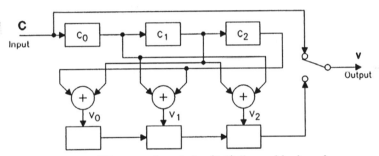

Figure 7.4: The encoder of the (6,3) linear block code.

7.2.2 Parity Check Matrix

Let us consider a systematic linear block code specified by its generator matrix in the form $\mathbf{G} = [\mathbf{P}\mathbf{I_k}]$. For any coder word $\mathbf{v}$, it is possible to show that

$$\mathbf{v} \cdot \mathbf{H^T} = \mathbf{0} \tag{7.5}$$

where

$$\mathbf{H} = \left[\mathbf{I_{n-k}}\mathbf{P^T}\right] \tag{7.6}$$

The $(n - k) \times k$ matrix $\mathbf{H}$ is called the *parity check matrix*. $\mathbf{I_{n-k}}$ is the $(n - k)$ identity matrix. The parity check matrix is used in error detection to test whether a received vector is a code word by checking (7.5).

7.2.3   Syndrome

Let us assume that the transmitted code word is $\mathbf{v}$ and the corresponding received vector is $\mathbf{r}$. In general, $\mathbf{v} \neq \mathbf{r}$ due to transmission errors. The component-wise modulo-2 sum of the transmitted and received vector gives the *error pattern* $\mathbf{e}$:

$$\mathbf{e} = \mathbf{v} + \mathbf{r}$$

or

$$
\begin{aligned}
\mathbf{e} &= (e_0, e_1, \ldots, e_{n-1}) \\
&= (r_0, r_1, \ldots, r_{n-1}) + (v_0, v_1, \ldots, v_{n-1}) \\
&= (r_0 + v_0,\ r_1 + v_1, \ldots, r_{n-1} + v_{n-1})
\end{aligned}
$$

If the value of e_j is 1, then this indicates that the j-th position of $\mathbf{r}$ has an error. There are a total 2^n possible error patterns. To check whether a received vector is a code word, we define the *syndrome* of $\mathbf{r}$ as an $(n-k)$-tuple

$$\mathbf{s} = (s_0, s_1, \ldots, s_{n-k-1}) = \mathbf{r} \cdot \mathbf{H}^{\mathrm{T}} \tag{7.7}$$

Then from the definition of a code word, expressed by (7.5), the received vector $\mathbf{r}$ is a code word in an (n, k) block code if and only if $\mathbf{s} = \mathbf{0}$. Error detection circuits compute the syndrome to check whether a valid code word is received. If $\mathbf{s} \neq \mathbf{0}$, the received vector $\mathbf{r}$ is not a code word. The error detector interprets it as evidence of transmission errors. When $\mathbf{s} = \mathbf{0}$, the detector assumes that the received vector $\mathbf{r}$ has no errors. A detection failure occurs if $\mathbf{r}$ is a code word but different from the actually transmitted one. It is undetectable by the code.

Exercise 7.2

Consider a (7,4) linear block code with the generator matrix

$$
\mathbf{G} = \begin{bmatrix}
1 & 1 & 0 & 1 & 0 & 0 & 0 \\
0 & 1 & 1 & 0 & 1 & 0 & 0 \\
1 & 1 & 1 & 0 & 0 & 1 & 0 \\
1 & 0 & 1 & 0 & 0 & 0 & 1
\end{bmatrix}
$$

Find the parity check matrix and an error detection circuit for this code.
Solution
The parity check matrix is obtained as

$$
\mathbf{H} = \begin{bmatrix}
0 & 0 & 1 & 0 & 1 & 1 & 1 \\
0 & 1 & 0 & 1 & 1 & 1 & 0 \\
1 & 0 & 0 & 1 & 0 & 1 & 1
\end{bmatrix}
$$

The syndrome for this code is computed as

$$s = \begin{bmatrix} r_0 & r_1 & r_2 & r_3 & r_4 & r_5 & r_6 \end{bmatrix} \begin{bmatrix} 1 & 0 & 0 \\ 0 & 1 & 0 \\ 0 & 0 & 1 \\ 1 & 1 & 0 \\ 0 & 1 & 1 \\ 1 & 1 & 1 \\ 1 & 0 & 1 \end{bmatrix}$$

The syndrome components are $s_0 = r_0 + r_3 + r_5 + r_6$, $s_1 = r_1 + r_3 + r_4 + r_5$, $s_2 = r_2 + r_4 + r_5 + r_6$. The syndrome circuit is illustrated in Figure 7.5.

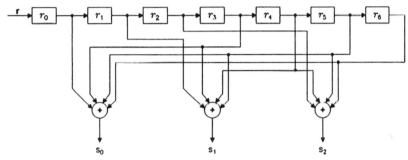

Figure 7.5: A syndrome circuit for the (6,3) linear block code.

Let us assume that the transmitted code word is $\mathbf{v} = (0000000)$ and the received vector is $\mathbf{r} = (1000000)$. The syndrome of $\mathbf{r}$ is

$$s = (100) \neq \mathbf{0}$$

Hence $\mathbf{r}$ is not a code word and a transmission error is detected.

7.2.4 Syndrome Decoding

Suppose that a code word $\mathbf{v}$ is transmitted and that $\mathbf{r}$ is the received vector. The role of the channel decoder is to estimate the transmitted code word $\mathbf{v}$ from the received vector $\mathbf{r}$.

The received vector can be expressed as

$$\mathbf{r} = \mathbf{v} + \mathbf{e} \tag{7.8}$$

where $\mathbf{e}$ is the error pattern.

Hence, if the decoder estimates the error pattern **e**, the actually transmitted code word **v** can be computed from (7.8). The syndrome has a property which is extremely useful to the decoder. That is, the syndrome is not dependent on which code word is transmitted, but depends only on the error pattern. From (7.7) the syndrome can be expressed as

$$\mathbf{s} = (\mathbf{v} + \mathbf{e}) \cdot \mathbf{H}^{\mathbf{T}} = \mathbf{e} \cdot \mathbf{H}^{\mathbf{T}} \tag{7.9}$$

since $\mathbf{v} \cdot \mathbf{H}^{\mathbf{T}} = \mathbf{0}$. Expanding (7.9) we obtain the following $(n - k)$ expressions for syndrome components of a linear systematic block code

$$
\begin{aligned}
s_0 &= e_0 + e_{n-k}p_{00} + e_{n-k+1}p_{10} + \cdots + e_{n-1}p_{k-1,0} \\
s_1 &= e_1 + e_{n-k}p_{01} + e_{n-k+1}p_{11} + \cdots + e_{n-1}p_{k-1,0} \\
&\ \ \vdots \\
s_{n-k-1} &= e_{n-k-1} + e_{n-k}p_{0,n-k-1} + e_{n-k+1}p_{1,n-k-1} + \cdots \\
&\quad + e_{n-1}p_{k-1,n-k-1}
\end{aligned}
\tag{7.10}
$$

These expressions form $(n - k)$ linear equations specifying n unknown error components. Since the number of equations is less than the number of unknowns, there is no unique solution. In fact, there are 2^k possible solutions for the error patterns.

A decoding process is a method of estimating a single error pattern out of 2^k possible error patterns. It is generally designed to minimize the probability of decoding error. The decoder will therefore select the most probable error pattern which satisfies Equation (7.10).

For a binary symmetric channel the error pattern that minimizes the probability of error is the vector with the smallest number of 1's. Thus, the decoder will assign to each syndrome the error pattern with the minimum number of 1's out of the 2^k possible error patterns. The syndromes and error patterns can be arranged in a look-up table. The look-up table contains 2^{n-k} syndromes and 2^{n-k} corresponding error patterns. This algorithm can be also implemented with combinatorial logic circuits.

7.2.5 The Minimum Distance of a Block Code

Let us consider an (n, k) linear block code. The *Hamming weight* (or weight) of a code word **v** is defined as the total number of nonzero components in the code word. The *Hamming distance*, $d(\mathbf{v}, \mathbf{u})$, between two code words **v** and **u** is defined as the number of places in which these two code words differ.

For example, if the two code words are $\mathbf{v} = (100111)$ and $\mathbf{u} = (010001)$ the Hamming distance between them is 4. The Hamming distance between two code words is identical to the weight of their modulo-2 sum.

The *minimum Hamming distance* (or minimum distance) of a code, d_{min}, is defined as the smallest Hamming distance between two different code words in the code. The linearity property of block codes requires that the modulo-2 sum of two code words is another code word. This implies that the minimum distance of a linear block code is the smallest weight of the nonzero code words in the code.

The importance of the minimum distance parameter is in the fact that it determines the error correcting and detecting capability of a code.

7.2.6 Error Detecting Capability of a Block Code

Let us assume that the transmitted code word is $\mathbf{v}$ and the received word is $\mathbf{r}$. A syndrome based error detection circuit can detect all error patterns which are not valid code words, since these error patterns will produce a nonzero syndrome. If the minimum distance of the code is d_{min} that implies that all error patterns with the weight of $(d_{min} - 1)$ and less can be detected.

However, not all error patterns with the weight of d_{min} can be detected since there is at least one error pattern of this weight that will change one code word into another code word. For this reason the error detecting capability of a block code is said to be $(d_{min} - 1)$.

7.2.7 Error Correcting Capability of a Block Code

For a BSC channel the most likely error patterns are those with minimum weight. A brute force decoding method for the BSC channel consists of comparing the received vector $\mathbf{r}$ with all code words in the code and selecting a code word that has the smallest distance from $\mathbf{r}$ as the estimate of the transmitted code word. This decoding method is called *minimum distance decoding*. Although this algorithm becomes impractical as the number of code words increases, it is instructive in determining the error correcting capability of a block code.

Let us imagine the received vector $\mathbf{r}$ and code words $\mathbf{v}$ and $\mathbf{u}$ as points in an n-dimensional space. Figure 7.6 shows two code words $\mathbf{v}$ and $\mathbf{u}$ at the distance equal to the minimum code distance $d_{min} = 3$.

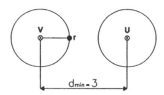

Figure 7.6: A code with minimum distance 3.

Let us assume that the transmitted code word is **v**. If the received vector **r** contains one error then it lies on the sphere of radius 1 around code word **v**. Clearly, vector **r** is closer to the transmitted code word **v** than to any other code word. If the minimum distance decoding rule is used, the error in **r** can be corrected.

Similarly, if $d_{min} = 2t + 1$, spheres of radius t around each code word will be disjoint, as shown in Figure 7.7. This code will be able then to correct up to

$$t = \lfloor \frac{d_{min} - 1}{2} \rfloor \tag{7.11}$$

errors where $\lfloor \cdot \rfloor$ denotes the largest integer no greater than the enclosed number. Therefore, if there are t or less transmission errors in the received vector, a minimum distance decoder will correct all of them.

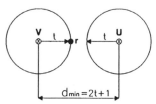

Figure 7.7: A code with minimum distance $d_{min} = 2t + 1$.

On the other hand, there exists at least one error pattern with the weight of $(t+1)$ that cannot be corrected and the decoder will deliver a wrong code word. For this reason the parameter t is called the *error correcting capability* of the code.

7.2.8   Maximum Likelihood Decoding of Block Codes for a BSC Channel

The decoding rule that provides the minimum probability of error is called *maximum likelihood decoding (MLD)* . For a BSC channel the demodulator makes hard decisions and the received sequence **r** is a binary sequence. Therefore, the decoder performs hard decision decoding.

The MLD rule for a hard decision decoder over a BSC channel can be stated as follows. For each code word $\mathbf{u}$ in an (n,k) block code and given $\mathbf{r}$, compute the conditional probability $P(\mathbf{r} \mid \mathbf{u})$. Then the code word with the maximum conditional probability is chosen as the estimate of the actually transmitted code word $\mathbf{v}$. For a BSC channel with the transition probability p the conditional probability is given by

$$P(\mathbf{r} \mid \mathbf{u}) = p^{d(\mathbf{r},\mathbf{u})}(1-p)^{n-d(\mathbf{r},\mathbf{u})} \qquad (7.12)$$

where $d(\mathbf{r},\mathbf{u})$ is the Hamming distance between the vectors $\mathbf{r}$ and $\mathbf{u}$.

Since for $p < 1/2$, $P(\mathbf{r} \mid \mathbf{u})$ is a monotonically decreasing function of $d(\mathbf{r},\mathbf{u})$, we will have

$$P(\mathbf{r} \mid \mathbf{u}) > P(\mathbf{r} \mid \mathbf{w}) \qquad (7.13)$$

if and only if

$$d(\mathbf{r},\mathbf{u}) < d(\mathbf{r},\mathbf{w}) \qquad (7.14)$$

Hence the MLD rule for a BSC channel can be obtained by comparing the received vector $\mathbf{r}$ with all code words and selecting the code word that has the smallest Hamming distance from $\mathbf{r}$ as the estimate of the actually transmitted code word. This decoding rule is known as *minimum distance decoding*.

7.2.9 Weight Distribution of Block Codes

The weight distribution of a block code is useful in computing probabilities of undetected or uncorrected errors. For an (n,k) block code we define A_i as the number of code words of weight i. Then the set

$$\{A_{d_{min}}, A_{d_{min}+1}, \cdots, A_i, \cdots, A_n\}$$

is called the *weight distribution* or the *weight spectrum* of the code.

There is a special class of codes with a binomial weight distribution. These codes are known as *perfect codes*.

Exercise 7.3
Find the weight distribution of the (6,3) block code given in Exercise 7.1.

Solution
$A_3 = 4$, $A_4 = 3$, $A_5 = 0$, $A_6 = 0$.

7.2.10   Probability of Undetected Errors

Let us consider an (n, k) block code which is used for error detection over a BSC channel with transition probability p. We have shown that the only error patterns that can transform one code word into another valid code word must themselves be code words. Hence, the undetectable error patterns are those which are code words in the code. For this reason the code weight distribution $\{A_{d_{min}}, A_{d_{min}+1}, \cdots, A_i, \cdots, A_n\}$ is the weight distribution of the undetectable error patterns. The probability of an undetected error is

$$P_u = \sum_{i=d_{min}}^{n} A_i p^i (1-p)^{(n-i)} \tag{7.15}$$

7.2.11   Probability of Erroneous Decoding for Hard Decision Decoders

Probability of erroneous decoding is generally used as the measure of performance of coded systems. It is expressed either as block or bit error probability. For transmission over a BSC channel with transition probability p the probability that a decoder with bounded distance decoding [5] commits an erroneous decision is given by

$$P_e = \sum_{t+1}^{n} A_j p^j (1-p)^{(n-j)} \tag{7.16}$$

where P_e is the probability that a block of n symbols contain at least one error and A_j is the number of code words with weight j.

If n is large the task of finding the weight distribution of a code is difficult. Hence an upper bound for block error probability is typically used. This is

$$P_e \leq \sum_{t+1}^{n} \binom{n}{j} p^j (1-p)^{n-j} \tag{7.17}$$

The bound becomes an accurate expression for perfect codes. Perfect codes are optimum in the sense that they minimize the probability of error among all codes with the same n and k. These codes can correct all errors of weight $\leq t$ but none of weight higher than that. An approximation to the bit error probability can be derived from the block error probability for high signal-to-noise ratios (SNR). At high SNR erroneous decoding arises predominantly from error patterns with $(t + 1)$ errors.

In this case the decoder will most likely make a false assumption that the error pattern contains t errors and will change the received vector in t places so that the total number of errors will be $(2t + 1)$. These errors are distributed all over the received vector. If the minimum distance of the code is $d_{min} = 2t + 1$, the bit error probability can be approximated by

$$P_b \simeq \frac{d_{min}}{n} P_e \tag{7.18}$$

Exercise 7.4
Compute the block error probability for the (7,4) code given in Exercise 7.2 if transmitted over a BSC with transition probability p.

Solution
For a perfect code the block error probability is given by

$$P_e = \sum_{t+1}^{n} \binom{n}{j} p^j (1 - p)^{n-j}$$

The computation can be simplified if the probability of erroneous decoding is expressed in terms of the probability of correct decoding. The decoder will make correct decisions if the error pattern contains no errors or only one error. Hence

$$P_e = 1 - (1 - p)^7 - 7p(1 - p)^6$$

7.2.12   Coding Gain

Coded systems typically require less SNR than uncoded systems to achieve the same output bit error probability. This reduction, expressed in decibels, in the required SNR to achieve a specific bit error probability, of a coded system over an uncoded system is called *coding gain*.

The *asymptotic coding gain* of a coded system with soft decision decoding over an uncoded reference system is defined as

$$G = 10 \log_{10} \frac{d_{Emin}^2}{d_u{}^2} \quad \text{(dB)} \tag{7.19}$$

where d_u is the Euclidean distance of the uncoded system. This is the coding gain when the system is operating with high SNR and error rates are low. However, to get a fair comparison of coded and uncoded systems we must take into account that coded systems require a higher bandwidth than uncoded systems resulting in different noise variances. Therefore, the comparison should not be performed on the basis of SNR but signal powers instead.

The asymptotic coding gain in this case will be

$$G = 10\log_{10}\frac{\sigma_u^2 d_{Emin}^2}{2d_u{}^2} = 10\log_{10}\frac{kd_{Emin}^2}{nd_u^2} \qquad (7.20)$$

where σ_c^2 and σ_u^2 are the variances of Gaussian noise for the coded and uncoded systems, respectively.

Exercise 7.5
Compute the asymptotic coding gain for the (7,4) block code given in Exercise 7.2 with soft decision decoding.

Solution
The minimum Euclidean distance of the uncoded system with binary antipodal modulation is $d_u = 2$. The Hamming distance of the (7,4) code is 3, while the minimum Euclidean distance of the code combined with binary antipodal modulation is $d_{Emin} = 2\sqrt{3}$.

The coded system with the (7,4) code requires a bandwidth $\sqrt[k]{7/4}$ times higher than the bandwidth of the uncoded system. Therefore, the variance of the white additive Gaussian noise of the coded system σ_c is related to the variance of the same noise in the uncoded system σ_u by

$$\sigma_c = \sqrt{7/4}\sigma_u$$

The asymptotic coding gain is then

$$G = 10\log_{10}\frac{12}{4}\frac{4}{7} = 2.34 \quad \text{dB}$$

7.3 CYCLIC BLOCK CODES

There is an important class of linear block codes, known as *cyclic codes*, for which implementation of encoding and decoding can be considerably simplified. The encoding and decoding operations of cyclic codes are based on their algebraic structure.

In this section we will briefly review some of their algebraic properties. See for example, Reference [3] for more details.

A cyclic code has the property that any cyclic shift of a code word is another code word. That is, if

$$\mathbf{v}^{(0)} = (v_0, v_1, v_2, \cdots, v_{n-1}) \qquad (7.21)$$

is a code word, then the vector obtained shifting $\mathbf{v}^{(0)}$ one place to the right is another code word $\mathbf{v}^{(1)}$ where

$$\mathbf{v}^{(1)} = (v_{n-1}, v_0, v_1, \cdots, v_{n-2}) \qquad (7.22)$$

while the vector obtained shifting $\mathbf{v}^{(0)}$ i places to the right is also a code word $\mathbf{v}^{(i)}$ in an (n, k) cyclic code, namely

$$\mathbf{v}^{(i)} = (v_{n-i}, v_{n-i+1}, v_{n-i+2}, \cdots, v_{n-i-1}) \qquad (7.23)$$

The cyclic property implies that a code word in an (n, k) code can be represented by a polynomial called the *code polynomial* of degree $(n-1)$ or less, with code word components as the polynomial coefficients, as follows

$$\mathbf{v}(X) = v_0 + v_1 X + v_2 X^2 + \cdots + v_{n-1} X^{n-1} \qquad (7.24)$$

where X is an arbitrary real variable. Multiplication of the code polynomial by X is equivalent to a cyclic shift to the right, subject to the constraint that $X^n = 1$.

In polynomial representation the cyclic property can be described as follows. If $\mathbf{v}(X)$ is a code polynomial, then the polynomial $X^i \mathbf{v}(X) mod(X^n - 1)$ is also a code polynomial, where $mod(X^n - 1)$ denotes multiplication modulo $X^n - 1$. This polynomial multiplication restores the polynomial $X^i \mathbf{v}(X)$ to order $(n - 1)$.

7.3.1 Generator Polynomial

Each cyclic code is characterized by a unique polynomial of degree $(n - k)$, called the *generator polynomial*. The generator polynomial, denoted $\mathbf{g}(X)$, plays the same role as the generator matrix in the case of block codes. The generator polynomial can be represented as

$$\mathbf{g}(X) = 1 + g_1 X + g_2 X^2 + \cdots + g_{n-k-1} X^{n-k-1} + X^{n-k} \qquad (7.25)$$

Every code polynomial of a cyclic code is divisible by the generator polynomial. Furthermore, every polynomial of degree $(n - 1)$ or less with binary coefficients that is divisible by the generator polynomial is a code polynomial.

7.3.2   Encoding of Cyclic Codes

Let us consider a message to be encoded

$$\mathbf{c} = (c_0, c_1, \cdots, c_{k-1}) \tag{7.26}$$

It can be represented by a message polynomial

$$\mathbf{c}(\mathbf{X}) = c_0 + c_1 X + \cdots + c_{k-1} X^{k-1} \tag{7.27}$$

Suppose that we require to encode the message $\mathbf{c}$ into an systematic (n, k) cyclic code. That is, the rightmost k components of the code word are unchanged message symbols. To accomodate the requirement of the systematic code format we multiply $\mathbf{c}$ by X^{n-k}

$$X^{n-k}\mathbf{c}(X) = c_0 X^{n-k} + c_1 X^{n-k+1} + \cdots + c_{n-1} X^{n-1} \tag{7.28}$$

Dividing the polynomial $X^{n-k}\mathbf{c}(X)$ by $\mathbf{g}(X)$ gives

$$X^{n-k}\mathbf{c}(X) = \mathbf{a}(X)\mathbf{g}(X) + \mathbf{b}(X) \tag{7.29}$$

where $\mathbf{b}(X)$ is the remainder of degree $(n - k - 1)$

$$\mathbf{b}(X) = b_0 + b_1 X + \cdots + b_{n-k-1} X^{n-k-1} \tag{7.30}$$

and $\mathbf{a}(X)$ is the quotient of degree $(k - 1)$

$$\mathbf{a}(X) = a_0 + a_1 X + \cdots + a_{k-1} X^{k-1} \tag{7.31}$$

Let us rearrange Equation (7.29) as

$$\mathbf{b}(X) + X^{n-k}\mathbf{c}(X) = \mathbf{a}(X)\mathbf{g}(X) \tag{7.32}$$

The polynomial on the left hand side of Equation (7.32) is a multiple of $\mathbf{g}(X)$ and its degree is $(n-1)$. Therefore, it is a code polynomial of the cyclic (n, k) code generated by $\mathbf{g}(X)$. The code polynomial in Equation (7.32) is in a systematic form with $\mathbf{b}(X)$ being the parity polynomial.

7.3.3   Important Cyclic Codes

Hamming Codes

The (7,4) code that has been used in examples throughout this section is a Hamming code.

Parameters of Hamming codes are given by

$$
\begin{aligned}
n &= 2^m - 1 \\
k &= 2^m - m - 1 \\
n - k &= m \\
d_{min} &= 3 \\
t &= 1
\end{aligned}
$$

where m is an integer. Hamming codes are perfect codes.

BCH Codes

BCH codes are an important class of cyclic codes named after their discoverers Bose, Chaudhuri and Hocquenghem. Their significance lies in the fact that there are efficient and practical decoding methods based on their algebraic structure [1], [2]. BCH codes are powerful in correcting random errors. The parameters of BCH codes are:

$$
\begin{aligned}
n &= 2^m - 1 \\
k &\geq n - mt \\
d_{min} &\geq 2t + 1
\end{aligned}
$$

where t is the error correcting capability.

BCH codes possess the cyclic property and therefore can be encoded by shift register circuits with feedback connections based on the generator polynomial. A (63,48) BCH code is used for error correction in transmission of signalling messages for the PAN European mobile radio network.

The generator polynomial for this code is

$$\mathbf{g}(X) = X^{15} + X^{14} + X^{13} + X^{11} + X^4 + X^2 + 1 \qquad (7.33)$$

An overall parity check digit is added to every code word to obtain code words of 64 bits in order to simplify DSP implementation.

Golay Code

The Golay code is a $(23, 12)$ triple error correcting binary BCH code with the generator polynomial

$$\mathbf{g}(X) = X^{11} + X^{10} + X^6 + X^5 + X^4 + X^2 + 1 \qquad (7.34)$$

or, alternatively

$$\mathbf{g}(X) = X^{11} + X^9 + X^7 + X^6 + X^5 + X + 1 \qquad (7.35)$$

The Golay code is one of a few nontrivial perfect codes. The Golay code is used in Inmarsat and other satellite mobile communication systems to protect coded speech signals.

The *extended Golay code* is obtained by appending an overall parity digit to the (23,12) Golay code. This operation gives a 1/2 rate code which is simpler for implementation than the original code. The minimum distance is increased from 7 to 8 so that the code can be designed to correct some but not all four error patterns.

Reed-Solomon Codes

Reed-Solomon (RS) codes are a subclass of BCH codes with code word symbols selected from a nonbinary alphabet. Each symbol consists of m bits. The code parameters are

$$
\begin{aligned}
n &= 2^m - 1 \\
k &= 1, 2, 3, ..., n-1 \\
n - k &= m \\
d_{min} &= n - k - 1 \\
t &= \lfloor \frac{d_{min}-1}{2} \rfloor
\end{aligned}
$$

The codes considered so far are efficient in correcting random isolated errors. Errors on many real channels tend to occur in groups rather than at random.

For example, in mobile communications, errors resulting from multipath propagation or shadowing, are not independent, but appear in *bursts*. RS codes are powerful in correcting multiple bursts of errors. An RS code can correct t symbols in a block of n symbols, or any single burst of length $m(t-1)+1$ digits.

RS codes have good distance properties and can be decoded by efficient hard decision decoding algorithms. The details of decoding algorithms can be found in [1], [3], [4], [5]. A (255,223) RS code is a standard code for NASA deep space communications.

This code is capable of correcting 16 symbol errors within a block of 255 symbols, where symbols are 8 digits long. It can also correct any single burst of 121 or less digits.

Figure 7.8 shows the bit error performance of a number of key cyclic codes of a similar code rate. Note that longer codes offer higher coding gains but their encoding/decoding algorithms are more complex.

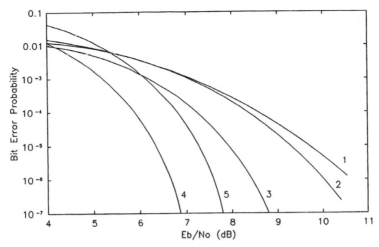

Figure 7.8: Bit error performance of linear block code on a Gaussian channel;
1: Uncoded BPSK; 2: The (7,4) Hamming code; 3: The (23,12)
Golay code; 4: The (127,71) BCH code; 5: The (31,17) RS code.

7.3.4 Interleaving

Interleaving is a useful technique for enhancing the burst error correcting capability of a code. Let us consider an (n, k) linear block code with correcting capability of t errors in any code word of n digits.

Assume that a transmitted code sequence consists of λ code words $\mathbf{v}_1$, $\mathbf{v}_2$, $\cdots$, $\mathbf{v}_\lambda$, arranged in a matrix

$$\mathbf{T} = \begin{bmatrix} v_{11} & v_{12} & \cdots & v_{1n} \\ v_{21} & v_{22} & \cdots & v_{2n} \\ \vdots & & & \vdots \\ v_{\lambda 1} & v_{\lambda 2} & \cdots & v_{\lambda n} \end{bmatrix} \qquad (7.36)$$

The matrix is transmitted serially column by column. It can be considered as a code word in an *interleaved* $(\lambda n, \lambda k)$ code. This code is able to correct bursts of λt digits or less. For example, if the original (n, k) code corrects single errors, the interleaved code can correct bursts of λ errors, where λ is called the *interleaving degree*.

At the receiver side the received sequence is stored into a matrix of the same format as matrix $\mathbf{T}$. The deinterleaving operation consists in reading the matrix row by row.

7.3.5   Concatenated Codes

Concatenated codes are a convenient method to design long codes with reasonable complexity. In order to correct burst errors on real channels, codes must be long enough. On the other hand, the decoding complexity of block codes increases exponentially with the code length.

This method aims to achieve both improved reliability and low implementation complexity. Concatenated codes are formed from two codes. An inner (n_1, k_1) codes handles random errors, presenting to an outer (n_2, k_2) code burst errors. RS codes are typically used as outer codes. The block diagram of a concatenated coding system is shown in Figure 7.9.

Figure 7.9: Block diagram of a concatenated error control coding system.

The encoding process consists of two encoding operations performed serially. The message sequence is divided into k_2 symbols of k_1 bits each. These k_2 symbols are applied to the input of the outer encoder. The output of the encoder is a code word **v** of n_2 symbols.

In the next encoding stage each $k_1 - tuple$ is encoded according to the inner code into n_1-tuple inner code words. This results in a binary code word of $n_2 n_1$ digits from a *concatenated* $(n_1 n_2, k_1 k_2)$ *code*. The decoding is accomplished in the reverse order from the encoding process.

Concatenated codes can correct combinations of random and burst errors. In general, the inner code is a random correcting and the outer code is a burst error correcting code. Concatenated codes are used as a standard means of providing high coding gains in deep-space and satellite communications.

7.4   CONVOLUTIONAL CODES

7.4.1   Introduction

In block coding described in the previous Section successive code words are not related in any way in the encoder. The decoder also operates on one block at a time. A convolutional coding scheme segments the input information stream into messages generally much shorter than messages for block codes. The messages are processed continuously in a serial manner producing longer output blocks at the output of the encoder. A distinguishing feature of this type of coding is that an output block depends on the history of a certain number of input messages.

In other words, the encoder is a finite memory system. Due to its memory the encoder effectively transforms long message sequences into even longer output code sequences. This property enables efficient use of statistical regularities of error sequences in the decoder. Probabilistic methods, typically used in decoding of convolutional codes, allow simple implementation of soft decision decoding.

7.4.2 Polynomial Representation

An (n, k, m) convolutional code can be generated by a k-input, n-output linear sequential circuit with input memory m. Typically, n and k are small integers. For simplicity, we will restrict our attention to a special class of rate $1/n$ convolutional codes. A general encoder for $(n, 1, m)$ convolutional codes is shown in Figure 7.10.

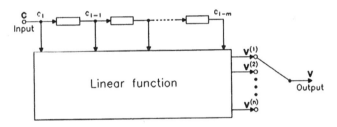

Figure 7.10: The encoder of a (n,1) convolutional code.

A single message sequence $\mathbf{c}$ is encoded into n output sequences $\mathbf{v}^{(1)}, \mathbf{v}^{(2)}, \cdots, \mathbf{v}^{(n)}$. The n output sequences are then multiplexed to form a single code sequence. If at time l, the input message bit is c_l, then the output is an n-digit sequence

$$\mathbf{v}_l = (v_l^{(1)}, v_l^{(2)}, \cdots, v_l^{(n)}) \qquad (7.37)$$

Since the encoder has m memory elements, the output $\mathbf{v}_l$ is a function of the input c_l and also $(m-1)$ previous input bits. A parameter $K = (m+1)$ is called the *constraint length* of the code. The input message sequence can be expressed as a polynomial of finite or infinite length

$$\mathbf{c}(X) = \ldots + c_0 + c_1 X + \ldots + c_l X^l + \ldots$$

where X is the delay operator.

The output code sequence is determined by connections from the m-stage shift register. Each of the n connections can be represented by a polynomial of degree m. Therefore, the encoder of an $(n, 1, m)$ convolutional code is specified by n polynomials called *generator polynomials* of the form

$$\mathbf{g}^{(j)}(X) = g_0^{(j)} + g_1^{(j)}X + g_2^{(j)}X^2 + ... + g_m^{(j)}X^m \quad j = 1, \cdots, n$$

The generator polynomials can be arranged in a matrix form as

$$\mathbf{G}(X) = \left[\mathbf{g}^{(1)}(X), \mathbf{g}^{(2)}(X), \cdots, \mathbf{g}^{(n)}(X) \right]$$

Matrix $\mathbf{G}(X)$ is called the *generator polynomial matrix*. Each j-th output code sequence can be expressed as a polynomial

$$\mathbf{v}^{(j)} = \cdots + v_0^{(j)} + v_1^{(j)}X + v_2(j)X^2 + \cdots + v_l(j)X^l \quad j = 1, \cdots, n$$

The entire output code sequence can be represented as a vector of n polynomials called the *code polynomial vector* as

$$\mathbf{v}(X) = \left[\mathbf{v}^{(1)}(X), \mathbf{v}^{(2)}(X), \cdots, \mathbf{v}^{(n)}(X) \right]$$

The encoding operation can be expressed as a matrix product

$$\mathbf{v}(X) = \mathbf{c} \cdot \mathbf{G} \tag{7.38}$$

Exercise 7.6
Consider the (2,1,2) convolutional code for which the generator polynomial matrix is $\mathbf{G}(X) = [1 + X^2, 1 + X + X^2]$.
If the message polynomial is $\mathbf{c}(X) = 1 + X^2 + X^3 + X^4$, *find the code word.*

Solution
The code polynomial vector is obtained as

$$
\begin{aligned}
\mathbf{v}(X) &= \mathbf{c}(X) \cdot \mathbf{G}(X) \\
&= \left(1 + X^2 + X^3 + X^4 \right) \cdot \left(1 + X^2, 1 + X + X^2 \right) \\
&= \left(1 + X^3 + X^5 + X^6, 1 + X + X^4 + X^6 \right)
\end{aligned}
$$

The encoded binary sequence corresponding to $\mathbf{v}(X)$ *is*

$$\mathbf{v} = (11, 01, 00, 10, 01, 10, 11, ...)$$

Similarly to block codes, an $(n, 1, m)$ convolutional code can be described by a *parity check polynomial matrix*, denoted by $\mathbf{H}(X)$. It is an $(n-1)$xn polynomial matrix that satisfies

$$\mathbf{v}(X) \cdot \mathbf{H}^{\mathrm{T}}(X) = \mathbf{0} \qquad (7.39)$$

for every code polynomial vector $\mathbf{v}(X)$. A systematic convolutional code can be described by the generator polynomial matrix in the form

$$\mathbf{G}(X) = [\mathbf{IP}(X)] \qquad (7.40)$$

where $\mathbf{I} = 1$ and $\mathbf{P}(X)$ is an $(n-1)$ polynomial vector. The parity check polynomial matrix for a systematic convolutional code can be obtained as

$$\mathbf{H}(X) = \left[\mathbf{P(X)}^{\mathrm{T}} \mathbf{I} \right] \qquad (7.41)$$

where $\mathbf{I}$ is an $(n-1)$x$(n-1)$ identity matrix. It is easy to verify that $\mathbf{H}(X)$ satisfies

$$\mathbf{G(X)} \cdot \mathbf{H}^{\mathrm{T}}(X) = \mathbf{0} \qquad (7.42)$$

Equation (7.42) enables the computation of $\mathbf{H}(X)$ from $\mathbf{G}(X)$ for nonsystematic codes.

Exercise 7.7
For the nonsystematic code in Exercise 7.6 the code polynomials are given by $\mathbf{v}^{(1)}(X) = \mathbf{c}(X) \cdot (1 + X^2)$ and $\mathbf{v}^{(2)}(X) = \mathbf{c}(X) (1 + X + X^2)$. Find the parity check polynomial matrix for the code.

Solution
If we multiply both sides of the first equation by $(1 + X + X^2)$ and of the second equation by $(1 + X^2)$, the right sides of the two equations are equal. Thus the left hand sides are equal as well

$$\mathbf{v}^{(1)}(X) \left(1 + X + X^2\right) = \mathbf{v}^{(2)}(X) \left(1 + X^2\right)$$

or

$$\mathbf{v}^{(1)}(X) \left(1 + X + X^2\right) + \mathbf{v}^{(2)}(X) \left(1 + X^2\right) = 0$$

The parity check polynomial matrix $\mathbf{H}(X)$ can be obtained using Equation (7.39) as

$$\mathbf{H}(X) = \left[1 + X + X^2, 1 + X^2 \right]$$

It is possible to build a systematic encoder for any convolutional code by using feedback polynomial division circuits. Figure 7.11 shows a systematic feedback encoder for the (2,1,2) code in Exercise 7.6.

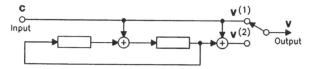

Figure 7.11: A systematic feedback encoder for the (2,1) convolutional code from Exercise 7.6.

The *syndrome polynomial* vector for an $(n, 1, m)$ convolutional code is defined by

$$\mathbf{s}(X) = \mathbf{r}(X)\mathbf{H}^{\mathrm{T}}(X) \qquad (7.43)$$

where $\mathbf{r}(X)$ is the received polynomial. The syndrome polynomial vector consists of $(n - 1)$ polynomials.

7.4.3 State Diagram

The operation of a convolutional encoder, as a finite state machine, can be described by a state diagram. Let us consider an (n, k, m) convolutional encoder. Its state is defined as the most recent message bits stored in the register at time l

$$\mathbf{S}_l = (c_{l-1}, c_{l-2}, ..., c_{l-m}) \qquad (7.44)$$

Whith a new information symbol shifted to the register, the encoder moves to a new state

$$\mathbf{S}_{l+1} = (c_l, c_{l-1}, ..., c_{l-m+1}) \qquad (7.45)$$

The current encoder state $\mathbf{S}_{l+1}$ is a function of the input message block c_l and the previous encoder state $\mathbf{S}_l$:

$$\mathbf{S}_{l+1} = f(\mathbf{c}_l, \mathbf{S}_l) \qquad (7.46)$$

There are 2^{mk} distinct states. The output block is a function of the current input message block $\mathbf{c}_l$ and the encoder state $\mathbf{S}_l$:

$$\mathbf{v}_l = g(\mathbf{c}_l, \mathbf{S}_l) \qquad (7.47)$$

Functions (7.46) and (7.47) can be graphicaly represented by a graph called a *state diagram*. The encoder states are depicted by nodes and state transitions by branches emanating from each node. Each branch is labeled with the corresponding message/output block. Given a current encoder state, the information sequence at the input determines the path through the state diagram, which gives the output code sequence.

For example, consider the (2,1,2) code given in Exercise 7.6. The state diagram for this code is shown in Figure 7.12. The encoder has four states, (00), (01), (10) and (11).

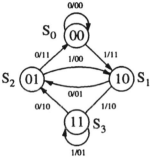

Figure 7.12: State diagram for the (2,1) convolutional code.

7.4.4 Trellis Diagram

A *trellis diagram* is derived from the state diagram by tracing all possible input/output sequences and state transitions. As an example, consider the trellis diagram for the (2,1,2) code specified by the state diagram shown in Figure 7.12. It is obtained by expanding in time every state transition and every path starting from the all zero state (00). The resulting trellis is illustrated in Figure 7.13.

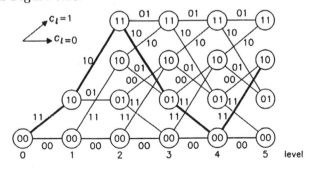

Figure 7.13: Trellis diagram for the (2,1) convolutional code.

When the first message is shifted into the register, the encoder can move either to state (10) if the input symbol is 1 or to state (00) if the input symbol is 0. When the next symbol enters the register the encoder can be in one of four possible states (00), (01), (10) and (11). The number of states is doubled upon every register shift.

The encoder reaches the maximum possible number of states, 2^{mk}, after m time units, where m is the encoder memory order. For $m > 2$, the structure of the trellis becomes repetitive. There are two branches emanating from each state corresponding to two different message symbols. There are also two branches merging into each state. Transitions caused by an input symbol 0 are indicated by lower branches while transitions caused by an input symbol 1 are indicated by upper branches. Each branch is labeled by the output block.

The output code sequence can be obtained by tracing a path in the trellis specified by the input information sequence. For example, if the input information sequence is $c = (11001)$, the output code sequence corresponds to the path indicated by the thick line in the trellis in Figure 7.13. That is $v = (1110101111)$. In general, a convolutional code is represented by a trellis diagram with 2^k branches leaving each state and 2^k branches entering each state. There are 2^{mk} possible states. The trellis diagram is used for decoding of convolutional codes.

Distance Properties of Convolutional Codes

The error probability performance of convolutional codes is determined by their distance properties. As with block codes, we consider two types of distances depending on the decoding algorithm. For hard decision decoding, the decoder operates with binary symbols and the code performance is measured by Hamming distance. A soft decision decoder receives quantized or analog signals from the demodulator and both the decoding operation and performance are based on Euclidean distance.

The *minimum free* distance, denoted d_{free}, of a convolutional code is defined as the minimum Hamming distance between any two code sequences in the code. Since convolutional codes are linear, the Hamming of distance between two code sequences is equal to the weight of their modulo-2 sum, which is another code sequence. Therefore, the minimum free distance is the minimum weight of all non-zero code sequences. In other words, the all-zero code sequence is used as the reference sequence in determining the minimum-free distance.

For example, for the code (2,1,2) represented by the trellis in Figure 7.13, it is clear that the path $v = (110111)$ is at the minimum Hamming distance from the all-zero path 0. Thus, the minimum free distance is $d_{free} = 5$.

The minimum free Euclidean distance, denoted by d_{Efree}, is defined as the minimum Euclidean distance between any two code sequences. The minimum Euclidean distance depends both on the convolutional code trellis structure and modulation type. For example, if the antipodal modulation is used to map each binary symbol into a signal from the $-1, 1$ modulation signal set, the minimum free Euclidean distance is $d_{Efree} = 2\sqrt{5}$.

Weight Distribution of Convolutional Codes

The weight distribution of a convolutional code is important for computing its error performance. It can be determined by using the transfer function approach [7]. We define A_i as the number of code sequences of weight i in the trellis that diverge from the all-zero path at a node and then remerge for the first time at a later node. Then the set

$$\left\{ A_{d_{free}}, A_{d_{free}+1}, ..., A_i, ... \right\} \qquad (7.48)$$

is called the *weight distribution* of a convolutional code. Methods of finding the weight distribution are described in more detailed coding texts.

7.4.5 Viterbi Decoding Algorithm with Soft Decisions

Consider transmission of convolutionally encoded data over an AWGN channel. The demodulator in this model makes soft decisions and presents this soft-decision sequence at the input of the decoder. The maximum likelihood rule for soft decisions over an AWGN channel consists in finding a code sequence with the minimum squared Euclidean distance from the received sequence. Note that hard decisions can also be used in a Viterbi decoder (with the distance metric being Hamming distance, just as in block codes). However, Viterbi decoders almost universally use soft-decisions due to its ease of incorporation into the algorithm, resulting in approximately 2dB more coding gain.

For a convolutional code, each code sequence **v** is a path in the trellis diagram of the code. The search for the received code sequence can, therefore, be achieved by comparing the received binary sequence with all possible paths in the trellis.

A brute force application of this rule would result in a prohibitively computationally complex algorithm. The Viterbi algorithm is a practical maximum likelihood decoding method with reduced computational complexity.

It is based on eliminating less likely paths at each node where paths merge and keeping only one path with the greatest likelihood. Hence, comparison need not be done between entire sequences but can be performed continously, level by level, along paths in the trellis, thereby reducing the number of computations and memory requirements.

Another important concept is that the comparison can be performed along relatively short sequences, approximately five times the code constraint length, without affecting the decoding performance. Suppose a message sequence c consists of message blocks of k bits each,

$$c = (c_0, c_1, ..., c_l, ...) \tag{7.49}$$

where the l-th message block is

$$c_l = \left(c_l^{(1)}, c_l^{(2)}, ..., c_l^{(k)} \right) \tag{7.50}$$

The code sequence v, at the output of an (n, k, m) convolutional encoder consists of code blocks of n digits each

$$v = (v_0, v_1, ..., v_l, ...) \tag{7.51}$$

where the l-th or code block is

$$v_l = \left(v_l^{(1)}, v_l^{(2)}, ..., v_l^{(n)} \right) \tag{7.52}$$

Each code block v_l is represented by a branch and each code sequence v by a path in the code trellis. Suppose code sequence v is transmitted. Let

$$r = (r_0, r_1, ..., r_l, ...) \tag{7.53}$$

be the received sequence where the l-th received block is

$$r_l = \left(r_l^{(1)}, r_l^{(2)}, ..., r_l^{(n)} \right) \tag{7.54}$$

Branch and Path Metrics

The *branch metric*, denoted by $\mu_l(r_l, v_l)$, is defined for hard decision decoding, as the Hamming distance between the received block r_l and a code block v_l in the trellis, written

$$\mu_l(r_l, v_l) = d(r_l, v_l) \tag{7.55}$$

For soft-decision decoding, the squared euclidean distance is used. The *path metric*, denoted $M_i^j(\mathbf{r}, \mathbf{v})$ is defined as the Hamming (or squared Euclidean) distance between the received sequence $\mathbf{r}$ and a path $\mathbf{v}$ in the trellis. Since the distance of a path is the sum of the branch distances, the path metric is computed as the sum of branch metrics

$$M_l^{(i)}(\mathbf{r}, \mathbf{v}) = \sum_{j=1}^{l} \mu_j(\mathbf{r}_j, \mathbf{v}_j) \quad i = 1, 2, ..., 2^l \qquad (7.56)$$

where l is the length of the received sequence and 2^l is the total number of paths of length l.

Viterbi Algorithm

The Viterbi decoding procedure consists of the following operations.

1. Generate the trellis for the code.

2. Assume that the optimum signal sequences from the infinite past to all trellis states at time l are known; their path metrics are denoted by $M_l^{(i)}$, $i = 1, 2^m$.

   Note that the number of optimum sequences is equal to the number of states in the trellis 2^m where m is the memory order of the code.

3. Increase time l by 1. At time $(l + 1)$, the decoder computes the path metrics $M_{l+1}^{(i)}$ for all the paths entering $(l+1)$st nodes by adding branch metrics to the path metrics of connecting survivors $M_l^{(i)}$.

4. Compare the path metrics of all paths entering each node and choose the path with the minimum path metric. These paths are called *survivors*. Store 2^{mk} survivors and discard all other paths.

5. Repeat the procedure iteratively.

6. Looking backwards in time survivors tend to merge into the same "history path" at some time $l - D$. Select the symbol from the common history path at time $(l - D)$ as the decoded output, where D is called the *decision depth* (approximately five times the code constraint length).

If the transmitted sequence is finite, m zero symbols are appended in the end to clear the encoder shift register.

Exercise 7.8

Consider the (2,1,2) convolutional code given in Exercise 7.6 whose trellis diagram is shown in Figure 7.4.4. Suppose the the all-zero sequence is transmitted over a BSC and that the received sequence is $\mathbf{r} = (01\ 00\ 00\ 10\ 00\ 00\ 00\ 00\ 00\ 00)$. *Describe the Viterbi decoding process.*

Solution

The decoding process starts at level l=2 by computing the path metrics of the four paths leading to the four nodes in the trellis. The received sequence at level 2 is

$$\mathbf{r} = (0100)$$

The four paths in the trellis represent the following code sequences

$$\mathbf{v}_1 = (0000)$$
$$\mathbf{v}_2 = (0011)$$
$$\mathbf{v}_3 = (1101)$$
$$\mathbf{v}_4 = (1110)$$

The path metrics for the four paths are

$$M_2^{(1)} = d(\mathbf{r}, \mathbf{v}_1) = 1$$
$$M_2^{(2)} = d(\mathbf{r}, \mathbf{v}_2) = 3$$
$$M_2^{(3)} = d(\mathbf{r}, \mathbf{v}_3) = 2$$
$$M_2^{(4)} = d(\mathbf{r}, \mathbf{v}_4) = 2$$

The path metrics are shown above each node at level 2 in Figure 7.4.5 (a). At level 3 the received sequence is

$$\mathbf{r} = (010000)$$

There are two paths entering each node at this level. The path metrics at this level are obtained by adding the branch metrics to the path metrics of the connecting paths $M_2^{(i)}$, *i=1,2,3,4. For each node only one survivor with the minimum path metric is retained. The survivors and their path metrics are shown in Figure 7.14 (b). The same procedure is repeated for levels 4 and 5. At level 5 there are four survivors with path metrics 3, 3, 3 and 2. The numbers above nodes indicate path metrics. At level 7 all the survivors merge into a single path over the first 3 branches. That is, all surviving paths lead to the decoded block at a level 7 levels back to the current one.*

The depth at which survivor merging occurs varies with the number and positions of errors in the received sequence. Let us assume that the decision depth is fixed to 10. After processing each new branch the decoder moves back 10 branches and decodes a message block on the path with the smallest metric. At level 10, in Figure 7.14, (j), the path with minimum metric is the all-zero path with metric 2 and the decoded message is the message block 0 at branch 1. In the next decoding step, at level 11, a new decoded message block released from the survivor is 0.

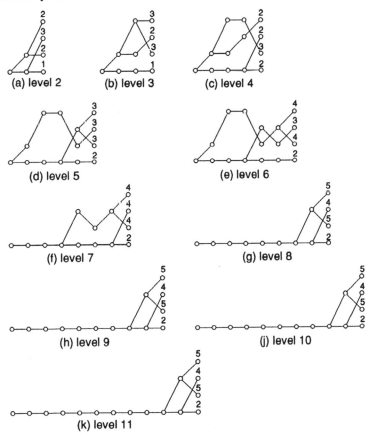

Figure 7.14: Viterbi decoding process for the (2,1) code in Exercise 7.8.

For a continous sequence the procedure repeats by processing a new branch and releasing a decoded message block. If the received sequence has a finite length, it is decoded by appending 2 zero's to the end of the message sequence to clear the register.

Note that Viterbi decoders are usually implemented in soft decision form since this provides approximately 2 dB improvement in performance over hard decision decoders.

As mentioned above, the soft decision decoder uses Euclidean distances as branch and path metrics instead of Hamming distances used by a hard decision decoder. This change does not require a substantial increase in decoder complexity.

7.4.6   Error Performance

Since convolutional codes are linear, their performance does not depend on the transmitted sequence. Hence, we assume that the all-zero code sequence is transmitted. This sequence is represented by the all-zero path in the trellis. In the process of decoding, we say that a *first-event error* is committed if the *correct* path (the all-zero path) is eliminated for the first time at a node in the trellis.

A measure of the error performance of a convolutional code is the *first-event error probability*, denoted $P(E)$. Another common measure of error performance is the decoded bit error rate (BER), denoted P_b. These two error probabilities can be approximated by union bounds based on the weight distribution of the code. We will consider only soft decision decoding.

Soft Decision Decoding:

Let us consider an (n,k) convolutional code followed by binary phase shift keying (BPSK) modulation over an AWGN channel with no output quantization. The first error probability P(E) is upperbounded by

$$P(E) \leq \sum_{d_{free}}^{\infty} A_d Q\left(d\sqrt{\frac{E_b}{N_0}\frac{R}{2}}\right) \tag{7.57}$$

where R is the code rate. The bit error probability is upperbounded as follows

$$P_b \leq \sum_{d_{free}}^{\infty} \frac{1}{k} B_d Q\left(d\sqrt{\frac{E_b}{E_0}\frac{R}{2}}\right) \tag{7.58}$$

At large E_b/N_0 the bit error probability can be approximated by

$$P_b \approx \frac{1}{k} B_{d_{free}} e^{-(Rd_{free})\cdot E_b/N_0} \tag{7.59}$$

Coding Gain

The asymptotic coding gain of a coded system with soft-decision Viterbi decoding over an uncoded BPSK is

$$G = 10\log_{10}(Rd_{free}) \quad \text{(dB)} \tag{7.60}$$

Asymptotically, a soft decision decoder requires 3dB less power than a hard decision decoder to achieve the same error probability. Over the entire range of E_b/N_0 ratios, the gain of soft-decision decoding over hard-decision decoding ranges between 2 and 3 dB.

Exercise 7.9

The most widely used convolutional code is the (2,1,6) Odenwalder code generated by the following generator sequences,

$$\mathbf{g}^{(1)} = (1101101)$$
$$\mathbf{g}^{(2)} = (1001111)$$

This code has $d_{free} = 10$. Find the coding gain.

Solution

With soft-decision decoding, the coding gain is

$$
\begin{aligned}
G &= 10\log_{10}(Rd_{free}) \\
&= 10\log_{10}(10/2) = 6.98\text{dB}
\end{aligned}
$$

7.4.7 Implementation Considerations for a Viterbi Decoder

Convolutional codes with Viterbi decoders have been used in various satellite communication systems. A number of VLSI (Very Large Scale Integration) circuits can provide implementation of high speed decoders up to 600 Mbps [8], [9] on a single chip. Low speed DSP decoders up to 64kbps [10] can be implemented.

The complexity of a Viterbi decoder mostly depends on the number of states in the trellis and the decision depth. A functional block diagram for the Viterbi decoder is shown in Figure 7.15. The input to the decoder is a continous binary or quantized analog sequence of demodulated signals.

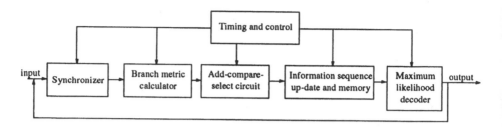

Figure 7.15: Block diagram of a Viterbi decoder.

The Viterbi decoder requires that in-phase and quadrature symbols for soft decisions or branch digits for hard decisions be delivered from the demodulator in the correct order.

The *synchronizer* is a block that determines the beginning of a branch in the received signal sequence. This operation can be performed without transmitting additional information since convolutional codes provide means for monitoring synchronization and channel condition.

There are two common methods for branch synchronization. One approach is based on monitoring bit error rate by reencoding the output decoded sequence and comparing it in an exclusive OR gate with the input sequence. The block diagram of an reencoding algorithm for bit error rate monitoring is shown in Figure 7.16.

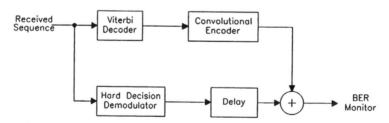

Figure 7.16: Block diagram of an re-encoding algorithm for synchronization in a Viterbi decoder.

If the bit error rate is high, a decision is made to skip or delay a symbol to establish synchronization. This function is also useful in monitoring modem carrier and timing recovery.

Another method relies on monitoring path metrics, which tend to increase rapidly when the decoderis out of synchronization. If the decoder is used with M-PSK or M-QAM modulation then as discussed in Chapter 4, there is an ambiguity problem in the signal phase.

It can be resolved by using a system of differential encoder/decoder and rotationally invariant codes. A convolutional code is rotationally invariant if the generator polynomials have odd weights. More details are given in reference [11].

The *branch metric calculator* in a Viterbi decoder computes either the Hamming distance, for hard decision decoders, or the Euclidean distance, for soft decision decoders. These are the distances between a received block and a code block corresponding to a branch in the code trellis. The complexity of branch metric calculations depends on the number of quantization levels and the type of metric assignment.

A common simplification for VLSI decoders is to quantize the input symbol value to 3 bits and use a look-up table to determine the branch metrics [11]. This reduces the size of the arithmetic logic units but results in a 0.25 dB loss relative to unquantized signals. In DSP based decoders with floating point arithmetic units, full precision can be maintained.

The *Add-Compare-Select (ACS)* block performs path metric updating and storage. This block is most computationally intensive part of the Viterbi decoder. Its functions are as follows.

1. Compute the path metric for each path leading to a node by adding the branch metric to the previous node survivor metric.

2. Compare path metrics of all paths leading to a node.

3. Determine the smallest path metric and store it in a path metric register.

If the code trellis has N_s states the ACS circuit can be either duplicated N_s times or be time shared for low speed applications.

The *path metric normalization* is performed periodically to make sure that path metric does not exceed the maximum metric range. On real channels the operation of adding branch metrics to path metrics causes the path metrics to increase. Growth can be avoided by periodically subtracting the shortest path metric from all others [11].

This is a significant problem with VLSI decoders because of the large computation effort required to monitor and prune this growth. It has been solved for fixed point DSP implementations by using 2's complement numbers to represent path metrics [12].

For 32 bit floating point accumulators the problem is further simplified. The high precision of path metric representation allows very infrequent pruning. For example, in transmission at 9600 bps over a poor channel, the period of pruning is greater than 30 minutes [10].

The *information sequence memory and update* block stores in the path memory a vector indicating the source state of the smallest path metric determined by the ACS block and performs its update for each received symbol. This operation can be handled by either *register exchange* or by *trace back* techniques. In the first method, N_s survivors are stored in buffer registers. These registers are interconnected according to the code trellis to allow that the information sequences on survivor paths be decoded at one end of each of the registers as a new symbol is shifted into each of them. In the *trace-back* technique the decoder does not memorize the actual survivors but only the trellis connections after each comparison. At the decision depth level the decoder moves back and traces the path by recalling trellis connections.

The *maximum likelihood decision* decoder determines the best path among N_s survivors. The optimum algorithm is to choose the path with the minimum path metric. This procedure requires N_s - 1 comparisons. The computational complexity can be reduced at slight performance sacrifice by selecting any path, since all survivors usually merge after sufficently high decoding delay. A minimum decision depth of approximately five times the code constraint length ensures the decoder achieves almost the theoretical coding gain.

7.5   TRELLIS CODES

Block and convolutional codes achieve coding gain by transmission of binary redundant symbols. The encoding operation results in a reduced data rate on channels with limited bandwidth and fixed symbol rate, or increased channelbandwidth if the data rate is maintained constant. This restricts the usefulness of block and convolutional codes to power limited systems and deep space communications where bandwidth is not a critical resource.

The major advantage of trellis codes is that coding gain can be obtained without bandwidth expansion or data rate reduction, which makes them attractive for bandwidth limited systems. Satellite communications have traditionally been power limited but bandwidth is becoming increasingly important.

Let us consider data transmission at the data rate r_b in a communication system with binary block or convolutional encoding. The data rate at the encoder output is $r_c = r_b/R$ where R is the code rate. The modulator maps b binary symbols at the output of the encoder into one of $M = 2^b$ possible transmit signals. Some typical modulation formats, called signal sets are M-PSK, M-AM, M-QAM and M-AMPM are shown in Figure 7.17.

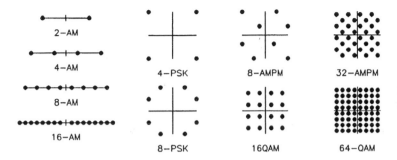

Figure 7.17: Typical modulation signal sets.

The symbol rate at the output of the modulator is

$$r_s = r_c/b = r_b/Rb$$

The required system bandwidth is r_s Hz to transmit signals at a modulator symbol rate of r_s symbols/sec (baud). The demodulator recovers the b bits by making an independent M-ary nearest neighbour decision on each signal received.

Spectral efficiency is defined as the ratio of the data rate over the system bandwidth

$$\eta = r_b/r_s = Rb$$

The spectral efficiency of the coded system is reduced by the code rate r relative to the spectral efficiency of an uncoded system. In bandwidth limited systems the spectral efficiency can be increased by using high level modulation schemes with large b. However, the performance of multilevel modulation schemes with binary and convolutional coding is poor if hard-decision decoding is used and codes are constructed using the Hamming distance construction criterion. An improvement can be achieved by soft-decision decoding technique based on Euclidean distance.

Further gain can be obtained by code design aimed at achieving large Euclidean distance. In a coded modulation system, message sequences are coded into signal sequences over a certain modulation signal set. The set of all possible signal sequences is called a modulation code. Modulation codes are generally decoded by soft decision decoding based on the code trellis and Euclidean distance metrics. Modulation codes with trellis structure are called trellis modulation codes. The Ungerboeck trellis codes [13] achieve coding gains of up to 6dB relative to uncoded systems with the same spectral efficiency. The principle of the trellis code design will be illustrated by an example.

Let us assume that it is required to improve the error performance of an uncoded QPSK system. A trellis code is constructed by using a rate-1/2 convolutional code with a 4-state trellis diagram and the 8-PSK modulation signal set. The trellis encoder block diagram is shown in Figure 7.18.

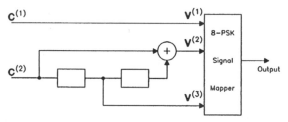

Figure 7.18:   Encoder for 4-state 8-PSK trellis code.

It consists of a binary encoder and an 8-PSK signal mapper. The input to the (2,1,2) binary convolutional encoder is a binary message sequence $\mathbf{c}^{(1)}$, while the output consists of two binary code sequences $\mathbf{v}^{(1)}$ and $\mathbf{v}^{(2)}$.

The message bit $c_l^{(1)}$ at the input to the encoder at time l is encoded into two bits, $v_l^{(1)}$ and $v_l^{(2)}$, based on the rate-1/2 convolutional code of memory order $m=2$. The encoder output digits are given by linear equations

$$v_l^{(1)} = c_{l-1}^{(1)}$$

$$v_l^{(2)} = c_l^{(1)} + c_{l-2}^{(1)}$$

The binary code is represented by a trellis diagram shown in Figure 7.19. By inspecting the trellis we can see that the free Hamming distance of the code is $d_{free} = 3$. A branch coming out from an l-th order node is labeled by an output pair $(v_l^{(1)}, v_l^{(2)})$ corresponding to an input $c_l^{(1)}$. At time l, the input to the 8-PSK mapper consists of the binary encoder outputs $v_l^{(1)}$ and $v_l^{(2)}$ and uncoded information bit $c_l^{(2)}$

$$v_l^{(3)} = c_l^{(2)}$$

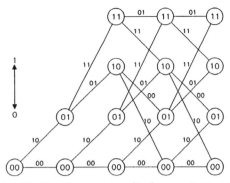

Figure 7.19:   Trellis diagram for (2,1) convolutional code.

Now we need to include the uncoded information bit $c_l^{(2)}$ in the binary code trellis diagram to make the trellis for the modulation code of rate 2/3. The binary code trellis diagram is shown in Figure 7.20. The uncoded input bit $c_l^{(2)}$ produces two parallel branches, one corresponding to $c_l^{(2)} = 0$ and the other corresponding to $c_l^{(2)} = 1$.

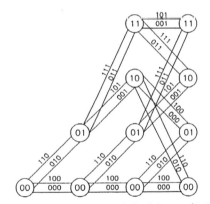

Figure 7.20:   Trellis diagram of the binary (3,2) code.

Each branch in the trellis is labeled with a triplet, $(v_l^{(3)}, v_l^{(2)}, v_l^{(1)})$. Two parallel branches are labeled with

$$(0, v_l^{(2)}, v_l^{(1)}) \quad \text{and} \quad (1, v_l^{(2)}, v_l^{(1)})$$

Note that parallel branches imply that single signal-error events can occur. This limits achievable free Euclidean distance to the minimum distance of the subsets of signals assigned to parallel transitions.

The overall trellis has the following properties

1. There are 4 states in the trellis.

2. There are two parallel branches connecting two adjacent nodes.

3. There are two groups of two parallel branches diverging from each node.

4. There are two groups of two parallel branches merging into a node.

The 8-PSK symbols can be represented as integers from the signal set $B = \{0, 1, 2, 3, 4, 5, 6, 7\}$. The signal set is shown in Figure 7.21.

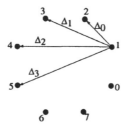

Figure 7.21: The 8-PSK signal set.

The squared Euclidean distance between two signal points in B is

$$\mathrm{d}^2(1,0) = \mathrm{d}^2(2,1) = \Delta_0^2 = 0.586$$

$$\mathrm{d}^2(2,0) = \mathrm{d}^2(5,3) = \Delta_1^2 = 2$$

$$\mathrm{d}^2(3,0) = \mathrm{d}^2(4,1) = \Delta_2^2 = 3.414$$

$$\mathrm{d}^2(4,0) = \mathrm{d}^2(5,1) = \Delta_3^2 = 4$$

Let B_i be a subset of B. Define $d^2[B_i] = min\{d^2(x, x') : x, x' \in X\}$.

Then $d^2[B_i]$ is the *minimum squared Euclidean* distance of B_i. It is also called the squared *intra-set distance*. Clearly, $d^2[B] = 0.586$. Let B_i and B_j be two subsets of B. The minimum squared Euclidean distancebetween B_i and B_j is defined as

$$\mathrm{d}^2[\mathrm{B_i}, \mathrm{B_j}] \equiv min\{d^2(b_i, b_j) :\ b_i \in B_i \ \text{and}\ b_j \in B_j\}$$

This distance is called the *separation* of two sets.

Let us form a chain of partitions of the signal set B with increasing number of subsets and maximizing the intra-set distances within subsets and separation between subsets. First we partition B into two subsets $B_0 = \{0, 2, 4, 6\}$ and $B_1 = \{1, 3, 5, 7\}$, with the intra-set distance $d^2[B_0] = d^2[B_1] = 2$ and the separation $d^2[B_0, B_1] = 0.586$. At the second step at partitioning, we obtain four subsets

$$
\begin{aligned}
B_{00} &= \{0, 4\} \\
B_{10} &= \{2, 6\} \\
B_{01} &= \{1, 5\} \\
B_{11} &= \{3, 7\}
\end{aligned}
$$

The intra-set distance of each subset is $d^2[B_{ji}] = 4$, $i=0,1$, $j=0,1$. The separations between subsets in this second partition are

$$
\begin{aligned}
d^2(B_{00}, B_{01}) &= d^2(B_{00}, B_{11}) = d^2(B_{01}, B_{10}) \\
&= 2
\end{aligned}
$$

At the last step of partitioning, we obtain single points

$$B_{000} = \{0\}, \ B_{100} = \{4\}, \ B_{010} = \{2\}, \ B_{110} = \{6\}$$

$$B_{001} = \{1\}, \ B_{101} = \{5\}, \ B_{011} = \{3\}, \ B_{111} = \{7\}$$

The set separations of this partitions are $d^2(B_{ij0}, \ B_{ij1}) = 0.586$, $i = 0, 1, j = 0, 1$ The partition sequence is shown in Figure 7.22.

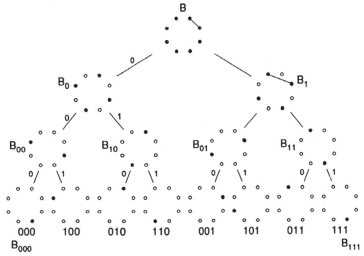

Figure 7.22: The partition sequence for 8-PSK trellis codes.

This set partitioning is used to map each triplet $(v_l^{(2)}, v_l^{(1)}, v_l^{(0)})$ into 8-PSK signal points. The mapping should be done in such a way that the resultant signal sequencesin the trellis have the largest possible free Euclidean distance.

Since single errors can occur due to parallel branches, the two triplets on two parallel branches should be mapped into two signal points in a set B_{ij}, $i=0,1$, $j=0,1$ with largest intra-set distance. For this set $d^2[B_{ij}] = 4$.

For example, two symbols on parallel branches are mapped into signal points from the subset $B_{00} = \{0, 4\}$. The two groups of two parallel branches diverging from or merging into a node should be assigned to signal points from two subsets, B_{0j} and B_{1j} with largest set separation, $d^2[B_{0j}, B_{1j}]$. In this case $d^2[B_{0j}, B_{1j}]=2$. For example, $B_{00} = \{0, 4\}$ and $B_{10} = \{2, 6\}$. Using the above mapping rules we obtain a 4-state trellis with signal symbols from the 8-PSK signal set as seen in Figure 7.23.

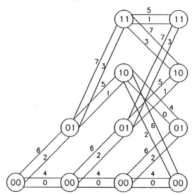

Figure 7.23:   Trellis diagram for the 4-state 8-PSK trellis code from Exercise 7.10.

The *squared free Euclidean distance* of the code is 4 and corresponds to two paths diverging and remerging through two parallel branches. The *spectral efficiency* is

$$\eta = \frac{r_b}{r} = 2\text{bit/s/Hz}$$

The *asymptotic coding gain* is measured relative to the reference uncoded QPSK system with the same spectral efficiency of 2bit/s/Hz and the minimum squared Euclidean distance of $d_u = 2$ and is given by

$$G = 10log_{10}\frac{d_{free}^2}{d_u^2} = 3\text{dB}$$

A general set partitioning procedure for design of trellis coded modulation is presented in [13].

The trellis codes that were presented by Ungerboeck [13] used two dimensional signal sets, e.g., 8PSK and 16QAM. This limits η to an integer number. In some situations, it may be desirable to have a non-integer η. For example, a specified data rate through a bandlimited channel cannot be achieved with η limited to integer values. The bandwidth efficiency may be too small (resulting in excess signal bandwidth, intersymbol interference, and adjacent channel interference) or too large (resulting in poor performance through reduced coding gain).

This limitation can be overcome through the use of multidimensional signal sets [1,2]. These coding schemes send k information bits into L 2-D signal sets, forming a $2L$ dimensional signal set. This gives a bandwidth efficiency of $\eta = k/L$. The signal sets are constructed in a similar fashion to trellis codes with 2-D signal sets. That is, the $2L - D$ signal set is partitioned so as to increase the distance between the multidimensional points in the subsets. Using multi-D signal sets also has other advantages, such as rotational invariance and reduced decoding complexity.

Note that to obtain rotational invariance for trellis codes with 2-D signal sets, it is necessary to use non-linear codes [17].

7.6 PROBLEMS

1. Design a feedback shift register encoder for an (8, 5) cyclic code with a generator polynomial $g(x) = 1 + X + X^2 + X^3$. Use the encoder to find the code word for the message 1 0 1 0 1 in systematic form.

2. A (15,7) cyclic code has a generator polynomial

$$g(X) = 1 + X^4 + X^6 + X^7 + X^8$$

   (a) Find the code word in systematic form for the message

$$\mathbf{c} = (1000001)$$

   (b) Draw block diagrams of an encoder and error detection circuit for this code.

   (c) Is $\mathbf{v}(X) = X^4 + X^6 + X^{10}$ a code polynomial?

   (d) For a channel bit error rate of $p = 10^{-2}$ find the word and bit error probabilities after decoding.

3. Consider a (24, 12) linear block code capable of double-error corrections. Assume that a coherently detected BPSK modulation format is used and that the received $E_b/N_0 = 14$ dB. Compute the coding gain.

4. Consider an (3,1,2) convolutional code generated by the generator sequences

$$g^1 = (1,1,0); g^2 = 1,0,1); g^3 = (1,1,1)$$

The code is used over a binary symmetric channel (BSC). Assume that the initial encoder state is the 00 state. The sequence $r = (1100000101101010...0)$ is received.

(a) Find the maximum likelihood path through the trellis diagram, and determine the first 5 decoded information bits. If a tie occurs between any two merged paths, choose the upper branch entering the particular state.

(b) Identify any channel bits in r that were inverted by the channel during transmission.

5. Consider the convolutional code (2,1,1) generated by the following two generator sequences:

$$g^{(1)} = (11)$$

$$g^{(2)} = (10)$$

(a) Find the free Hamming distance of the code.

(b) Give the expression for the bit error probability on a BSC channel, for high E_b/N_0 at the output of a hard decision Viterbi decoder. Assume that the channel transition probability is p.

(c) Assume a binary phase shift keying modulator with the symbols (-1,1) following this convolutional encoder. Find the free Euclidean distance of the code.

(d) Give the expression for the bit error probability on a Gaussian channel at high E_b/N_0 at the output of a soft decision Viterbi decoder. Assume that the transmitted energy per noise power spectral density is E_b/N_0.

(e) Estimate the coding gain obtained in d).

6. Design an interleaver for communication system operating over a bursty noise channel at a transmission rate 14,400 coded symbols/s. A noise burst typically lasts for 2 ms. The system is encoded by a (64,48) BCH code with $d_{min} = 5$ while the modulation format is BPSK. The end-to-end delay is not to exceed 100 ms.

7. Consider the 4-state code whose trellis diagram is shown in Figure 7.23. Suppose that the all zero binary message sequence is transmitted. Every three digits at the output of the encoder are mapped into an 8-PSK symbol. The transmitted sequence consists of signals from the two-dimensional 8-PSK signal set. The received sequence is

$$\mathbf{r} = (1, -0.4, 1.1, -0.3, 1, 0.1, 0.8, -0.4, 1.2, -0.2)$$

Compute the decoded sequence.

REFERENCES

1. Berlekamp, E.R., *Algebraic Coding Theory*, McGraw-Hill, 1968.

2. Sugiyama, Y., Kasahara, M., Hirasawa, S. and Namekawa, T., "A Method for Solving Key Equation for Decoding Goppa Codes", *Inf. Control*, 27, pp. 87-99, January 1975.

3. Lin, S. and Costello, D.J. Jr., *Error Control Coding: Fundamentals and Applications*, Prentice-Hall, 1983.

4. Blahut, R., *Theory and Practice of Error Control Codes*, Adison-Wesley Publishing Company, 1983.

5. Imai, H., *Essentials of Error-Control Coding Techniques*, Academic Press, 1990.

6. Chase, D., "A Class of Algorithms for Decoding Block Codes with Channel Measurement Information", *IEEE Trans. Inf. Theory*, IT-18, 170-182, January 1972.

7. Viterbi, A.J. and Omura, J.K., *Principles of Digital Communication and Coding* , McGraw-Hill, New York, 1979.

8. Ishitani, T. et al, "A Scarce-State-Transition VLSI Viterbi Decoder for Bit Error Correction", *IEEE Journak for Solid State Circuits*, Vol. SC-22, No. 4, pp. 575-582, Aug. 1987.

9. Fettweis, G., Dawid, H. and Meyer, H., "Minimized Method Viterbi Decoding: 600 Mb/s Per Chip", *Conf. Rec., Global Commun.*, pp. 1712-1716, 1990.

10. McDonald, P. and Vucetic, B., "A Viterbi Codec for the AT&T DSP32C Signal Processor ",*Proc. IREECON'91*, Sydney, p. 464-467, Sept. 1991.

11. Clark, G.C. and Cain, J.B., *Error-Correction Coding for Digital Communication* , Plenum Press, 1982.

12. Hekstra, A.P., "An Alternative to Metric Rescaling in Viterbi Decoders", *IEEE Trans. Commun.* Vol. 37, No. 11, pp. 1220-1222, Nov. 1989.

13. Ungerboeck, G., "Channel Coding with Multilevel/Phase signals", *IEEE Trans. Inform. Theory*, Vol. IT-28, pp. 55-67, Jan. 1982.

14. Biglieri, E., Divsalar, D., McLane, P. and Simon, M., *Introduction to Trellis-Coded Modulation*, Macmilan Publishing Company, 1991.

15. Wei, L.F., "Trellis-coded modulation with multi-dimensional constellations," *IEEE Trans. Inform. Theory*, Vol. IT-33, pp. 483-501, July 1987.

16. Pietrobon, S.S., Deng, R.H., Lafanechére, A., Ungerboeck, G. and Costello, D.J. Jr., "Trellis-coded multidimensional phase modulation," *IEEE Trans. Inform. Theory*, Vol. 36, pp. 63-89, Jan. 1990.

17. Wei, L.F., "Rotationally invariant convolutional channel coding with expanded signal space - Part II: Nonlinear codes," *IEEE J. Select. Areas Commun.*, Vol. SAC-2, pp. 672-686, Sep. 1984.

Chapter 8

SIGNALLING SYSTEMS

by Jean-Luc Thibault
Bond University, Australia

A typical mobile satellite communications system can be characterized by four major elements which interact with each other

- A Network Management Station (NMS) controls, monitors and allocates the overall communication resources (bandwidth and power) of the system;

- The Base Stations (BSs) interface with the space segment and with the Public Switched Telephone Network (PSTN) for communications with the Mobile Terminals;

- The Mobile Terminals (MTs) interface with the space segment for communications with the Network Management Station and the Base Stations;

- A space segment transponds all upward channels to downward channels with or without any switching or processing.

The functions of these four system elements are combined to form two major subsystems: a *communications subsystem* which provides the demand-assigned satellite communications links between MTs and BSs, and a *signalling subsystem* which provides the signalling links between MTs, BSs and the NMS. This Chapter analyses the characteristics and the performance of the signalling subsystem.

8.1   INTRODUCTION

A mobile satellite system may support different types of service (circuit-switched, packet-switched, mobile radio) and each of them relies on the signalling subsystem for their realisation. The performance required from the signalling subsystem varies from one service to another. Some services are time critical in their realisation (for example, the channel establishment of a voice call). Others are less sensitive to delay (for example the transfer of messages for the packet-switched service).

The most important service of a mobile satellite system, from the traffic as well as revenue point of view, is the circuit-switched service (voice and data). Therefore, the signalling subsystem, on which the channel establishment procedure relies, must be carefully designed to guarantee a quality of service which will meet the users' expectation of this service.

A typical design specification of a signalling subsystem for the circuit-switched service could be stated as follows

- less than 1% of all channel establishment requests are lost;

- the 99 percentile of the channel establishment delay is less than 5 seconds.

To assess the performance of the signalling subsystem, a detailed description of the channel establishment procedure must be established with a modelling of the system components involved in the procedure. In addition, some external factors affecting this channel establishment procedure need to be considered. Then their impact on the channel establishment delay can be estimated from mathematical models developed in queueing theory.

Two factors are analysed in this Chapter. The first one deals with fading on the radio links which affect the reliability of transmission channels. A transmission failure on a signalling channel significantly disturbs the channel establishment procedure due to the long propagation delay on the satellite link ($\simeq 0.5$ s for a round trip). The second factor analysed in this Chapter deals with the traffic generated by the MTs. A large number of MTs operating simultaneously in the system may lead to a severe deterioration of the signalling system performance.

Two cases are studied: one where the access to the signalling channel is random (which leads to collision of channel requests), the other where the access to the channel is controlled (which leads to possible queueing of channel requests).

8.2 SIGNALLING SUBSYSTEM CONFIGURATION

A typical configuration of a mobile satellite signalling subsystem is illustrated in Figure 8.1. Note that the satellite is not shown specifically in this figure because it does not perform any call processing function. It acts only as a reflector in the sky transponding all upward channels to downward channels.

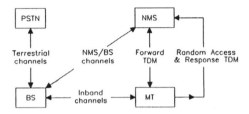

Figure 8.1: Typical configuration of a satellite signalling system.

The configuration of a signaling subsystem depends on how the call processing functions are distributed between the system elements. In a centralized system (for example, Mobilesat), all the functions allocating the system resources (bandwidth and power) are located at the NMS. However, in a more distributed system (for example, Inmarsat-B), both the NMS and the BSs can allocate these communications resources.

The configuration studied in this chapter assumes that the system resource allocation functions are centralized at the NMS and uses the Mobilesat system specifications to illustrate the procedures. The reader should bear in mind that slight variations in the signalling subsystem configuration can arise from one specific system to another.

8.2.1 Signalling Channels

Each signalling channel shown in Figure 8.1 performs specific functions in relation to the call establishment procedure for the circuit-switched service. These functions are now described and analysed.

Forward TDM Channel

The Forward TDM channel, also called the outbound TDM channel, is a time-division multiplex channel transmitted at Ku-band by the NMS and monitored in the L- band by all active MTs.

The functions of the Forward TDM channel are to:

- Provide general system information to the MTs such as the number of Forward TDM and Random Access channels in the system respectively, and also, their frequencies, the maximum number of attempts for a channel request and the number of repeats per attempt;

- Acknowledge correct reception of all information sent by MTs;

- Assign circuit-switched channels to MTs;

- Provide the frequency reference for all MTs;

- Transport messaging traffic generated by the packet-switched service.

The Forward TDM channel usually has a highly structured format with fixed length frame. A frame consists of a synchronization word followed by a certain number of fixed length time slots which are used to carry Signalling Units (SUs). An SU is a short word of fixed length encapsulating the information to be sent. Only one SU is sent per time slot but it is assumed that the majority of transactions in the system can be completed with a single SU. A queueing model of the Forward TDM channel is depicted in Figure 8.2.

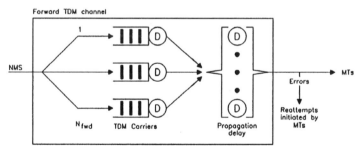

Figure 8.2:   Model of the Forward TDM channel.

In the Mobilesat system, as illustrated in Figure 3.4 , the frame of the Forward TDM channel has a length of 110 ms and contains 8 time slots of 13.33 ms. Its transmission capacity is therefore of 72.7 SUs per second. Additional carriers can be allocated to this channel when the traffic generated by the MTs becomes too excessive. MTs are then divided into different groups (based on their serial number) and each group is assigned a specific carrier. Another important characteristic of the Forward TDM channel is that its access is controlled by the NMS.

When the channel is busy, the NMS puts all new incoming SUs in a queue according to a pre-established queueing policy. The waiting time that an SU has to spend in the queue adds to the delay in the channel establishment procedure and its contribution will be estimated in Section 8.3.

Response TDM Channel

The Response TDM channel is used by the MTs to carry acknowledgements/responses for messages received from the NMS. For each SU transmitted on the Forward TDM channel, a slot is reserved on the Response TDM channel to return a response or acknowledgement.

In the Mobilesat design, this channel is slotted at 110 *ms* and eight Response TDM channels are allocated for each Forward TDM channel. This implementation provides the appropriate capacity to acknowledge every SU transmitted on the Forward TDM channel (8 SUs per frame of 110 *ms*). SUs have a dedicated access to the Response TDM channel and are not affected by collision or queueing delays.

Random Access Channel

The random access channel is a slotted Aloha channel used in the return direction (MTs to NMS) to carry signalling information, status information and user messages. A model of the random access channel is depicted in Figure 8.3.

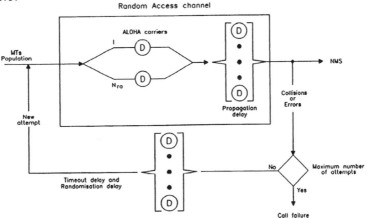

Figure 8.3: Model of the random access channel.

Depending on the number of MTs in the system, up to N_{ra} carriers can be assigned to this channel. When an MT initiates a call, it selects randomly one of the N_{ra} carriers to transmit its SU. When two MTs choose the same time slot and the same carrier to transmit their SUs, a collision occurs resulting in a loss of information for both MTs. A collision resolution algorithm is then required to resolve this deadlock situation. The random access channel is slotted in time and in the Mobilesat system, the slot has a duration of 110 *ms*.

BS/NMS Signalling Channels
The NMS to BS signalling channel performs the following functions:

- Pass on information on circuit-switched channels requested by MTs;

- Provide system updates and command information;

- Acknowledge correct information received from the BSs;

- Provide frequency offset information of the satellite transponders;

- Poll BSs for status information.

The BS to NMS signalling channel functions are to

- Initiate channel requests orignated in the PSTN and destined to a MT;

- Acknowledge correct information received from the NMS;

- Provide circuit-switched call termination information to the NMS;

- Report circuit-switched call status to the NMS.

The physical connection between the NMS and the BSs could be via a radio link or via a line-based data network (like X.25 or Ethernet). The terrestrial solution eliminates the long propagation delay of a satellite link and improves the performance of the signalling system.

BS/MT Signalling Channels
The BS to MT signalling channel can be defined as an inband signalling channel. BSs and MTs exchange information through the communication channel allocated by the NMS. This signalling channel is required to control the set up and release of the communication channel at the start and the end of a call.

8.2.2 Channel Establishment Procedure

The channel establishment procedure defines the entire set of interactive signals necessary to establish, maintain and release a connection. Three different channel establishment procedures can be identified:

- MTs to PSTN (outgoing calls);

- PSTN to MTs (incoming calls);

- MTs to MTs;

The majority of calls in a mobile communications system are usually expected to be outgoing calls (60% to 80 %), while incoming calls count only for 20% to 30%. MTs to MTs calls are technically possible but are not frequent and will not be considered in this Section.

MT Originated Call

When a mobile subscriber initiates a call, the MT sends a *channel request SU* to the NMS on the random access channel. This SU contains the call priority, the type of service and the resources (power, bandwidth) required, the MT identification and the called destination address.

On receipt of the request, the NMS checks the authorisation of the MT and determines if the network has the necessary resources to provide the requested service. If all the requirements are fulfilled, the NMS sends a *call announcement SU* to the BS and the MT. The SU contains the frequencies of the communication channels, their operating power level and bandwidth, and the MT identifier.

On receipt of the channel assignment, the BS starts to send a *channel assignment confirmation poll SU* with its identifier and the calling MT identifier. In the meantime, the MT receives a response to its request from the NMS and tunes to the allocated carrier. It then synchronizes to the BS transmitted signal after which it starts to transmit the called address.

Upon reception of the address, the BS informs the NMS that the circuit has been established. The BS starts to transmit the *call in progress* tone to the MT and dials the called party number. When the called party answers, the BS signals to the NMS that the call has been established.

At the end of the call, a party going on hook causes a *call clear SU* to be sent. Once this has happened, both parties remove their carriers and the BS signals to the NMS that the call has been cleared down. The NMS acknowledges the *call idle SU* and deallocates the channel frequencies. The MT on clearing the call retunes to the Forward TDM channel. The sequence of events of this channel establishment procedure is illustrated in Figure 8.4.

The sequence of events described above characterizes a typical channel establishment procedure. In some cases however, some unwanted events modify this sequence and further actions are required to complete the channel establishment procedure. Some mechanisms must be implemented to resolve deadlock situations which arise when SUs are corrupted. for example, an MT is locked into a waiting state indefinitely if its *channel request SU* isn't received correctly by the NMS.

To avoid such situations, timers are used in the channel establishment procedure. If an MT doesn't receive a response after a prefixed timeout period, it will initialize a new attempt. These attempts are repeated until the channel establishment procedure is successful or until a maximum number of attempts N_a is reached, in which case a call failure indication is sent to the calling party. On each new attempt, a randomization delay is added to reduce the probability of collision with other MTs involved in the same channel establishment procedure. The maximum number of attempts N_a is a parameter set dynamically by the NMS.

PSTN Originated Call

A call initiated in the PSTN is first routed to the BS which acts as the gateway between the PSTN and the mobile satellite system. The BS translates the dialled telephone number to the MT address and sends an *access request SU* to the NMS. The NMS looks up its database to verify that the MT is in attended mode, not busy and authorised for the requested communication service. The NMS also verifies that the resources required (power, bandwidth) are available before polling the MT with a *call announcement SU*.

On receipt of the *call announcement SU* from the NMS, the MT sends a *response to call announcement SU* on the Response TDM channel. Upon reception of this SU, the NMS sends a *channel assignment SU* to the MT and the BS. The BS on receipt of this SU transmits a *channel assignment confirmation poll SU* on the carrier allocated to the MT. The MT tunes to the allocated carrier and synchronises to the BS transmitted signal. It checks its validity and then acknowledges the BS signal. The BS then sends a *ring command SU* to the MT and informs the NMS that the circuit has been established. When the called party answers the call, the MT sends an *off-hook SU* to the BS.

The remainder of the call establishment procedure and clear down procedure is identical to that described for the mobile originated call procedure. Figure 8.4 illustrates the sequence of events for this channel establishment procedure.

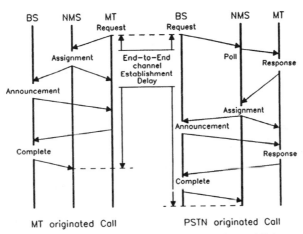

Figure 8.4: Sequence of events of a typical channel establishment procedure.

8.3 EFFECT OF FADING

The mobile satellite system operates under a harsh environment and as discussed in Chapter 2, transmission channels suffer from severe fading. This reduces the capability to transmit information reliably. To increase the probability of a successful transmission, each SU is duplicated N_r times before being sent.

As described in Chapter 3, the repeats are then transmitted at regular time intervals reducing therefore the probability of failure of a transmission. This time diversity strategy results in a reduction of the channel throughput. Its implementation must address two fundamental questions: How many repeats are required and at what time intervals should they be sent? These questions can be answered by analysing the fading process on the satellite link.

8.3.1 Markov Fading Model

The fade duration and depth on a radio link depends on the environment surrounding the MTs (forests, cities, plains , and so on.) and the power level of the signal transmitted. As discussed in Chapter 2, the fading characteristic of a channel can be modelled by observing the evolution in time of the power level of the signal. The time axis is first divided into a contiguous set of equal length intervals.

The length of an interval is chosen according to the slot duration on the signalling channel. A channel state is then associated with each time slot according to the following rule: if the power level of the signal stays above a preset threshold during the entire time slot, the channel is said to be in a *Good* state, otherwise it is in a *Bad* state.

With this simple model, the error process on the signalling channel can be described by a two-state Markov chain built at the time slot level. When the Markov chain is in a *Good* state, the SU is transmitted correctly and when it is in the *Bad* state, the SU is corrupted. After each time slot, the Markov chain makes a state transition. The channel state changes from a *Good* state to a *Bad* state with probability p and goes from a *Bad* state to a *Good* state with probability q. This is illustrated in Figure 8.5.

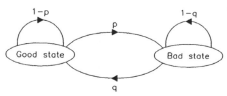

Figure 8.5: Markov chain description of errors on a link.

The state transition probabilities can be expressed in a matrix form as follows

$$\Psi = \left(\begin{array}{cc} \psi_{G,G} & \psi_{G,B} \\ \psi_{B,G} & \psi_{B,B} \end{array} \right) = \left(\begin{array}{cc} 1-p & p \\ q & 1-q \end{array} \right) \tag{8.1}$$

The parameters p and q can be derived from experimental observations by estimating the mean time spent by the channel in a *Good* and a *Bad* state.

The mean times and the transition parameters p and q of the Markov model are related by the following geometric distribution

$$\begin{aligned} E[T_{Good}] &= 0p + 1(1-p)p + 2(1-p)^2 p + 3(1-p)^3 p + \ldots \\ &= p \sum_{i=1}^{\infty} i(1-p)^i = \frac{1-p}{p} \end{aligned} \tag{8.2}$$

$$E[T_{Bad}] = 0q + 1(1-q)q + 2(1-q)^2 q + 3(1-q)^3 q + \dots$$

$$= q \sum_{i=1}^{\infty} i(1-q)^i = \frac{1-q}{q} \qquad (8.3)$$

where $E[T_{Good}]$ and $E[T_{Bad}]$ are expressed in terms of the number of time slots. The probability of being in a *Good* state (π_G) or in a *Bad* state (π_B) can be calculated assuming that the observed process is under stationary conditions. This assumption leads to the set of following equations:

$$p\pi_G = q\pi_B \qquad \text{and} \qquad \pi_G + \pi_B = 1 \qquad (8.4)$$

which when resolved leads to the following state probabilities:

$$\pi_G = \frac{q}{p+q} \qquad \text{and} \qquad \pi_B = \frac{p}{p+q} \qquad (8.5)$$

Exercise 8.1
A statistical analysis of a signal received on a heavily faded satellite link shows that the average duration of a fade is 19.2 ms and they occur at an average rate of 5.7 fades per second.
With this information, develop a Markov model describing the error process on the radio link for a signalling-channel slotted at 15 ms. Calculate the probability that the channel is in a Good state.

Solution
The average duration of a fade can be translated into a number of slots from which the parameter q can be deduced. $E[T_{Bad}] = 1.28$ time slots which leads to a parameter q of 0.44. The average time spent in a Good state can be deduced from the rate at which fades occur. This leads to an average time spent in a Good state of 156.2 ms. Using the same approach as earlier, one can find that the value of p is 0.088 and the probability of finding the channel in a Good state is therefore 0.833.

Some experimental values of the parameters p and q of the Markov model are shown in Table 8.1. These experimental values were obtained for different fading environments (light, medium, heavy), different signal levels expressed in E_b/N_0 values (8 dB and 12 dB) and different slot durations (110 ms and 13.75 ms). Table 8.1 shows that the value of the transition parameter p varies with the fading environment whilst the value of the transition parameter q is rather insensitive to it.

The fading environment of an MT depends on its specific geographical position and different MTs see different error processes on the same global channel. For a specific MT, a unique fading environment can be assumed for all its signalling channels (random access, Forward/Response TDM and inband channels).

	Time Slot Duration							
	110 ms				13.75 ms			
	Signal Level				Signal Level			
	8 dB		12 dB		8 dB		12 dB	
Fading	p	q	p	q	p	q	p	q
Light	0.009	0.24	0.002	0.40	0.005	0.18	0.003	0.27
Medium	0.022	0.28	0.007	0.38	0.010	0.22	0.005	0.28
Heavy	0.052	0.22	0.030	0.37	0.031	0.17	0.019	0.27

Table 8.1: Experimental values of the Markov model parameters.

8.3.2   Delay Distribution for a Successful Transmission

With the Markov model, the delay distribution for a successful transmission can be calculated given the number of repeats N_r of an SU and the delay in transmission between repeats expressed in terms of number of slots N_s. In this analysis, we are interested in the first SU received successfully at the NMS since this first SU determines the delay of the channel establishment procedure. The outcomes of a transmission can then be seen as the distribution of first success which can occur at the initial SU sent, at the first repeat, at the second repeat and so on.

The probability that the first success occurs with the initial SU is the probability that the channel is in a *Good* state (π_G). The probability that the success occurs with the first repeat is the probability that the channel is in a *Bad* state for the first SU (π_B) and that the channel makes a transition from a *Bad* state to a *Good* state after a delay of N_s time slots. This probability denoted as $\psi_{B,G}^{(N_s)}$ can be calculated from the n-step transition probabilities given by:

$$
\Psi^n = \begin{pmatrix} \psi_{G,G}^{(n)} & \psi_{G,B}^{(n)} \\ \psi_{B,G}^{(n)} & \psi_{B,B}^{(n)} \end{pmatrix}
$$

$$
= \frac{1}{p+q} \begin{pmatrix} q + p(1-p-q)^n & p - p(1-p-q)^n \\ q - q(1-p-q)^n & p + q(1-p-q)^n \end{pmatrix} \qquad (8.6)
$$

where $\psi_{i,j}^{(n)}$ is the probability of moving from a state i to a state j after n transitions.

The outcomes of a transmission can be expressed as $(s_0, s_1, ... s_{N_r}, \ell)$ where s_i is the probability that the first successful transmission of an SU occurs at the ith repeat and ℓ is the probability that the transmission fails (all the $(N_r + 1)$ SUs are corrupted). The probability of failure of a transmission can therefore be expressed as

$$\ell = 1 - \sum_{i=0}^{N_r} s_i \tag{8.7}$$

The outcomes of a transmission can be calculated by defining a Bernoulli random variable X_i which describes the channel state for the ith repeat of an SU. The possible outcomes for X_i are

$$X_i(\omega) = \begin{cases} 1 & \text{if } \omega = \textit{Good} \text{ state} \\ 0 & \text{if } \omega = \textit{Bad} \text{ state} \end{cases} \tag{8.8}$$

The state probabilities of the channel for the first SU sent is given by:

$$P\{X_0\} = \begin{cases} \pi_G & X_0 = 1 \\ \pi_B & X_0 = 0 \end{cases} \tag{8.9}$$

and the probability that the first successful SU occurs at the ith repeat is given by:

$$
\begin{aligned}
s_i &= \begin{cases} P\{X_0 = 1\} & i = 0 \\ P\{X_i = 1, X_{i-1} = 0, X_{i-2} = 0, ..., X_0 = 0\} & i > 0 \end{cases} \\
&= \begin{cases} P\{X_0 = 1\} & i = 0 \\ P\{X_i = 1 | X_{i-1} = 0\} P\{X_{i-1} = 0, ..., X_0 = 0\} & i > 0 \end{cases} \\
&= \begin{cases} \pi_G & i = 0 \\ \psi_{B,G}^{(N_s)} \left(\psi_{B,B}^{(N_s)} \right)^{i-1} \pi_B & i > 0 \end{cases}
\end{aligned}
\tag{8.10}
$$

The delay distribution of a successful transmission can be calculated by normalizing the s_i's with $1 - \ell$.

Exercise 8.2

A strategy of 2 repeats delayed by 20 time slots is to be implemented on a signalling channel slotted at 13.75 ms. Calculate the delay distribution of a successful transmission and the probability of failure of a transmission if the MT operates in a heavy fading environment with a signal level of 12 dB.

Solution

From Table 8.1, p equals 0.019 and q equals 0.27. The probability that the channel is in a Good state is $\pi_G = 93.43\%$ and in a Bad state $\pi_B = 6.57\%$. The 20-step transition probability from a Bad to Good transition is equal to $\psi_{B,G}^{(20)} = 0.933$ and from a Good to Bad state $\psi_{B,B}^{(20)} = 0.067$. From which $s_0 = 93.43\%$, $s_1 = 6.13\%$ and $s_2 = 0.41\%$ and $\ell = 0.03\%$.

Note that the redundancy strategy adds 6.54 % to the probability of success of a transmission. The delay distribution of a successful transmission is as follows: 93.46% of successes occur with the initial SU (no delay), 6.13% of successes occur after a delay of 275 ms and 0.41% of successes occur after a delay of 550 ms.

The Markovian model can also be used to determine the value of the parameters N_r and N_s to guarantee a certain level of reliability of the transmission channels. The probability of failure (ℓ) of a 8 dB power level signal on a medium faded channel slotted at 13.75 ms is illustrated in Figure 8.6 for different repeat strategies.

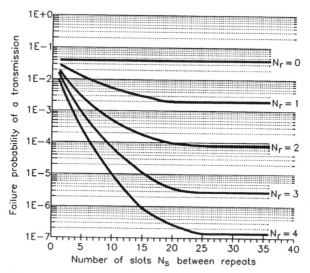

Figure 8.6 Failure probability on a medium faded channel with different repeat scenarios.

The Figure shows that there is not much to gain by delaying a repeat by more than 25 slots ($\simeq$ 350 ms) since the probability of failure doesn't decrease significantly after such delay.

It is interesting to note that the actual Mobilesat system Specification is based on a strategy of 2 repeats $N_r = 2$ and a delay of 24 slots $N_s = 24 (\simeq 330\ ms)$. With such a strategy, the probability of failure of a transmission on the Forward TDM channel in a medium faded environment would be in the order of 0.009%.

8.4 EFFECT OF TRAFFIC LOAD

Traffic in the system is another important factor influencing the call establishment delay . Its influence depends on the type of access that SUs have to the transmission channel.

Two cases need to be considered. In the first, SUs have a direct access to the transmission channel and many SUs coming from different MTs can be sent simultaneously on the same channel. This situation may result in collisions between SUs so a collision resolution algorithm is required to avoid deadlocks. In the second case, the access to the transmission channel is regulated by a controller. An SU is granted access to the transmission channel only if no other SU is being, or waiting to be transmitted.

If the transmission channel is busy, the SU is put in a queue according to a pre-established queueing policy. The Forward TDM channel is an example of a controlled access channel in which SUs are queued and served on a First Come First Serve (FCFS) policy.

8.4.1 Delay on a Random Access Channel

Different multiaccess algorithms can be implemented on the random access channel. The one discussed in this section is called *slotted Aloha* . The basic idea of the algorithm is that each MT simply transmits its SU in the first slot after its arrival, thus risking occasional collisions but achieving very small delay if collisions are rare.

The probability of collisions can be estimated if the offered traffic to the channel is known. The offered traffic depends on many factors such as the rate of SUs generated per MT, the number of MTs in the system, the number of carriers allocated to the random access channel. To estimate its value, let the normalized offered traffic per carrier be defined:

$$\rho_N = \frac{\lambda_{ra} d_{ra}}{N_c} \qquad \text{with} \quad \lambda_{ra} = (1 + N_r)\lambda_{mt} N_{mt} \qquad (8.11)$$

where λ_{mt} is the arrival rate of SUs per MT (including timeout reattempts), N_r the number of repeats of an SU, N_{mt} the number of active MTs, d_{ra} the duration of a time slot on the random access channel and N_c the number of carriers.

Figure 8.7 shows values of the normalized offered traffic for the following set of parameters: $N_r = 1$ repeat, $\lambda_{mt} = 0.50$ SUs/hour and $d_{ra} = 110$ ms. The traffic load on the random access channel can be set at any desirable level by choosing the appropriate number of carriers. The probability of collision on the random access channel can be calculated if the distribution of interarrival times between SUs is known.

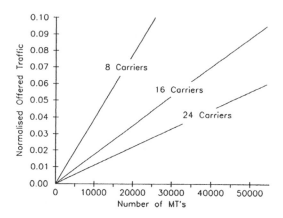

Figure 8.7: Normalized offered traffic on the random access channel.

When the population of traffic sources is very large and the traffic generated by each source is small, the arrival process can be described by a Poisson distribution. However, the implementation of redundancy and reattempt mechanisms distorts the arrival distribution by introducing correlation between arrivals. More complex models are then necessary to describe the arrival process. These models are beyond the scope of this Chapter and won't be considered here.

Assuming a Poisson process of intensity λ_{ra} on the random access channel, the number N of SUs arriving within an interval d_{ra} is given by:

$$P(N = k) = \frac{e^{-\lambda_{ra}d_{ra}}(\lambda_{ra}d_{ra})^k}{k!} \qquad (8.12)$$

The probability to have no collision on the random access channel is the probability of having no SU arrival during a time slot period d_{ra}.

The probability of a collision p_c is given by:

$$p_c = P(N = 0) = 1 - e^{-\lambda_{ra} d_{ra}} \qquad (8.13)$$

Given the probability of collision on the random access channel, the delay distribution of a successful transmission of an SU can be calculated. Let Y_i be the random variable describing the possible outcomes for the ith repeat on the random access channels, as follows

$$Y_i(\omega) = \begin{cases} 1 & ; \omega = \text{Good channel and no collision} \\ 2 & ; \omega = \text{Bad channel} \\ 3 & ; \omega = \text{Good channel and collision} \end{cases} \qquad (8.14)$$

The probability that the first success occurs at the ith repeat is given by:

$$s_i = \begin{cases} P\{Y_0 = 1\} & i = 0 \\ P\{Y_i = 1, Y_{i-1} \neq 1, ..., Y_0 \neq 1\} & i > 0 \end{cases} \qquad (8.15)$$

with $P\{Y_i = 1, Y_{i-1} \neq 1, ..., Y_0 \neq 1\}$ equals to

$$\begin{aligned} = \quad & P\{Y_i = 1 | Y_{i-1} = 2\} P\{Y_{i-1} = 2, Y_{i-2} \neq 1, ..., Y_0 \neq 1\} \\ & + P\{Y_i = 1 | Y_{i-1} = 3\} P\{Y_{i-1} = 3, Y_{i-2} \neq 1, ..., Y_0 \neq 1\} \end{aligned} \qquad (8.16)$$

where

$$P\{Y_i = 1 | Y_{i-1} = 2\} = (1 - p_c) \psi_{B,G}^{(N_s)}$$
$$P\{Y_i = 1 | Y_{i-1} = 3\} = (1 - p_c) \psi_{G,G}^{(N_s)}$$
$$P\{Y_{i-1} = 2, ...\} = \psi_{B,B}^{(N_s)} P\{Y_{i-2} = 2, ...\} + \psi_{G,B}^{(N_s)} P\{Y_{i-2} = 3, ...\}$$
$$P\{Y_{i-1} = 3, ...\} = p_c \psi_{B,G}^{(N_s)} P\{Y_{i-1} = 2, ...\} + p_c \psi_{G,G}^{(N_s)} P\{Y_{i-1} = 3, ...\}$$

These relations applied to a strategy of one repeat give the following expressions:

$$s_i = \begin{cases} (1 - p_c)\pi_G & i = 0 \\ (1 - p_c)[\psi_{B,G}^{(N_s)}\pi_B + p_c\pi_G\psi_{G,G}^{(N_s)}] & i = 1 \end{cases} \qquad (8.17)$$

Figure 8.8 shows the probability of failure of this strategy for different fading environments and normalized offered traffic. The value of N_s is set at 5 and the parameter q is assumed to be constant and equal to 0.250.

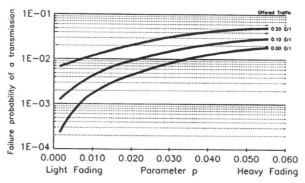

Figure 8.8: Transmission failure probability on a random access channel.

This figure shows that the traffic offered to a random access channel can reduce significantly the probability of success of a transmission. Therefore, the traffic on this channel must be kept as low as possible in order to insure proper channel establishment procedures.

8.4.2 Delay on a Controlled Access Channel

On a controlled access channel, SUs are queued when transmission contention occurs. The queueing delay (or waiting time) depends on various factors such as the traffic in the system, the transmission capacity of the channel and the queueing policies implemented.

This Section does not attempt intend to study all possible queueing disciplines. It focuses on a particular example, namely the Forward TDM channel of the Mobilesat system to illustrate how analytical models can be used to calculate the average queueing time of an SU.

The frame structure of the Forward TDM channel consists of a leading synchronization word followed by 8 TDM slots. Nine consecutive TDM frames form a TDM Super Frame. This is illustrated in Figure 8.9. Out of the 72 slots of a Super Frame, 6 are reserved by the NMS to transmit general information to the MTs.

This information (called Bulletin Board information) is updated periodically and transmitted in the first 2 slots of Frame 1 (slots 1.1 and 1.2), Frame 4 (slots 4.1 and 4.2) and Frame 7 (slots 7.1 and 7.2). A strategy of 2 repeats delayed by 24 time slots is used to combat fading on the radio link.

The transmission channel can be modelled as a server with a service time equal to the duration of a time slot. Neglecting the synchronization word, the service time is approximately equal to 13.75 *ms*. The reservation of slots for the Bulletin Board and the repeats interrupts the availability of the server and therefore increases the qucueing delay of an SU.

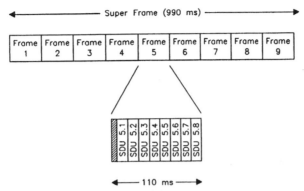

Figure 8.9: TDM frame structure.

An accurate analysis of such a system can be obtained by using a cyclic Markov chain model but such analysis goes beyond the scope of this study. In our analysis, it will be assumed that the server is always available and additional traffic is used to represent the slots reserved for the Bulletin Board and the repeats. The traffic generated by the Bulletin Board (λ_{BB}) is six SUs per Super Frame or 6.06 SUs per second and the traffic generated by the repeats is N_r times the traffic offered to the Forward TDM channel(λ_{fwd}).

The Forward TDM Channel can then be approximated by an M/D/1 discrete-time queueing system assuming that arrivals are described by a Poisson process. The probability α of having k SU arrivals during a time slot of length d_{fwd} is given by:

$$\alpha_k = P\{N = k\} = \frac{(\lambda_T d_{fwd})^k}{k!} e^{-\lambda_T d_{fwd}} \qquad (8.18)$$

where λ_T is the total arrival rate of SUs $(\lambda_{BB} + (1 + N_r)\lambda_{fwd})$. The queue length transitions q_{ij} observed at the beginning of each slot can then be derived from the α_k as follows

$$q_{ij} = \begin{cases} \alpha_0 + \alpha_1 & i = j = 0 \\ 0 & i > j + 1 \\ \alpha_{j-i+1} & \text{otherwise} \end{cases} \qquad (8.19)$$

The resulting queueing system with its state transition is illustrated in Figure 8.10. In this Figure, a state i represents a queue length of i SUs and a transition occurs at every beginning of a new time slot.

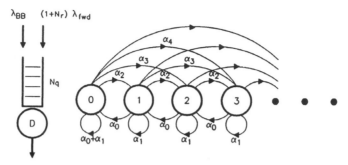

Figure 8.10: Discrete time $M/D/1$ queueing system.

The queue length probabilities π_i can be expressed as a function of the α_i as follows:

$$\pi_1 = \pi_0(1 - \alpha_0 - \alpha_1)/\alpha_0$$

$$\vdots$$

$$\pi_i = \left[\pi_{i-1} - \sum_{j=0}^{i-1} \alpha_{i-j}\pi_j\right]/\alpha_0$$

$$\vdots$$

The numerical value of the queue length probabilities can then be calculated recursively starting with the probability of having an empty queue.

The normalisation condition $\sum_{i=0}^{\infty} \pi_i = 1$ is used to calculate the final solution. From the queue length distribution at a beginning of slot, we want to obtain the mean waiting time of a random arrival SU. As the arrival process is Poisson, the number of SUs in the queue can be observed at any random points, because a Poisson arrival sees the same distribution as a random observer of the process.

The mean number of SUs in the queue N_q at a random observation point is the sum of the mean number of SUs in the queue at the beginning of the time slot and the mean number of arrivals between the beginning of the slot and the random observation point. This is expressed mathematically as follows:

$$N_q = \sum_{i=0}^{\infty} i\pi_i + \frac{\lambda_T d_{fwd}}{2} \tag{8.20}$$

The average waiting time of an SU in queue W_q can then be calculated by applying Little's theorem:

$$W_q = \frac{N_q}{\lambda_T} \tag{8.21}$$

8.5 END-TO-END ESTABLISHMENT

The end-to-end channel establishment delay is the sum of a large number of delays which are caused by various factors such as the transmission and processing time, the reliability of the transmission linksand the traffic in the system. Each factor contributing to the delay can be represented as a random variable which has a specific probability distribution.

The delay distribution of a successful channel establishment can be obtained by convolution of the probability distributions of all these variables. From this distribution, the performance of the channel establishment procedure can be assessed.

Let us define a random variable Y as the sum of two other random variables Y_1 and Y_2. The probability distribution f_Y of the random variable Y can be obtained by convolution of the probability distribution f_{Y_1} of Y_1 and f_{Y_2} of Y_2. This is usually expressed as

$$f_Y(y) = f_{Y_1}(y_1) * f_{Y_2}(y_2) \qquad (8.22)$$

The convolution operation is commutative, associative and can be applied to any number of variables n. If Y_1 and Y_2 are discrete mutually independent variables, the probability distribution of Y can then be calculated as follows:

$$f_Y(y) = \sum_{y_1} f_{Y_1}(y_1) f_{Y_2}(y - y_1) \qquad (8.23)$$

In this analysis, all random variables are assumed to be mutually independent and the end-to-end performance is evaluated for a channel establishment procedure initiated by an MT. An MT disposes of up to N_a attempts to succesfully establish a channel. An attempt is said to be successful if the transmission on the random access channel and the transmission on the Forward TDM channel are both successful.

A transmission is successful if at least one of the SUs sent on the channel is received correctly at the other end. Let us define s_a as the success probability of an attempt, N_a as the maximum number of attempts and ℓ as the probability of failure of a channel establishment procedure. Assuming that the probabilities of a successful attempt are mutually independent, the probability of failure of a channel establishment procedure can be expressed as follows:

$$\ell = (1 - s_a)^{N_a} \qquad \text{with} \quad s_a = s_{ra} s_{fwd} \qquad (8.24)$$

where s_{ra} and s_{fwd} are the probabilities of a successful transmission on the random access channel and the Forward TDM channel respectively.

These probabilities are assumed to be mutually independent. The mean number of transmissions on the random access channel (N_{ra}) for a successful channel establishment can be derived from s_a as follows:

$$N_{ra} = \sum_{i=1}^{N_a} i s_a (1 - s_a)^{i-1} \tag{8.25}$$

and the mean number of transmissions on the Forward TDM channel can be calculated as follows:

$$N_{fwd} = s_{ra} N_{ra} \tag{8.26}$$

Given the probability of failure ℓ of a channel establishment and the probability of success of an attempt s_a, the delay distribution of a successful channel establishment procedure $f(k)$ can be expressed as follows:

$$f(k) = \frac{1}{1-\ell} \sum_{i=1}^{N_a} s_a (1 - s_a)^i f_i(k) \tag{8.27}$$

where $f_i(k)$ is the delay distribution of a successful channel establishment occuring at the ith attempt. This distribution is calculated in discrete-time and the variable k refers to the number of slots on the Forward TDM channel.

The delay distribution $f_i(k)$ can be obtained as follows:

$$f_i(k) = f_{ra,i}(k) * f_{fwd}(k) * \Omega_i(k) \tag{8.28}$$

where $f_{ra,i}(k)$ is the overall delay distribution of a successful transmission on the random access channel, $f_{fwd}(k)$ the overall delay distribution of a successful transmission on the Forward TDM channel and $\Omega_i(k)$ is the delay distribution related to the deterministic components of the channel establishment procedure.

On the random access channel, a successful transmission is delayed because of fading, collisions and the waiting time imposed by the collision resolution algorithm (CRA). At the ith attempt, the CRA delay is randomly chosen over a discrete set varying from 0 to $2^{i-1} - 1$ RA time slots. One RA time slot is equal to eight Forward TDM time slots.

To illustrate the application of the convolution theorem, let us assume a strategy of one repeat delayed of one time slot. A transmission is successful with probability $s_{ra} = s_0 + s_1$.

The delay distribution of a successful transmission $f_{ra,D}$ can be found by normalising the s_i by s_{ra}:

$$a_0 = \frac{s_0}{s_{ra}} = \frac{s_0}{s_0 + s1} \tag{8.29}$$

$$a_1 = \frac{s_1}{s_{ra}} = \frac{s_1}{s_0 + s1} \tag{8.30}$$

This distribution convoluted with the delay distribution of the CRA gives the overall delay distribution of a successful transmission on the random access channel $f_{ra,i}(k)$. Figure 8.11 illustrates these distributions for the first three attempts. The delay is expressed as the number of Forward TDM time slots with one time slot being equal to 13.75 ms.

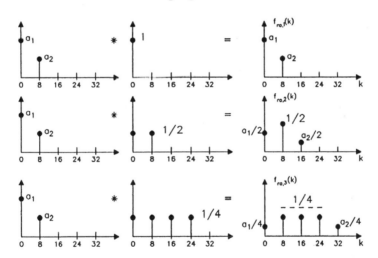

Figure 8.11: Delay distribution of a successful transmission on the random access channel.

On the Forward TDM channel, delay of a success may be caused by fading on the radio link or the waiting time in front of the channel. The overall delay distribution of a successful transmission $f_{fwd}(k)$ can be calculated as follows:

$$f_{fwd}(k) = f_{fwd,D}(k) * f_{fwd,W}(k) \tag{8.31}$$

where $f_{fwd,D}(k)$ represents the delay distribution of a successful transmission on the radio link and $f_{fwd,W}(k)$ represents the queueing time distribution of

an SU. Let us assume a strategy of 2 repeats delayed by 24 time slots on the
Forward TDM channel. The delay distribution of a successful transmission
can then be represented as follows:

$$b_0 = \frac{s_0}{s_{fwd}} = \frac{s_0}{s_0 + s_1 + s_2} \tag{8.32}$$

$$b_1 = \frac{s_1}{s_{fwd}} = \frac{s_1}{s_0 + s_1 + s_2} \tag{8.33}$$

$$b_2 = \frac{s_2}{s_{fwd}} = \frac{s_2}{s_0 + s_1 + s_2} \tag{8.34}$$

The waiting time distribution can be obtained from the analysis presented in
the previous section. Assuming the following distribution (70% no delay, 20%
1 time slot delay and 10% 2 time slots delay), the overall delay distribution
of a successful transmission on the Forward TDM channel can be calculated
and is illustrated in Figure 8.12.

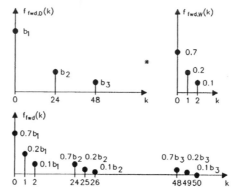

Figure 8.12:   Delay distribution of a successful transmission on the Forward
              TDM channel.

The last component $\Omega_i(k)$ groups the deterministic delay components of
the channel establishment procedure which includes the propagation delay
on the satellite links and the processing time at the NMS. Assuming a total
deterministic delay of 1100 ms (80 time slots), a distribution $PT(k)$ can be
defined:

$$PT(k) \;=\; \begin{cases} 1 & \text{if } k = 80 \\ 0 & \text{otherwise} \end{cases}$$

$$\tag{8.35}$$

In addition to this distribution, a second one is required to take into account the time out period that an MT has to wait before initiating a new attempt. Assuming a time out period of 3 seconds (218 time slots) between each attempt, the delay distribution at the ith attempt $(TO_i(k))$ can be expressed as follows:

$$TO_i(k) = \begin{cases} 1 & \text{if } k = 218(i\text{-}1) \\ 0 & \text{otherwise} \end{cases}$$

(8.36)

Note that this delay distribution is calculated from the time the call has been initiated and changes with the attempt being considered. The $\Omega_i(k)$ distribution can be obtained by convolution of $PT(k)$ and $TO_i(k)$.

From the convolution of $f_{ra,i}(k)$, $f_{fwd}(k)$ and $\Omega_i(k)$, the delay distribution of a successful attempt $f_i(k)$ is obtained. The weighted sum of these successful attempt distributions gives the delay distribution of a successful channel establishment procedure $f(k)$. From this distribution, the mean channel establishment delay $(\bar{k})$ and its 99 percentile $k_{99\%}$ can be calculated. The mean delay is given by

$$\bar{k} = \sum_{k=0}^{\infty} k f(k)$$

(8.37)

and the 99 percentile is given by the smallest value $k_{99\%}$ for which

$$\sum_{k=0}^{k_{99\%}} f(k) \geq 0.99$$

(8.38)

The accuracy of the analytical method presented in this Section depends largely on the modelling assumptions made. This Chapter has presented simple teletraffic models which can be improved with more complex mathematical analysis. The validity of the analytical models can also be verified experimentally through simulation.

8.6 PROBLEMS

1. A mobile terminal (MT) located in a heavy faded environment sends a *channel request signalling unit SU* at a power level of 8 dB. The random access channel is slotted at 110 *ms*. To combat fading, the SU

is repeated once and delayed by 2 RA slots. Upon receipt of the SU, the NMS allocates the required transmission resources and sends to the MT a *Call announcement SU* on the Forward TDM channel which is slotted at 13.75 *ms*. This SU is repeated twice and each repeat is delayed by 10 Forward TDM slots. If the MT does not receive any response from the NMS after a prefixed period of 5.5 s, it initiates a new attempt. At the ith attempt, the MT waits for a period of time randomly chosen over a discrete set of $\{0, 2^{i-1} - 1\}$ RA time slots. The deterministic delay due to propagation time on the satellite links and the processing time at the NMS is estimated at 0.8 seconds.

(a) Calculate the delay distribution of a successful transmission on the RA channel $f_{ra,D}(k)$. Plot the delay distribution. Express time in number of Forward TDM slots.

(b) Calculate the delay distribution of a successful transmission on the Forward TDM channel $f_{fwd,D}(k)$. Plot the delay distribution.

(c) Calculate the probability of success of an attempt s_a. Calculate also the mean number of transmission on the random access channel N_{ra} and on the Forward TDM Channel N_{fwd} for a successful channel establishment procedure. Calculate the probability of success of a channel establishment procedure if the maximum number of attempts is set to 3.

(d) Calculate values for the overall delay distribution of a successful transmission on the RA channel $f_{ra,i}(k)$ for each of the following cases:

   i. 1st attempt, b) 2nd attempt, c) 3rd attempt and

   ii. 4th attempt. For each case, plot the delay distribution.

(e) Assuming the following waiting distribution in queue for an SU (60 % no delay, 25% 1 slot delay, 10 % 2 slots delay and 5 % 3 slots delay), calculate the overall delay distribution of a successful transmission on the Forward TDM channel $f_{fwd}(k)$. Plot the delay distribution.

(f) Calculate the delay distribution of a successful attempt $f_i(k)$ for success at each of the following cases: a) 1st attempt, b) 2nd attempt and c) 3rd attempt. For each case, plot the delay distribution.

(g) Assuming a maximum of two attempts $N_a = 2$ for the channel establishment procedure, calculate the average delay of a successful channel establishment and the 99 percentile of the distribution.

Chapter 9

SATELLITES IN NON-GEOSTATIONARY ORBITS

by Zoran M. Markovic
OPTUS Communications, Sydney, Australia

This Chapter examines recent trends towards the use of communications satellites in low earth orbits and in other non-geostationary orbits. It is shown why the use of such orbits could offer the possibility of satellite communications from miniature hand-held earth terminals. The various classes of non-geostationary orbits are first described. If satellites in low non-geostationary orbits are to provide continuous coverage of a region, then a constellation of several satellites is required. We examine the inter-relationships between orbital parameters and ground coverage for such satellites.

In systems using satellite constellations in low orbits, a limited set of frequencies is available and the same frequencies must be reused for simultaneous transmission to different regions of the earth. To facilitate frequency reuse, it may be desirable to divide the total coverage area from a given satellite into a number of adjacent cells in a similar fashion to terrestrial cellular mobile radio systems. That is, each satellite incorporates an array of antennas, one for each cell within its coverage area. Factors affecting the choice of the number of satellites and the number of cells covered by each will be considered.

Issues involved in the design of satellites for low orbit use are then described. Link budget analysis techniqes for such systems are reviewed. Doppler shift effects can be significant for these systems since the satellites are moving with respect to the terminals. Metods of predicting and compensating for these effects are described.

Finally, some examples are given of systems involving the use of satellites in low orbits for data transmission using a store-and-forward technique, and of cellular mobile systems for speech communications.

9.1 WHY NON-GEOSTATIONARY ORBITS?

The most frequently chosen orbit for communication satellites is the *geostationary earth orbit* (GEO), a unique circular orbit in the equatorial plane at an altitude of 35,786 km. Geostationary satellites can cover large areas of the globe and allow fixed pointing ground antennas. However, they require high transmitter powers and large antenna apertures. They cannot cover high latitude regions of earth. They produce long communication delays, and they require high cost and high risk satellite launches.

Table 9.1 shows a summary of a link budget for a typical geostationary mobile satellite system such as was described in Chapter 3. Here the system is assumed to be a single-beam (non-cellular), L-band system, using non-coded QPSK for regional mobile communications.

	MobileTerminal-Satellite	Satellite-MobileTerminal
SATELLITE	receiver G/T=-2.0 dB/K	transmit EIRP=24.5 dBW
-Fade margin - Boltzmann's constant - required C/N_0 (for 6600bps voice)=		
- 5.0 dB + 228.6 dB - 47.7 dBHz = +175.9 dB		
Path-loss (L-band)	-188.7 dB	
Req. TERMINAL	transmit EIRP=14.8 dBW	receiver G/T=-11.7 dB/K

Table 9.1: A mobile satellite system link budget.

Note that in the Table, the figures in the last row specify the earth terminal requirements. They are equal but opposite in sign to the sum of the figures in the first three rows. Most GEO mobile satellite systems would have a link budget similar to the one in this Table.

In essence, the design of any satellite communications link involves consideration of the division of relative performance requirements between the satellite repeater and the ground terminals respectively. Path-loss is virtually a constant. The important design parameters are the G/T ratios and the EIRP values of the satellite and the terminals. These are determined by the antenna gains, receiver noise and transmitter powers which become the key design variables.

Each communication satellite system design represents one solution to the cost/performance trade-offs for the system taking into account the expected population of terminals. As discussed in Chapter 1, the history of satellite communications began with small, low performance satellites and a few very large earth stations. Over the years, satellites have grown in function and size. Earth terminals have decreased in size but increased in numbers. These trends are due to the fact that the savings made on a large population of terminals can offset the cost of satellite improvements.

Recently, the concept of a *Personal Communication System* (PCS) has been introduced. PCS terminals will be required to have small antennas, low output power and high equivalent system noise temperature. That is, they will be hand-held terminals with very low EIRP and G/T of up to 8dBW and -18dB/K respectively. Such a requirement presents a new challenge in system design. There are three possible solutions to the problem of utilizing miniature terminals in satellite communications.

The first solution is to improve further the performance of an L-band GEO satellite in order to compensate for small terminals. Yet, some limitations exist to this approach. The satellite receiver noise performance is limited by the noise temperature of the antenna looking at the "warm earth". The satellite transmitter output power is limited by the satellite power supply capability and its mass.

If a cellular division of satellite coverage area is employed to reuse frequencies, decreasing the cell size and increasing the satellite antenna gain (that is, its size), would improve both G/T and EIRP. However, the antenna size is restricted by the launch vehicle fairing envelope, and by the mass, complexity and cost. In summary, the increase in the GEO satellite receiver G/T and transmitter EIRP, with the associated build-up in satellite mass, may not be the most cost effective solution to PCS design.

A second solution is to use a frequency band at which even the small terminals would have high enough antenna gains. For example, at Ka-band (26-40 GHz), antenna gains of 20-25 dB can be achieved with small portable antennas. The use of such frequencies would also enable a GEO satellite cellular system to utilize a high number of cells with a satellite antenna array of moderate size and mass. Note that the path loss is somewhat increased with increasing frequency. The current availability of higher frequency bands mitigates in favor of this solution.

However, several problems present serious impediments for such an approach. These include the problem of the low power efficiency of Ka-band transmitters, the dependence of link availability on weather, the lack of technology, incompatibility with the existing terrestrial PCS standards, and especially, personal radiation safety.

A third, radically different, solution is to decrease the path loss (represented in the third row in Table 9.1) by bringing the satellite closer to the terminals. This can be achieved by placing the satellites into a lower orbit than geostationary. Besides the advantages of lower path loss and shorter communications delay, this technique brings several disadvantages.

Many such non-geostationary satellites are required in a constellation in order to provide coverage equivalent to the geostationary satellite coverage. Associated control telemetry and network management systems become complex. In addition, high Doppler frequency shifts are introduced into the satellite-terminal links. Furthermore, intersatellite cross links may be required for coverage extension. Still, such a solution has some potential for reducing the overall cost of a satellite personal communication system.

9.2   SATELLITE ORBITAL ANALYSIS

9.2.1   Satellite Orbits

The non-geostationary orbits are grouped as either *low earth orbits* (LEO), *medium earth orbits* (MEO) or *highly elliptical orbits* (HEO).

Low earth orbits are those in the range of 200-3000 km between the so-called constant atmospheric density altitude and the Van Allen radiation belts.

Medium earth orbits start above 3,000 km and extend up to GEO. Satellites in these orbits always pass through the radiation belts, and thus encounter high radiation levels.

Highly elliptical orbits approach the earth's surface within several hundred kilometers, and then reach and surpass the GEO distance.

The choice of optimum satellite orbit is usually determined by the earth coverage required. Satellite orbits may be circular or may be elliptic with different altitudes. Let the inclination of an orbit i be the angle that the orbital plane makes with the plane of the equator. All orbits can then be divided according to their inclination into two classes, namely

1. equatorial orbits $i = 0^0$, and

2. inclined orbits $i \neq 0^0$.

Inclined orbits can further be divided into: retrograde orbits ($i < 90^0$), polar orbits ($i = 90^0$) and posigrade (prograde) orbits ($i > 90^0$).

All stable orbits may be divided into four classes, as illustrated in Figure 9.1. These are as follows:

- inclined circular geo-synchronous orbits (Class i),

- inclined elliptical geo-synchronous orbits (Class ii),

- non-synchronous orbits of the above two and the equatorial type (Class iii),

- geostationary orbits (Class iv).

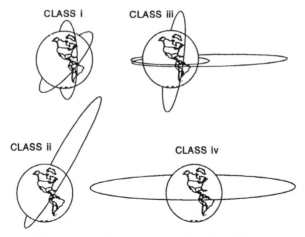

Figure 9.1:   Four classes of earth orbits.

An important term used to specify the location of an orbiting satellite is its *subsatellite point*. The subsatellite point is the point where a line from the centre of the earth to the satellite passes through the earth's surface.

In a *geo-synchronous* orbit (Classes i and ii), a satellite completes an integral number of orbits per day. Its subsatellite point tracks the same shape on the ground regularly, for example, every 24h, or every 12h, or every 8h, and so forth.

The *geostationary orbit* (Class iv), is the non-inclined circular geo-synchronous orbit that lies in the earth's equatorial plane. In this case, the subsatellite point remains in virtually the same place on the earth.

Note that not all orbits are suitable for communications. Non-synchronous orbits (Class iii) are seldom used for commercial communications, but may be preferred for defence or scientific applications. For example, nonsynchronous random LEO defence communications satellite systems have been under consideration for some time [3].

Class i and Class ii orbits are the subject of this Chapter. Class i circular LEO and MEO orbits are particularly suitable for cellular mobile communication systems since the cell sizes remain constant throughout the orbit. Constellations of geo-synchronous satellites in polar and/or inclined circular orbits can be combined to provide optimum continuous earth coverage.

All elliptical orbits have two characteristic points. They are the *perigee* (the point of closest approach to the earth), and the *apogee* (the point farthest from earth). A general problem with Class ii low-perigee non-equatorial orbits is that the earth's oblateness causes rotation of the orbiting plane. This instability would constantly move the coverage area. The larger the eccentricity, the smaller is the perigee height, the shorter is the period, and the greater is the instability of the orbit.

The only way to stabilize a low-perigee elliptical orbit is to use a particular value of inclination of $i = 63.435^0$. One such stable inclined elliptical geosynchronous orbit is, the so called, *Molniya* orbit. It has a period of one half a sidereal day (approx. 12h), a perigee height of 500 km and an apogee height of 39,800 km. The apogee of this orbit always remains at a constant geographic latitude. Another orbit from this class with lower eccentricity is called the *Tundra* orbit. Its period is one full sidereal day (86164 s). In both cases the most useful part of the orbit is its apogee. In the region of the apogee, the satellite appears to move slowly and stays close to the observer's zenith for a long period.

Table 9.2 shows the values of several important orbital and communication parameters for different values of the altitude of any circular earth orbit. Note that maximum Doppler values are given in ppm, that is, parts-per-million of the carrier frequency. Also the zero values in the last row of Table 9.2 disregard the Doppler shift from geostationary satellite drift (caused by the fact that the earth is not a perfect sphere with symmetric distribution of mass).

Orbital altitude (km)	Orbital radius (km)	Orbital period	Tangential velocity (km/s)	Maximum Doppler (ppm)	Max. Doppler $f_0 = 1.6GHz$ (kHz)	Coverage area radius (km)
200	6,556	1h28m	7.797	24.8	39.7	808
400	6,756	1h32m	7.681	23.7	38.0	1333
600	6,956	1h36m	7.569	22.7	36.4	1737
800	7,156	1h40m	7.463	21.8	34.9	2068
1,000	7,356	1h45m	7.361	20.9	33.5	2349
1,200	7,556	1h49m	7.263	20.1	32.2	2592
1,400	7,756	1h54m	7.168	19.3	30.9	2805
2,000	8,356	2h07m	6.906	17.3	27.7	3320
3,000	9,356	2h30m	6.527	14.6	23.4	3914
5,000	11,356	3h21m	5.924	11.0	17.6	4615
10,000	16,356	5h47m	4.936	6.5	10.4	5361
20,000	26,356	11h50m	3.889	3.4	5.5	5819
35,786	42,164	23h56m	3.075	0	0	6027

Table 9.2: Parameters of circular orbits (elevation look angle $\gamma_X = 10°$).

Note that the higher the altitude of the satellite, the greater will be the coverage area, the lower the relative velocity of the satellite (Doppler shift), and the longer the satellite stays above the horizon. These advantages must be balanced against the fact that the propagation conditions deteriorate with an increase in the distance between satellite and earth terminals.

9.2.2   LEO Satellite Constellations

Because of their limited field of view, LEO communications satellites for a given network have to be launched in significant numbers and organized into a constellation. There are two basic types of satellite constellations, namely *phased* and *random* constellations.

Phased constellations have satellites in constant positions relative to each other. Such a constellation necessitates good control of satellite orbit. On the other hand, random constellations avoid the need for complex orbit control systems and can therefore use simpler satellites. They can achieve an equivalent earth coverage but only if a higher number of satellites is employed. Some of the better known circular orbit satellite constellations are the *Walker, Beste* and the *Ballard* constellations [4]. The *Draim* constellations consist of satellites in moderately inclined elliptical orbits. The *Molniya* and *Tundra* HEO orbits also require 3 or 4 satellites for an uninterrupted 24h coverage.

The Inclined Walker Delta Arrays/indexinclined Walker delta arrays [4] are a special set of inclined circular orbits. Each array is designated by three numbers: $T/P/F$, where T is the total number of satellites, P the number of equispaced orbital planes and F the phasing factor for satellites in adjacent planes. The separation between satellites in adjacent planes is: $F(360°/T)$. As an example of a Walker constellation, a 7 satellite $7/7/2$ constellation of 61.81° inclination and 20,000 km altitude provides continuous dual global coverage.

Elliptic orbit constellations have two more degrees of freedom than circular orbits, namely eccentricity and argument of perigee. Thus, they are more easily adjusted to provide the desired coverage.

Circular polar constellations, have both advantages and disadvantages in communications. Their advantages are

- system synchronism

- constancy of coverage patterns

- simplified satellite management

- simplified communication parameter prediction

- easier satellite position prediction

Their main disadvantages are the relatively high number of satellites needed, and the unproductive coverage overlap that occurs in the vicinity of the earth poles. This coverage overlap results from the congestion occuring as the satellites pass over the poles. An more efficient LEO constellation would have maximum coverage overlap in the area of the maximum expected communication traffic. For a global system, that would be at approximately 45°N latitude.

Two-level and multi-level constellations are also possible. They offer an additional degree of freedom to improve the efficiency of coverage. Particularly effective could be a combination of LEO and GEO satellites.

Depending on the number of satellites visible to a terminal at any time, a non-GEO satellite system can be described as a *single coverage* system, *dual coverage* systems, and so on. A dual coverage system, for example, is one in which each user on the ground is in the view of at least two satellites.

The placement of the satellites into their required orbits represents a large portion of the total launch cost. Launch vehicles deliver payloads to LEO. If a satellite is destined for GEO, then a powerful rocket motor is used to propel it from that so-called *low transfer orbit* into its final orbit. LEO satellites are generally much smaller than GEO satellites. As a result, it is usually possible for several LEO satellites to be launched on the same launch vehicle, a smaller, less expensive launch vehicle being required than for the launch of GEO satellites.

Almost all launch vehicles used for standard GEO satellite launches may be modified for the clustered LEO satellite launches. Though such powerful launch vehicles may be used for the initial system placement, they may not be the optimum choice for satellite replenishment if required during the lifetime of the constellation.

Two strategies are possible. The system could either have in orbit spares (for example, one per each orbital plane), or have on-ground spares for rapid deployment. In the latter case a single or dual satellite launch vehicle has advantages. The changing of a satellite's orbital plane requires a significant fuel expenditure. Therefore, the most efficient way to deploy the satellites in a multi-satellite launch is on a per-orbital-plane basis. In-orbit spares could also be implemented on per plane basis.

9.3 LEO COMMUNICATION SATELLITES

9.3.1 Basic LEO System Parameters

In order to provide a basis for the further analysis, let us define some key orbital and coverage parameters. Under the limiting condition that the eccentricity $e = 0$, the orbit is a circle with the earth at its centre.

In any orbit of *semimajor axis a* (that is, half the length of the major axis of the ellipse), and *instantaneous altitude h*, the satellite linear and angular velocities are given by

$$v_s = \sqrt{\mu \left(\frac{2}{r_E + h} - \frac{1}{a} \right)} \qquad (9.1)$$

$$\omega_s = (r_E + h)v_s \qquad (9.2)$$

where r_E is the radius of the earth and $\mu = 398,600.5 \times 10^9 m^3/s^2$ is the gravitational constant. The value of earth radius used in Equations (9.1) and (9.2) may be either the polar value, $r_{Ep} = 6356.75$ km, or the equatorial value, $r_{Ee} = 6378.14$ km.

The distance between the center of the earth and the center of the orbital ellipse is equal to $a \cdot e$ where e is the eccentricity of the orbit.

Let the instantaneous flight path angle η be defined as the angle subtended at the earth's center in the orbital plane between the semimajor axis and the direction of the satellite. Then the instantaneous height of the satellite is given by

$$h = \frac{a(1 - e^2)}{1 + e \cdot \cos \eta} - r_E \qquad (9.3)$$

The satellite orbital period is

$$T_s = 2\pi \sqrt{\frac{a^3}{\mu}} \qquad (9.4)$$

Now consider a circular orbit, as illustrated in Figure 9.2. For this case, the velocities and period are given by

$$v_s = \sqrt{\frac{\mu}{r_E + h}} \quad \omega_s = \sqrt{\frac{\mu}{(r_E + h)^3}} \quad T_s = 2\pi \sqrt{\frac{(r_E + h)^3}{\mu}} \qquad (9.5)$$

The coordinates to which a directional earth terminal antenna must be pointed to communicate with a satellite are called the *look angles*. These are specified as the *azimuth look angle* α (the angle between north and the path to the satellite in the horizontal plane), and the *elevation look angle* γ (the angle above the horizon at which a terminal sees the satellite).

By examination of Figure 9.2, it can be seen that the *coverage angle* (angular separation between satellite and terminal) is given by

$$\phi = \arccos \left[\frac{r_E}{r_E + h} \cos\gamma \right] - \gamma \qquad (9.6)$$

Expressed in longitude between satellite and terminal meridians (L), sub-satellite point latitude (l_s), and terminal latitude (l_t), we have

$$\cos \phi = \cos L \cos l_s \cos l_t \sin l_s \sin l_t \qquad (9.7)$$

The elevation look angle is

$$\gamma = \arccos \frac{(r_E + h) \sin \phi}{s} \qquad (9.8)$$

The azimuth look angle is

$$\alpha = \arcsin \left[\frac{\sin L \cos l_s}{\sin \phi} \right] \qquad (9.9)$$

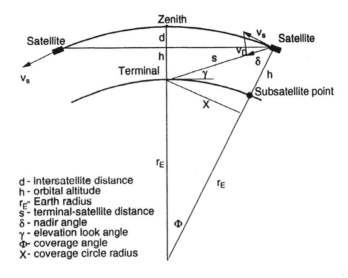

d - intersatellite distance
h - orbital altitude
r_E- Earth radius
s - terminal-satellite distance
δ - nadir angle
γ - elevation look angle
Φ- coverage angle
X- coverage circle radius

Figure 9.2: A circular LEO cross-section.

Let γ_X be some specified minimum look angle, where the index X is used to represent the edge of coverage. (Note that in Figure 9.4 X is the radius of the coverage circles). Then, the coverage angle, that is the maximum angular separation between a satellite and a terminal, is

$$\phi_X = \arccos \left[\frac{r_E}{r_E + h} \cos \gamma_X \right] - \gamma_X \qquad (9.10)$$

The minimum elevation look angle γ_X is a characteristic angle for the system. It is the angle at which the terminal at the edge of the satellite coverage sees the satellite. It is this angle that determines the availability of the communication link and must be taken into account in considering call handoffs from satellite to satellite.

From Figure 9.2, the terminal to satellite distance is

$$s = \sqrt{r_E^2 + (r_E + h)^2 - 2r_E(r_E + h) \cos \phi} \qquad (9.11)$$

In the design of a LEO communication system, a trade-off must be made between the number of satellites in the constellation, the signal propagation delay and the path loss. While the coverage area increases and the required number of satellites decreases with an increase in orbital altitude, both communication parameters (delay and loss) become worse.

For a satellite constellation utilizing circular polar orbits, the number of equispaced satellites per orbital plane and the number of equispaced planes, for the complete coverage without overlap are [2], respectively

$$N_s = \frac{\pi}{\phi - \varepsilon} \qquad N_p = \frac{\pi - 2\arccos[cos\phi/cos(\phi - \varepsilon)]}{\phi + \arccos[cos\phi/cos(\phi - \varepsilon)]} + 1 \qquad (9.12)$$

with

$$\phi = \arccos\left[\frac{r_E \cos\gamma}{r_E + h}\right] - \gamma \qquad (9.13)$$

and

$$\varepsilon \cong \frac{(-\phi + \sqrt{\phi^2 + 6\phi\tan\phi})^2}{18\tan\phi} \qquad (9.14)$$

The total number of LEO satellites in the constellation will be $N_t = N_p N_s$. Values of N_t are plotted versus orbital altitude in Figure 9.3 for three different values of elevation look angles. Note that the use of a large look angle reduces the chances for communication path blockage by trees or other obstacles. However, increasing the look angle sharply increases the number of satellites required.

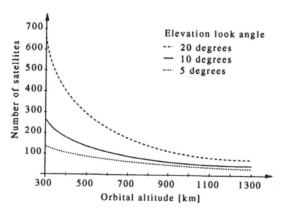

Figure 9.3: The required number of satellites for earth coverage.

As explained in Chapter 2, mobile satellite communication links are particularly sensitive to blockage by roadside trees, hills and buildings. The links require look angles as high as possible to maintain good link margins. Alternatively, they require very high link margins if it is required to maintain communications over paths with low look angles.

One deficiency of GEO satellites is the low elevation look angle at higher latitude areas such as in Scandinavia, North Asia, Alaska and North Canada. These areas require link margins of 20-30 dB to overcome blockages. Also, most of the Arctic and the Antarctic regions cannot have any satellite coverage at all from GEO satellites. Inclined elliptical and circular polar orbits, on the other hand, can maintain good coverage and high look angles for these regions.

9.3.2 LEO Cellular Systems

In a LEO satellite system, to facilitate frequency reuse the coverage area would normally be divided into cells. Since LEO satellites are moving with respect to any point on the earth, the cell sizes cannot be chosen according to the user population distribution (as would be possible in GEO satellite mobile systems).

Now consider the coverage circles and constituent cells for a cellular system as illustrated in Figure 9.4.

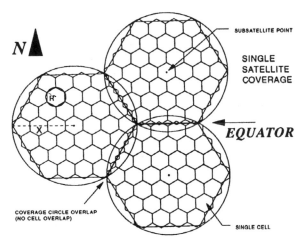

Figure 9.4: Coverage circles and hexagonal cells.

The radius of the projected coverage circle will be

$$X = r_E \sin \phi_X \qquad (9.15)$$

the number of hexagonal calls within one circulat coverage zone and the cell radius are, respectively

$$N = 1 + 3n(n+1) \qquad R \cong \frac{2\pi r_E}{N_s(4n - \sqrt{3})} \cong \frac{2X}{(2n+1)\sqrt{3}} \qquad (9.16)$$

where n is the number of rings of cells per coverage area.

Table 9.3 compares the number of cells N required within one coverage zone, for circular and rectangular aperture antennas respectively, for different values of the number n.

n	0	1	2	3	4	5	6	7
N (circ. aperture)	1	7	19	37	61	91	127	169
N (rect. aperture)	1	9	25	49	81	121	169	225

Table 9.3: The number of cells within one coverage area.

For the case where $n = 37$ cells per coverage area, Figure 9.5 shows how the cell radius R varies with satellite orbital altitude. The Figure also shows values of the anticipated satellite antenna array dimension versus altitude.

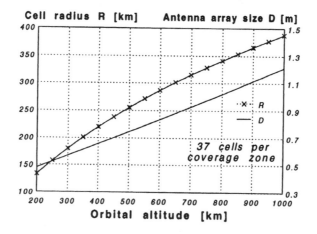

Figure 9.5: Cell and antenna size versus altitude.

In a cellular system, a *call handoff* is performed every time the terminal changes from one cell to another, or from one satellite to another. That is, the terminal connection is changed from one assigned channel frequency in a given cell to a new frequency in an adjacent cell. Each handoff involves a significant amount of control processing and signalling. In contrast to terrestrial cellular systems, a LEO satellite cellular system has very rapidly moving cells and slowly moving users. Handoff will occur relatively often (every few minutes). However, the handoff cell is always known in advance.

Each cell is covered by a separate antenna beam. The more cells there are within one coverage zone, the more efficient the frequency reuse. On the other hand, antenna design is more difficult, and the call handoff from cell to cell is more frequent. Also, the higher the orbital altitude, the fewer the number of satellites needed to achieve coverage. However, with an increase in distance between satellites, higher power and greater traffic handling capacity is required of each satellite, increasing the per satellite cost. As the satellite altitude is increased, the antenna gain must be increased to maintain the same cell size, thus increasing its complexity and mass.

Clearly, LEO satellite system parameter values must be chosen based on a complicated trade-off between many factors including the number of satellites, the number of cells per coverage zone, the cell size (antenna and transponder complexity), the number of channels per cell, the launch vehicle capability, and so on.

9.3.3 Effects of the LEO Environment

All satellites in space experience some detrimental effects, such as thermal stress, outgassing in vacuum, and so forth. Still, the environment of LEO is somewhat different from the environment of GEO. Ionizing radiation is a major hazard, which exists at all orbital altitudes. Two main sources are solar flares and cosmic rays. However, the earth's magnetic field traps most of the charged particles into the two concentric toroidal belts centered on the equator, called *the Van Allen radiation belts*. The lower one consists predominantly of high speed protons, the higher one of electrons. Radiation level in the belts varies with the time of the year, geographical latitude, and geo-magnetic and solar activity. On the average, in the plane of the equator the radiation peaks at around 2,200 and 18,500 km above the earth.

Due to the existence of the belts, all orbits below them are somewhat protected from space radiation. The satellites approaching or passing through them, though, require special protection measures such as radiation hardening of electronic components. In HEO, the charged particles of the solar wind may cause electrostatic damage of electronic components.

At very low altitudes there is an effect of atomic oxygen erosion, which is more pronounced on the "leading edge" of LEO spacecraft; coating of composites and polymeric materials is required to prevent this effect on them. Solar cells, in particular, are vulnerable to radiation and oxidization. A thicker transparent polymer or glass coating must be used to protect them against those two effects.

Various space debris, decommissioned satellites, natural meteoroids, and even spent fuel add to a moderately hostile environment of LEO. For example, because of the earth's gravitational attraction, more natural meteoroids are found at lower altitudes. Also, the debris flux is increased between 600-1100 km [14].

The received radiation dose versus shielding thickness on the three equatorial circular orbits and one elliptic orbit, over a five years mission, is shown in Figure 9.6.

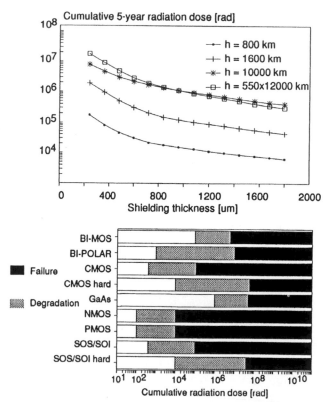

Figure 9.6: Cumulative radiation dose and its effect on semiconductors.

9.3.4 Reliability and Availability

Reliability and availability measures can be defined for a satellite communication system, for a satellite, or for a communication link.

The *system reliability* is the probability that it will remain operational over some period. It is determined by the reliability of the vital system components. To eliminate *single points of failure* and increase the system reliability vital system components are usually duplicated and arranged in redundancy configurations.

The *system availability* is the percentage of time, over some period, that the system can perform its function at the specified level of performance.

Satellite reliability is the dominant factor in the overall satellite communication system availability. A failed satellite, a subsystem, or a unit cannot be repaired. Satellite failure can either be at random, by component wearout, or by fuel exhaustion.

Communication system availability and link availability are related, but different characteristics. The availability of the system, A_s, depends on the terminal availability, A_{term}, satellite availability, A_{sat}, and ground station availability, A_{gs}, and can be expressed as

$$A_s = A_{term} \cdot A_{sat} \cdot A_{gs} \tag{9.17}$$

The most critical part of communication system availability is the satellite availability. Assuming an exponential model for probability of satellite failure, the availability of a satellite in a single orbital position and the number n of satellites required in a period T_L are given respectively by

$$A_{sat} = 1 - \frac{T_R}{pT_F} \qquad n = \frac{T_L}{T_F(1 - \exp^{-\frac{T_M}{T_F}})} \tag{9.18}$$

For a system of k in orbit spares per orbital position this becomes

$$A_{sat} = 1 - k\frac{T_R}{pT_F} \qquad n = k\frac{T_L}{T_F(1 - \exp^{-\frac{T_M}{T_F}})} \tag{9.19}$$

where T_L is the system duration, T_R the time to replace the satellite in orbit, T_M the satellite mission (fuel) life, T_F the satellite mean time to failure (MTTF) and p denotes the probability of launch success.

When properly designed, the satellite MTTF can be considerably greater than its mission life.

The availability of a single-hop communication link A_L depends on the equipment availability, A_{equip}, and propagation path availability, A_{prop}, as

$$A_L = A_{equip} \cdot A_{prop} \tag{9.20}$$

The propagation path availability (A_{prop}) is influenced by atmospheric loss, shadowing by objects, multipath effects, and interference. Equipment availability (A_{equip}) depends on the reliability of all equipment in the chain, from terminal transmitter and receiver, through to satellite transponder, and ground station transmitter and receiver.

For a multi-satellite system, with or without intersatellite links, the link availability also depends on the number of satellites in view at any instance. Double, or multiple coverage systems (refer to Section 9.2.2) will have higher availability than single coverage systems.

The main difference between LEO satellites and GEO satellites in the domain of reliability is that, due to lower weight constraints, a higher level of equipment redundancy can be provided. This enables the use of lower cost units, and simplifies system testing.

9.4   THE SPACE SEGMENT

One GEO satellite may achieve regional coverage, while global, excluding high latitude, coverage may be achieved with three GEO or HEO satellites. In LEO the coverage area is reduced. The space segment, then, must consist of a high number of small, identical satellites. Such satellites are small due to the lower communication traffic they handle. Because the satellites can be identical almost a serial production of satellites is possible, which further reduces costs.

Each satellite can be divided into the so called bus and payload. The satellite payload is the communication equipment, while the bus provides mechanical support, power, control and temperature stabilization of the spacecraft. For example, a LEO satellite payload may be only 1/4 of the total satellite dry-mass.

9.4.1   Satellite Bus

A satellite platform or bus is made of the following subsystems: mechanical structure, orbit and attitude control, propulsion, telemetry, tracking, command and ranging, thermal control, and electric power. Communication satellites in LEO have a bus of a similar design to other types of nongeostationary satellites: meteorological, defense and earth resources. However, it is somewhat different from the bus of GEO satellites. Let us examine some of the reasons for this.

Orbit and attitude control subsystem

Non-disturbed satellite orbits are governed by Newton's and Kepler's laws. But, while on-station, the satellite experiences a multitude of perturbations caused by the sun, moon and planet gravitation, asphericity of the earth, solar wind pressure, magnetic forces, and so forth. These disturb its orbit and attitude. In LEO, the earth's asphericity is the dominant source of perturbations. For inclined circular orbits it causes the precession of the satellite orbital plane (the so called *nodal regression*), and pendulation of the major axis in the direction of motion for inclined elliptical orbits (*line-of-apsides rotation*). This second effect is minimized in an orbital inclination of 63.435°.

One way to counter the effects of perturbations is to select a satellite orbit less influenced by them, that is, *an inherently stable orbit* (Molniya, or polar circular, for example). Satellite thruster firing must be used also to periodically correct orbit. The task of the satellite attitude control is to achieve stabilization, while pointing antennas at the desired coverage area and maintaining a satisfactory sun aspect angle for solar arrays. The satellite is kept aligned to a three axis coordinate system: the yaw axis, the pitch axis and the roll axis. The attitude control is, usually, semi-autonomous.

The attitude control subsystem has three essential parts: sensors, control electronics, and actuators. The sensors commonly used on earth orbiting satellites are: sun sensors, earth sensors and rate gyros.

To maintain the desired attitude there are several means: thruster firing (propulsion subsystem), mass spinning, gravity gradient compensation, magnet-torquing, and solar-wind-sailing. In general, the first is used to correct for large perturbations, and it has high mass cost. The remaining three are, alternatively or in combination, used against smaller perturbations. Magnet-torquing uses an electromagnet and the earth's magnetic field to apply force on the body of the spacecraft. Gimbaled or rotating solar panels may be used to "sail" on the flux of photons and charged particles from the sun.

Mass spinning maintains the position of the satellite axis through the law of angular momentum preservation. For mass spinning there are two choices: to spin as much of the spacecraft mass as possible, that is all parts for which the pointing is not required (the *spin-stabilized spacecraft*), or to spin a number of momentum-wheels within the body (the body or the *three-axis stabilized spacecraft*).

There are, also, three spacecraft pointing options: axis-parallel orientation, nadir pointing, and sun-pointing orientation. The axis-parallel option is the classical GEO satellite orientation. The spacecraft body axis is parallel to the orbit normal. In nadir orientation one satellite's body axis and the nadir facing antennas are pointed toward the center of the coverage area. The solar arrays have to be gimbaled and pointed toward the sun. In the sun-pointing orientation satellite body and fixed solar arrays are pointed toward the sun. The antennas are gimbaled and pointed toward the coverage area. Any of the three pointing methods can be used for LEO communication satellites. The choice is based on the antenna and solar array requirements.

Electric power subsystem

The function of the power subsystem is to generate and store electric power for use by other spacecraft subsystems. The power source for the spacecraft is chosen based on power/mass ratio. Solar arrays are still the most efficient means of providing power for all earth orbiting communication satellites. They have to be larger because of the shorter illumination time and to make up for wearout and radiation damage losses during satellite life in LEO. GaAs solar cells have double the power/mass ratio of Silicon cells, better resistance to radiation, and lower output power drop with an increase in temperature, but they are more expensive.

Satellites with solar panels as the main source of power require batteries to deliver power during eclipses. Batteries are a major contributor to satellite mass. LEO satellites stay in the shade for a shorter period than geostationary satellites. Therefore, their batteries can be smaller. Yet, they also have a shorter period to recharge, and a shorter life due to a much larger number of charge-discharge cycles. For a satellite in circular orbit the solar eclipse time is

$$T_e = \frac{T_s}{\pi} \arcsin \sqrt{\left(\frac{r_E}{r_E + h}\right)^2 - \sin^2 \rho}, \quad 0 \le |\rho| \ge \arcsin\left(\frac{r_E}{r_E + h}\right)$$

$$T_e = 0, \quad |\rho| > \arcsin\left(\frac{r_E}{r_E + h}\right) \tag{9.21}$$

where T_s is the orbital period and ρ is the angle between the sun-earth vector and the satellite orbital plane.

The maximum eclipse time is

$$T_e = \frac{T_s}{\pi} \arcsin\left(\frac{r_E}{r_E + h}\right) \quad \text{for} \rho = 0 \qquad (9.22)$$

The required solar array size and battery capacity can be determined from this.

Telemetry, tracking, command and ranging

These functions are shared between the space and the ground segment. The telemetry subsystem constantly acquires important parameters of satellite subsystems and transmits this information down to earth. Tracking allows for satellite antennas to track base stations on the ground, and other satellites on orbit if intersatellite links are necessary. The command subsystem receives commands from earth and decodes and verifies them before execution.

Thermal control subsystem. The function of the thermal control subsystem is to maintain the satellite temperatures within the operating limits of its units and components. Passive thermal control techniques are preferred: insulation by coatings and blankets, and optical reflectors/radiators. Electric heaters are commonly used to replace heat of the switched-off units, or to keep transponders and propulsion above the minimum operating temperature during eclipses. Thermal control can be thermostatic, by command, or a combination of both. LEO satellites experience much higher number of thermal cycles and their thermal design is more difficult.

9.4.2   Satellite Payload

The satellite payload consists of the transponder electronics, the signal processing circuitry, and the antennas.

The favored choice for the LEO satellite antennas is a multiple beam array (MBA), or even an adaptive MBA. Multiple beam arrays are the phased arrays that can be designed to have the beam nulls directed at co-frequency cells to minimize interference. VLSI beam-forming circuits operated under computer control, together with arrays of solid-state power amplifiers, make beam-forming, beam-switching and beam-scanning possible.

In a multibeam system a major source of interference are signals from the adjacent beams. The co-channel reduction factor is the number of cell radii before the same frequency may be reused. With a well designed array antenna it could be as low as 4. This means that any non-adjacent cells could use the same frequencies. The disadvantages of the array antenna are its complexity and mass.

In a very advanced solution for inclined elliptical orbits, the variable beamwidth antenna may be used to maintain the constant cell size, independently of orbital altitude.

The transmitter high power amplifiers can be low weight, high efficiency solid state power amplifiers (SSPAs). For higher frequency intersatellite links and satellite - base station links, travelling wave tube amplifiers (TWTAs) may be a more power efficient choice. Each beam/cell should have its own amplifier to avoid power saturation. Monolithic microwave integrated circuit (MMIC) front-ends and distributed output amplifiers offer a chance for performance enhancements, mass reduction, and cost decrease in relatively large series production needed for LEO systems.

9.5   THE GROUND SEGMENT

The ground segment consists of terminals, communication traffic concentrators (*hubs*) and control and monitoring (TTC&R) facilities. Hubbing and TTC&R are usually integrated into one facility, known as a major earth station. LEO satellites need to have at least one ground station always in view, or enough intersatellite link (ISL) capacity to pass the traffic to the nearest station. The expected number of long distance users determines the minimum ISL capacity.

9.5.1   Base Stations

Base stations consist of one or several ground antennas and associated communication equipment. All base stations in LEO and HEO satellite systems must have tracking capability, but some may perform the communication function only. Those are called minor base stations. Telemetry, commanding, Doppler and position determination functions are performed in the major base stations. The ground stations through which the communication system is linked with other networks are called *gateways*.

With GEO satellites the ground station antennas could be fixed, high gain antennas. With LEO system, however, they have to be position tracking antennas, and polarization tracking antennas in case linear polarization is used on the uplinks and downlinks (*feeder links*). Still, the acquisition of satellites is easier if their orbits are synchronous.

A multi-satellite system requires an elaborate network monitoring and control system. Massively parallel architectures, and distributed control algorithms in an environment of networked base stations also need to be considered. Fully automatic monitoring and control, even with the application of artificial intelligence principles, may be necessary for complicated satellite constellations.

9.5.2 Terminals

Since the satellites are moving fast in the sky, the terminal antennas should be of a broad beam - low gain type. The lower path loss between LEO and the ground allows for less directive, thus smaller user terminal antennas. Still, with the future technology advancements a low cost and simple phase steerable narrowbeam tracking antenna may be used to improve the performance (link availability) and reduce the size (output power) of terminals.

The path loss (both the freespace and atmospheric component) increases with an increase in signal frequency. When omnidirectional antennas are chosen for user terminals the optimum frequency band for communication moves toward lower frequencies (UHF and VHF). Low capacity messaging systems operate better at those frequencies. However, voice communication systems require wider channels available only at L- and higher bands. This again illustrates the interdependence of many system factors and the complexity of the system design.

9.6 LINK ANALYSIS

In any satellite communications system satellites either perform a role of simple repeaters only, or they attain additional roles in traffic concentration and processing. In the first case the following radio links are used: terminal to satellite uplink, satellite to terminal downlink, base station to satellite uplink, satellite to base station downlink. In the second case intersatellite and interorbit links may be required.

9.6.1 Link Parameters

For a LEO satellite system, the terminal to satellite distance S has amximum value of

$$s_{max} = (r_E + h)\frac{\sin\phi}{\cos\gamma} = \sqrt{(r_E + h)^2 - r_E^2\cos^2\gamma} - r_E\sin\gamma \qquad (9.23)$$

where the parameters are defined in Section 9.3.1. The propagation (free-space) loss and propagation delay (half-hop) are respectively given by

$$L_f = 20 \log \frac{4 \pi s f_0}{c} \tag{9.24}$$

$$\tau_p = s/c \tag{9.25}$$

The total link loss consists of the freespace loss (the major component) plus a number of minor link losses. As discussed in Chapter 1, other sources of link loss are

- atmospheric propagation loss

- atmospheric refraction loss

- atmospheric scintillation loss

- atmospheric precipitation loss

- Faraday polarization rotation loss (only for linear polarization)

- antenna pointing error loss (including satellite station-keeping errors)

The total communication delay consists of propagation delay plus signal processing delay in the satellite digital, or analog, circuitry. Intersatellite links can be an additional source of propagation delay. They also add signal processing and switching delay.

Time delay may be a problem in voice communication, especially if echo is present due to reflections. In GEO communication the maximum half-hop propagation delay is 138.9 ms, and the minimum is 119.3 ms. For LEO systems the range of possible communication distances, propagation losses, propagation delays, and satellite view time versus orbital altitude is shown in Figure 9.7. The figure illustrates the maximum satellite view time, that is, the maximum time a single satellite can stay in the view of a user. The maximum is achieved for zenith passes.

It can be noticed that the view time is strongly dependent on orbital altitude. This time determines the frequency of the satellite handoff, which involves intersatellite signalling.

A universal measure of link quality, convenient for comparing analog and digital links, is the carrier to noise density ratio C/N_0. It is given by

$$\frac{C}{N_0} = EIRP - 20 \log \frac{4 \pi s f_0}{c} + 10 \log \frac{G}{T} + 10 \log L_d - 10 \log k \tag{9.26}$$

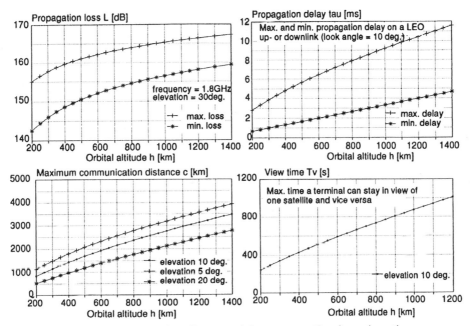

Figure 9.7: Communication distance,delay,propagation loss,view time.

As an example, in Table 9.4, a typical LEO satellite system link budgets are shown. The system uses digital QPSK without error correction coding. The table shows link budgets between terminals and satellites, as well as link budgets between ground stations and satellites. Reciprocal antennas for transmit and receive are assumed. The transmitter power and receiver noise temperature, used to calculate the EIRP and G/T, are also shown.

Worst-case links	Uplink $T \to S$ at 1.6GHz	Downlink $S \to T$ at 1.6GHz	Uplink $B \to S$ at 29.4GHZ	Downlink $S \to B$ at 20GHz
Antenna & T_X	1dBi, 3.7W	24.3dBi, 3.5W	56.3dBi, 20W	26.9dBi, 1W
Circuit loss	0.7dB	2.1dB	1.3dB	3.7dB
T_x EIRP(dBW)	6	27.7	68	23.2
Prop. loss(dB)	164.5	164.5	189.1	185.8
Athmosph. loss	0.7	0.7	1.7	1.7
Antenna & R_x	23.9dBi, 500K	3dBi, 250K	29.3dBi, 1295K	53dBi, 731K
R_x G/T (dB/K)	-3.1	-23.0	-1.8	+24.3
Boltzzmann's(dB)	228.6	228.6	228.6	228.6
C/N_0 (dBHz)	53.1	53.1	75.7	75.7
Margin [dB]	13.2	15.0	28.3	18.0

Table 9.4: Terminal to satellite and hub to satellite links.

In the design of a LEO satellite system, the choice of frequency band is very important for the system performance and cost. Lower communication frequencies are subject to scintillation caused by the ionosphere, especially in the VHF and UHF bands. Higher frequencies suffer from rain loss, particularly bands including and above X-band.

For mobile communication from terminal to satellite and from satellite to terminal, the preferred frequency bands are VHF, UHF, L-band, and S-band. If terminals have omnidirectional antennas lower frequencies have an advantage because the required output power is lower than at higher frequencies. This is due to the reduced propagation loss. For other applications, higher frequency bands may be considered.

For voice communications, L-band seems to offer the best compromise between all the important parameters: antenna size, available capacity, low cost transmitter power, and so forth. The only disadvantage is the crowding of many other radio systems in this band and, thus, lack of available spectrum.

In LEO mobile satellite systems, user terminals are limited in both their antenna aperture (gain) and receiver noise (non-cooled, low cost front-end), so they need maximum possible power from the satellite. The links between user terminals and the satellite are the most critical. For the uplink, performance is limited in gain (terminal aperture), RF power (terminal power consumption and size) and noise (satellite antenna looking at "warm earth"). It is this link that sets the overall system performance and terminal size limits. That is, the system is uplink limited.

Both intersatellite links and downlinks in LEO suffer from an additional source of noise interference - the sun. Clearly, the impact is worse on the satellite to terminal downlinks, due to their already low link margins.

Links between terminals and the satellite, especially downlinks, suffer from scatter and multipath effects. This prohibits linear antenna beam polarization and circular polarization must be used. Then, the frequency reuse through orthogonal polarization is not possible. However, for feeder links, namely intersatellite and interorbital links, linear beam polarization with frequency reuse can be employed.

9.6.2 Onboard Processing

Digital satellite communication links can be classified as either regenerative and non-regenerative. Regenerative operation involves demodulation, decoding, processing at baseband, recoding and remodulation of the carrier. When compared with the conventional nonregenerative operation it offers some advantages through decoupling of the uplink and downlink.

These advantages are

- uplink and downlink data rates can be individually optimized,

- different modulation schemes may be employed on uplink and downlink,

- forward error correction (FEC) may be used on the satellite.

Further, onboard processing may include buffering, error detection, data compression, transponder and beam switching and routing, timing and synchronization, digital linearization and equalization of channels, interference rejection, and so forth.

Regeneration requires signal processing hardware onboard the satellite. The contents and complexity of this hardware is determined by the type of information, channel access methods, transmission protocols, modulation type, and so forth. Regeneration is particularly important for the processing of intersatellite traffic.

In any communication satellite system the two primary resources are transmitter power and channel bandwidth. Sophistication of signal encoding and modulation type determine spectral efficiency and output power. Modulation selection can be used to trade information transmission rate for power. However, linearity of the channel amplifier limits the spectral efficiency by constraining the modulation type and order. In power limited channels coding can be used to trade signal bandwidth for power. Coding increases system noise bandwidth, but, ultimately, it improves C/N.

As mentioned before, frequency reuse is crucial for LEO systems with large numbers of users. One way to improve frequency reuse is to implement a code division scheme. A combination of Spread-Spectrum Multiple Access and Code Division Multiple Access (SSMA/CDMA) may offer high spectrum efficiency with graceful performance degradation under a traffic load increase [9]. Another advantage of SSMA/CDMA is that it can facilitate spectrum sharing with other L-band ground mobile and GEO mobile systems. The spread spectrum techniques are

- Direct Sequence (DS) or Pseudo Noise (PN) operation,

- Non-coherent Frequency Hopping (FH),

- Time Hopping (TH), and

- Hybrid FH/DS.

Packetized or burst transmission of signal can improve LEO communications network efficiency. It has strong advantages over continuous emission of carrier on satellite downlinks and crosslinks in the reduced output power and supply power. A packetized communication satellite network can be of two basic types of architectures: synchronous or non-synchronous.

Synchronous architectures are

- space switched time division multiple access (SS-TDMA)

- pseudorandom packet scheduling (PRS)

Non-synchronous architectures are

- multiple frequency cellular,

- time domain random access (TDRA).

Refer to Chapter 4 for a more detailed description of channel access and traffic handling methods.

9.6.3   Doppler Shift and Compensation

Both LEO and HEO satellite communication links suffer from relatively large Doppler frequency shift. Doppler shift is a result of movement of a satellite relative to the terminal, base station, or another satellite. Its effect on communication depends on the type of modulation, multiplexing technique and satellite access method. Doppler shift is more harmful to digital communication links, which require highly coherent demodulation.

In **LEO** satellite systems, the satellites move at around 7.5 km/s relative to the surface of the earth, and the velocity of mobile terminals, including most aircraft, can be neglected in the first approximation. Because of the earth's rotation, the points on the surface, and thus terminals, move at between 0 km/s at the poles and 0.46 km/s at the equator.

For users on the ground, the Doppler frequency shift changes as the satellite passes overhead. Its value depends on the terminal latitude and position within the moving satellite coverage zone. If the satellite is on the ascending part of the orbit, the user on the ground experiences a maximum positive value of frequency shift as the satellite appears on the south horizon. The Doppler frequency then goes through a minimum absolute value when the satellite is around zenith and attains a maximum negative value at the lowest elevation angle before the satellite disappears on the north horizon.

In a circular orbit the Doppler shift is given by

$$\Delta f_D = f_0 \frac{v_D}{c}$$

$$v_D = \left[\sqrt{\frac{\mu r_E^2}{(r_E + h)^3}} \cos\gamma \sin\psi - \frac{2\pi}{86164} \cdot r_E \cos l_t \cos\gamma \cos\psi \right] (9.27)$$

where l_t is the terminal geographical latitude, and ψ is the angle between the projection of the satellite-terminal line onto the tangential plane at the subsatellite point and the latitudinal tangent at the subsatellite point. A general expression for Doppler shift in elliptical orbits cannot be derived in a closed form.

Doppler shift, though a problem that has to be compensated for in LEO communications, can be a useful source of information if used in position location systems. If the position of the satellite is accurately known, then based on the Doppler frequency shift and cell number (to remove ambiguity) a relatively accurate active/passive ground position determination is possible.

The Doppler shift in Equation (9.27) reaches a minimum value (maximum absolute value) for a terminal on the equator, at the edge of the coverage zone and with

$$\psi = \psi_X = \arctan\left[-\frac{43082}{\pi} \sqrt{\frac{\mu}{(r_E + h)^3}} \right] \qquad (9.28)$$

Table 9.5 shows absolute values of the maximum possible Doppler shift for terminals on the equator ($l_t = 0°$) for satellites at different orbital altitudes.

	$T \leftrightarrow S$	$f_0 = 1.5 GHz$	$B \leftrightarrow S$	$f_0 = 11 GHz$
MOLNIYA	8.7ppm	13.1kHz	87ppm	95.7kHz
TUNDRA	4ppm	6.0kHz	40ppm	44kHz

Table 9.5: The maximum Doppler shift in HEO.

Figure 9.8 presents a 3D view of the Doppler shift and its pendulation within the coverage zone of one LEO right above the equator. Within one coverage zone there is an S-like curve (marked "ZERO" and dashed in the figure), passing through the subsatellite point, on which all points have zero Doppler. It is crossed under an angle of 90° by a curve containing all points of maximum positive and negative Doppler shift.

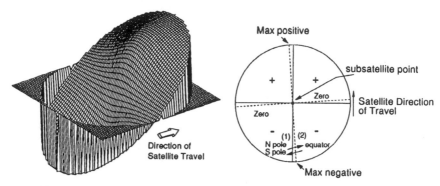

Figure 9.8: Doppler shift within one zone and its pendulation.

For a satellite above the South or North Pole, the position of maximum, minimum and zero Doppler curves is (1). For a satellite above the equator the position is (2). The maximum offset between (2) and (1) is around 4°. Basic methods for Doppler compensation are

- closed terminal-satellite frequency control loop,

- on-board satellite Doppler correction,

- pre-correction on the receive side of the link, and

- pre-correction on the transmit side of the link.

The first method offers a very precise control over the frequency shift for every user, but requires complicated equipment, either on-board or within terminals. The second is independent of terminals (does not involve them), and requires simpler equipment, but introduces receive frequency errors within a coverage cell.

In **HEO** the Doppler effect is smaller, due to lower satellite relative velocity in the apogee. It is most pronounced around the time of satellite handoff, when both satellites are away from their apogee. Due to the smaller frequency shift the methods of compensation are different. Often, only a wider guard band between channels is used.

9.6.4   Frequency Spectrum, Spectrum Sharing and Interference

Radio frequency spectrum is a very limited and costly resource. Its availability is crucial for any new radio communications system. Every radio communication system with a large user base must achieve high spectral efficiency and maximum frequency reuse. LEO systems have a disadvantage of a large Doppler frequency shift of up to 80 kHz peak-to-peak. Each communication channel would require a broad frequency allocation. The methods that can be used to improve frequency reuse are: narrowband and SSB operation, time division, cellular division, and code division in spread spectrum operation.

At present there exist world-wide frequency allocations to MSS uplinks and downlinks in the VHF/UHF band and the L/S band. Allocations are on the primary (P), or on the secondary (S) basis. The frequencies allocated exclusively to NON-GEO, or to be shared in coordination between GEO (G) and NON-GEO (NG) systems are shown in Table 9.6.

Bandwidth MHz	uplink	Bandwidth MHz	downlink
148–149.9	(P, NG)	137–137.025	(P, NG)
149.9–150.05	(S, G + NG)	137.025–137.175	(S, NG)
387–390	(S, G + NG)	137.175–137.825	(P, NG)
1610–1646.5	(P, G + NG)	137.825–138	(S, NG)
1656.5–1660	(P, G + NG)	312–315	(S, G + NG)
1980–2010	(P, G + NG)	400.15–401	(P, NG)
2670–2690	(P, G + NG)	1525–1544	(P, G + NG)
		1613.8–1626.5	(S, G + NG)
		2170–2200	(P, G + NG)
		2483.5–2520	(P, G + NG)

Table 9.6: Frequency allocations for moblie satellite systems.

The problem of interference with terrestrial radio services operating in the same radio band is pronounced by the requirement to communicate while the satellite is low over the horizon. We need a different approach from the deterministic model of GEO interference. In the case of LEO, mathematical models for interference have to be statistical. The interference power from a LEO satellite into a co-frequency ground receiver can be expressed as a time function

$$I = 20 \log \frac{c}{4\pi f_0 s(t)} + L_a(t) + L_p(t) + G_T(t) + P_t + G_R(t) \qquad (9.29)$$

where $L_p(t)$ is the polarization discrimination, $G_T(t)$ satellite gain in the direction of receiver, $G_R(t)$ ground receiver gain in the direction of the satellite, $L_a(t)$ atmospheric and fade loss, and P_t satellite transmit power.

After determining all the time functions involved in (9.29) and accounting for the multisatellite operation, the probability distribution of the interference for a specific ground location can be determined.

9.6.5   A Comparison of GEO, LEO and Fibre optic Links

The parameters of LEO and GEO mobile systems can be compared in many different ways, to the advantage of one or the other. We will present here a comparison in terms of two basic communication parameters: loss and delay. Table 9.7 summarizes comparative values for GEO, LEO and fibre optic links in configurations illustrated in Figure 9.9.

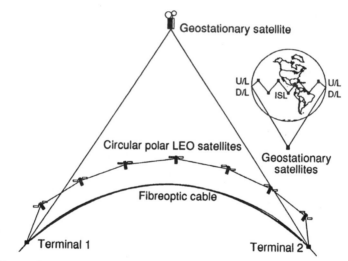

Figure 9.9: A comparison of LEO, GEO, and fibre optic links.

In the table it is assumed that two terminals are at the maximum distance that one GEO satellite can link for a chosen minimum look angle. That angle is set to be the same $\gamma = 10°$ for both GEO and LEO systems. A fictional fibre optic link is assumed, made of a group index 1.466 monomode fiber with pure silica cladding and core doped with germanium, at $0.85 \mu m$.

No signal processing delays were included in the total delay calculation, though they can be significant in regenerative digital links. For LEO satellites and the fibre optic cable this delay should be of a similar order of magnitude. Modern fibre optic cables can have repeaters at 150 km or more, but not all need to be regenerative.

		half-hop distance	half-hop link freespace loss	total communication distance	total propagation delay
distant terminals	GEO	40,585 km	188.7 dB	81,170 km	271 ms
	LEO	2,367 km	164 dB	23,568 km	78.5ms
close terminals	GEO	35,786 km	187.6 dB	71,572 km	239 ms
	LEO	800 km	154.6 dB	1,600 km	5.3 ms
fibre optic cable		N/A	N/A	15,903 km	77.7 ms

Table 9.7: A comparison of communication links (The LEO system in the table is a circular orbit, $N_s N_p = 77$, $\gamma_X = 10°$, $f_0 = 1.6$ GHz system.)

Table 9.7 shows that LEO links have losses 24.7-33 dB smaller and total communication delay 190-230 ms smaller than GEO. The propagation delay of the LEO system is comparable with the propagation delay of the fibre optic system. This is due to the propagation through denser media (glass) in the latter case. For geographically close terminals (for example most domestic communications) LEO links have a strong advantage over equivalent GEO links in communication delay and loss, with similar processing delays. For long distance communications the relation depends on processing delays.

9.7 NETWORK CONCEPTS

There are two broad types of non-geostationary communication satellite systems

- real-time communication systems, and

- store & forward systems.

Real-time systems
Real-time systems provide almost instantaneous transfer of information (audio, video, data). This kind of transfer is important for interactive communications, like telephony, conferencing, and so on. Packet switching networks also belong to this type, and are often the most efficient way of information transmission. However, real-time communications require constant satellite coverage of both the information source and destination. They also require ground crosslinks or intersatellite links.

Store & forward systems
Store & forward systems receive and store the information while linked to a source, to transmit it when they establish a link with the destination

terminal. Periodic simultaneous transmission and reception is also possible. This type of system is suitable for non-interactive information: messages, computer data, images, digitized voice, and so on. Its main advantage is in the much lower number of satellites required. But, the links are not constantly available and satellites need additional onboard processing hardware and large memory.

A switched communication system can provide one-way or two way calling. *Outcalling*, or placing calls from the terminals is relatively simple. In order to enable *incalling*, however, the system needs terminal position maps and complicated call processing and routing. There are three options for call routing: centralized adaptive, distributed adaptive, and flooding.

Flooding is the simplest and the fastest but the least efficient method of routing. A message is sent down all the available routes between the source and the destination. It is suitable for random satellite constellations and military LEO systems.

Centralized routing implies centralized terminal position maps and call routing decisions. This method requires a high number of signalling channels, but it is fairly simple and fast. However, due to centralization the system reliability may be reduced. It is preferred for GEO fixed and mobile networks. In distributed or localized routing, position maps are kept in ground or satellite network nodes. This routing method creates lower signalling traffic on the network. It can adapt to a failure on the network to maintain services. This is the most suitable method for LEO communication systems because of their dynamic structure.

9.7.1 Crosslinks

In a mobile satellite communication system the satellites serve as nodes of traffic concentration and distribution. In geostationary systems these nodes are fixed relative to the ground, but in non-geostationary they are moving. For large area and global coverage satellites have to be arranged into systems.

To complete a multi-satellite system network for real-time communications the crosslinks between nodes are needed. The crosslinks may be established either *through ground based links*, or *through satellites* directly, as intersatellite links (ISL). In case the system involves satellites in orbits of different types, for example, LEO satellites and a GEO satellite, the interorbit links (IOL) may be needed also.

Ground based crosslinks can be established through fibre optic and coaxial lines, or microwave relay links. Intersatellite links can be microwave or millimeter wave (mm wave) radio links, or laser-optical links. Crosslinks through satellites have some advantages over ground based crosslinks

- reduction in communication delay (no double hop)

- minimisation of secondary traffic distribution problems

- reduction, or elimination, of the need for on-board storage (digital systems)

- potential reduction in the total system cost.

9.7.2 Intersatellite and Interorbit Links

Intersatellite and interorbit links are, fundamentally, tools for improving the connectivity, the coverage, the flexibility, and the economics of multisatellite networks. Though not exclusive to LEO or HEO satellite networks, it is for those that ISL and IOL can realize their full potential. For example, they can reduce the number of ground stations needed. This is particularly important for LEO satellites during passes over large ocean surfaces with no ground stations in the view. Also, the IOL and ISL may be used as a primary, or backup means of satellite telemetry and command transmission.

An important factor in ISL is the clearance angle between ISL and the earth. The intersatellite link is not shaded by earth if

$$(r_E + h) \cos \phi - r_E > 0 \qquad (9.30)$$

The intersatellite link does not pass through the atmosphere of height h_A if

$$(r_E + h) \cos \phi - (r_E + h_A) > 0 \qquad (9.31)$$

In circular polar orbits the maximum intersatellite distance is

$$d = 2(r_E + h) \sin(\frac{\pi}{N_s}) \qquad (9.32)$$

Intersatellite links should be designed for that distance. For co-plane satellites the intersatellite distance remains constant. For co-rotating satellites, interplane satellite distance varies between this value and one half this value.

The ISL topology depends on the amount of traffic. For the minimum time delay and traffic congestion each satellite must have at least four ISL:

two to the two neighboring satellites in the same orbital plane and two to the adjacent orbital planes, for the system with more than two orbital planes. The intraplane ISLs carry most of the handoff traffic and signalling. The interplane ISL distances are never constant, and interplane links will always have some Doppler frequency shift. For non-circular orbits this frequency shift will exist in the intraplane links also.

Intersatellite and interorbit links are not handicapped by atmospheric absorption and operating frequencies are freely chosen from these bands. Yet, a good choice of frequency should provide for high atmospheric absorption to prevent mutual interference with ground radio systems.

Also, while the uplinks and downlinks are subject to unpredictable atmospheric fades, intersatellite links/indexintersatellite link are subject to predictable noise increases due to sun interference. However, the design of ISL and IOL is not simple. Due to large distances they require high transmitter powers, which increase total power consumption and satellite mass, and makes isolation in the duplex operation difficult.

In the digital satellite systems with ISL, regeneration and on-board processing must be utilized to separate the downlink and the crosslink traffic. Intersatellite data links may require even broader bandwidth than individual up/down links, because the traffic from several other satellites may be passing through them. On the complete terminal to ground link of non-regenerative type, including one or more intersatellite links, the overall link performance is determined by the worst amongst the uplink, ISLs, and the downlink.

Three types of technology are available for ISL: microwave, millimeter-wave, and optical. The preferred choice depends on several system factors: communication capacity, attitude control precision, available power, and so on.

Optical and millimeter wave intersatellite links have some advantages and disadvantages compared with microwave ISL. The advantages are

- small antenna (or telescope), that is, savings on size and mass

- low output power and compact transmitters bring lower power consumption and size and mass reduction

- high operating frequency means inherently large communication bandwidth, that is capacity, and

- there is no interference with other radio systems (optical).

But, the disadvantages are

- the requirement for high quality attitude control and antenna pointing, which implies an increase in mass, and

- most laser transmitters and receivers require active cooling.

The advantages of the microwave technology are

- high reliability of components and proven designs

- low receiver noise figure and high transmitter output power

- smaller pointing and acquisition problems, and

- easier filtering of low data rate channels.

Radio frequencies assigned to ISL on the primary (P) and secondary (S) bases, and present levels of receiver low noise amplifier (LNA) noise figure and transmitter tube power amplifier (TWTA) output power, are shown in Table 9.8.

Freq.[GHz]	Band	Wavelength	LNA NF	TWTA power
22.55-23.55 (p) 25.25-27.50 (s) 32-33 (p)	23/32 GHz band	Microwaves	3-4 dB	150 W
54.25-58.2 (p) 59-64 (P)	60 GHz band	Microwaves	4-5 dB	75 W
116-134 (P)	120 GHz band	Millimeter waves	8 dB	30 W
170-182 (P) 185-190 (P)	180 GHz band	Millimeter waves	-	-

Table 9.8: Comparative link budget values for microwave and fibre optic intersatellite links.

9.7.3 Optical Intersatellite and Interorbit Links

Optical links are more sensitive to mechanical vibrations than other. Vibrations are produced by movable parts of the spacecraft (solar-wings, mechanically steerable antennas, and so forth.). Also, optical links are subject to contamination of the spacecraft environment (outgassing and propellant), and interference from other optical sources in the sky.

The maximum telescope gain, that is diameter, is limited by the RMS value of the statistical mispointing angle (σ) and the probability of burst errors (P_{BE})

$$G_{OTmax} = (\frac{1.53}{\sigma\sqrt{\ln(1/P_{BE}^2)}})^2 \tag{9.33}$$

The optical acquisition may be impaired by the attitude uncertainty, which may be in the order of several hundred beamwidths. One way to ease the acquisition is to broaden - defocus the optical transmit beam and use a high gain, wide field of view receivers. The other is to scan a narrow optical beam. The stabilization and tracking could be achieved by moving a coarse and fine pointing mirrors until the acquisition is achieved. Then, a fine pointing mirror would be used for communication. Optical ISL frequencies are at least 1000 times higher than the highest mm wave ISL frequencies. Thus, the gain of the antenna of the same size is at least 60 dB greater, and antenna diameter can be significantly reduced. This is illustrated by the link budget in Table 9.9.

Digital links (d=4300km)	Microwave ISL	Optical ISL
Frequency/wavelength	32 GHz	0.85 μm
Average transmit power	10 dBW (10 W)	-17dBW (50 mW)
Transmit loss [dB]	-2.4 dB	-7.3 dB
Antenna/telescope t_X gain	55 dBi($\phi = 1m$)	119.5dBi($\phi = 0.25m$)
Free space loss [dB]	-195.2 dB	-276.1 dB
Antenna/telescope R_X gain	55 dBi($\phi = 1m$)	119.5dBi($\phi = 0.25m$)
Pointing loss [dB]	-0.1 dB	-3.5 dB
Receiver loss [dB]	-2.0 dB	-4.0 dB
Required receiver power for 65 Mb/s link of BER=10^{-6}	-87.5 dBW	-87.5 dBW
Link margin [dB]	7.8 dB	18.6 dB

Table 9.9: Microwave versus optical ISL budget.

Due to high antenna gain, the transmitter output power, and its power consumption, can be reduced. Unlike microwave systems which use metal waveguides and coaxial interconnections, optical transponders use fibre optic guides. This brings mass savings.

Another difference between the radio and optical ISL is that the radio systems usually need upconversion after the carrier modulation. The optical carrier source is directly modulated by the baseband, channel multiplexed signal.

Due to propagation delays, the direction of ISL transmission must be toward the point where the other spacecraft will be at the moment of signal arrival. This offset, or *look ahead angle*, can be up to 0.003° for satellites in non-geostationary orbits. For most microwave ISLs this is much below the standard beamwidth of 0.5°. But, for optical ISLs, where the typical beamwidths are of the order of 0.0003°, some measures have to be taken to compensate for this offset.

9.8 NON-GEO SYSTEM DESIGN

A satellite communication system consists of the space segment and the ground segment. The division of roles between the ground and space segment determines the structure of the system. Because of their distributed nature, the LEO systems should not play the role of bent-pipe repeaters only. They should perform a part of the networking function too.

The design of the space and ground segments cannot be done independently. The design objective is to get the desired performances levels with a minimum total cost of all segments, the simplest and the most reliable satellite and ground segment, and the maximum frequency efficiency (minimum spectrum, maximum reuse).

To achieve these goals, the ground and space segment must be designed concurrently. The satellites must be designed to satisfy the four basic constraints of

1. minimum mass

2. minimum electric power consumption

3. minimum cost, and

4. high reliability and availability.

The cost per communication satellite channel is proportional to the required output power per channel. The lower power consumption and higher power efficiency lead to multitudinous mass reduction: savings on power generating and power distribution circuitry, and savings on attitude control.

A LEO system design begins with the definition of the basic requirements: the type of traffic, the amount of traffic, locations of sources and so forth. Based on these requirements the orbit type and satellite configuration are chosen.

The next step in the system design is to determine the minimum look angle γ_X from the specified minimum link availability. Then, the orbital altitude is selected, and, thus, the number of satellites. Orbital altitude is a crucial parameter in the trade-off between communication performance and system cost. Costing analysis is performed next to estimate the system effectiveness. The selection of orbital altitude may require several iterations.

In summary, the main advantage of LEO satellites is that the decreased terminal to satellite distance allows lower power spacecraft payloads and smaller ground user terminals, due to lower free space loss. An additional benefit is the shorter propagation delay. Also, cellular frequency reuse is made easier due to smaller required satellite antenna size. Satellites in LEO can be smaller and lower cost due to modest output power and antenna gain requirements and smaller communication capacity (limited coverage area). However, a number of satellites must be used to provide continuous coverage of larger areas of the earth.

9.8.1 A LEO System Example

All LEO systems can be divided into ultra-narrowband systems (messaging and other very low speed data) and narrowband systems (voice and narrowband data). Based on this a system is called "small" or "large" LEO system. Here, one large, the well known IRIDIUM system will be discribed.

The space segment will consist of 66 3-axis staibilized satellites in 6 orbital planes of 86° inclination. The satellite orbital altitude is 780km. Co-rotating orbital planes are spaced 31.6° apart, which leaves about 22° for the spacing between counter-rotating planes 1 and 6. There are 11 satellites in one plane, spaced 32.7° apart. Satellites wet-mass (including station keeping fuel) is around 700kg and total power consumption around 600W. The IRIDIUM system employs FDMA/TDMA with QPSK modulation.

For terminal-satellite uplinks and downlinks the frequency band of operation is 1616–1626.5MHz (10.5MHz). Polarization is right-hand-circular. Satellite antennas project 48 beams onto the earth of 16.8–24.3dBi gain at the edge ot the cell. The cell size increases as the nadir angle increases, that is, outer cells are larger than inner cells. The total number of active cells at any one time is close to 2,150 and the frequency reuse is about 180 times. Satellite transmitter power is up to 3.5W per channel in bursts. For simplified IRIDIUM link budget refer to Table 9.4. The communications channel bandwidth is 31.5kHz, channel spacing is 41.67kHz. There are 3840 uplink and the same number of downlink channels per satellite. On up and downlinks multiple rate forward error correction (FEC) is used with a coded data of 50kb/s.

For base station to satellite links the frequency band is 29.35–29.45GHz (100MHz) for uplinks and 19.95–20.05GHz (100MHz) for downlinks. Polarization is also right-hand-circular. Channel bandwidth is 4.375MHz and spacing is 7.5MHz. An FEC rate of 1/2 is used and the coded data rate is 25Mb/s.

Terminal units will be hand-held and compatible with terrestrial cellular systems. The user will get an IRIDIUM dial tone wherever his local cellular system is not available. Services to be provided are: 4.8kb/s voice, and 1.2–9.6kb/s data, position finding (GSM). Services will be available on the country-by-country switched basis as negotiated with individual governments and carriers.

IRIDIUM is a world-wide digital cellular system for commercial mobile communication with low traffic density but high terminal population. It is designed for minimum size terminals and almost negligible communications delay, when compared with other proposed large LEO systems. However, it is a high implemetation cost system that requires a very large user base up-front to be economically viable.

9.9 THIRD GENERATION NETWORKS

Analog cellular mobile phones, wireless phones and pagers have experienced tremendous growth in the past decade. Initially developed as car phones, both the terminal units and cells of mobile phones are decreasing in size, in the trend toward pocket sized hand-held personal units for operation in microcells. All analog mobile systems belong to the *first generation systems*.

Due to the high demand in urban areas, and relative spectrum inefficiency of the analog system, congestion became a limiting factor in the wireless personal phone system development. *Second generation systems* are currently replacing analogue systems. The second generation standards are digital (for example, the European GSM and North American IS-54), and facilitate somewhat standardized signalling, switching and network management. Interoperation and roaming within all countries deploying the system is another advantage in comparison with the first generation. However, incompatibility still exists, particularly between mobile and other wireless systems. Services are still not integrated technically or commercially. There is as yet no single world-wide standard. Even after full deployment of the second generation mobile and personal networks there will remain areas in which the demand exists but the traffic level is not large enough to make the cellular implementation economical.

A possible *third generation network* architecture is illustrated in Figure 9.10. Third generation wireless communication systems are often called "tetherless" to symbolize their universal and seamless nature [12]. They will be characterized by the confluence of personal and business communications, and integration of all wireless networks, standards, and services. The network management and control should be intelligent, adaptive and distributed.

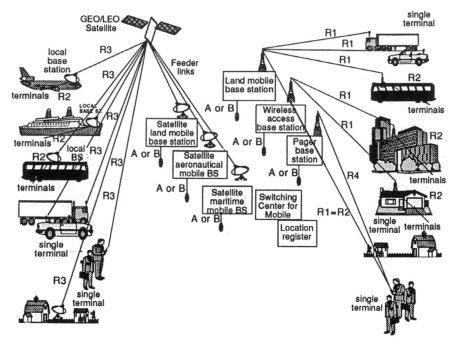

Figure 9.10:   Architecture of third generation networks.

Apart from personal voice terminals, facsimile and information terminals will be able to access the same network: palm-top computers and data entry terminals. All systems, from microcellular to geostationary cellular should be fully compatible. Modular terminals will be usable at all network levels. Phone numbers will be personal and perambulatory. All personal terminals should be able to reach the public switched telecommunications network (PSTN), wireless, at any time and from any place in the world.

Basic building blocks of the third generation networks are: personal and mobile terminals, mobile earth stations, mobile base stations, small rural exchanges, and PABXs. There are four types of radio interfaces designated R1-R4, and two types of PSTN interfaces: (A)-directly into the network, and (B)-through an independent mobile switching center.

As LEO satellites allow communication with small terminals, a wide range of old and new applications is possible. Services that third generation systems can provide are listed in Table 9.10.

stwtt - two way terminal to terminal (private voice and data lines)	
pstn - interconnect with public switched telephone network	
em - environment monitoring (rivers, dams)	
www - world weather watch (around 10,000 terminals world-wide)	
mat - mobile asset tracking (cattle, trucks, carriages, and so forth)	
twm - two way messaging	vrm - vehicle road monitoring
pd - position determination (gps, rdss)	prm - pollution road monitoring
ea - emergency alert (fire, police)	bsm - bio-sensor monitoring
ats - anti-theft security	jl - judicial location
rc - remote control	sr - search and rescue

Table 9.10: Third generation system services.

Non-geostationary systems can provide mobile satellite voice and data services (MSS). They may, also, be applied to navigation and Ground Positioning (GPS), Radio Determination (RDSS), paging, store and forward data and messaging, satellite sound broadcasting, alarm monitoring, road, river and sea traffic monitoring, carriage and truck tracking, weather data collection and distribution, ATM, credit card and check transactions, environmental monitoring, and many other. Integration of several of these services is possible in one very small aperture mobile (for example car, plane, train, and so forth), or personal terminal.

LEO and other non-geostationary satellites, can be organized into stand alone but compatible systems, or they can provide one type of local loop for network access.

In the second case they could be fully integrated into the third generation ubiquitous network, to serve as a means of network access in low traffic density areas.

In both cases, they should complement other wireless systems to achieve the full range of cell sizes, from miniature in PCS (for highly urbanized areas), to very large in GEO satellite mobile (rural and remote areas). Likely relative costs are illustrated in Figure 9.11.

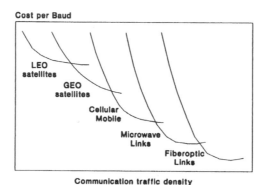

Figure 9.11:   Cost comparison for different communication systems.

9.10   PROBLEMS

1. Which are the possible orbital types for communication satellites? Suggest some reasons why non-synchronous orbits are not suitable for communications.

2. List advantages and disadvantages of placing a communication repeater in a low earth orbit. In general, which satellite system parameters improve with an increase in orbital altitude and which become worse?

3. Draw the ground track of the 24h, elliptical orbit called Tundra (the answer is available in ref. [5]). Which coverage is optimized with this type of orbit? Could the same be done for the southern hemisphere? What is the main advantage of HEO for mobile communications?

4. How do LEO satellite systems compare with HEO satellite systems? Compare benefits, complexity and costs of each. What advantages would a combination of LEO and GEO satellites bring?

5. How does Doppler shift depend on frequency, orbital type and altitude? Which types of compensation are possible?

6. An L-band terminal is communicating through a satellite in circular LEO at 780 km. Determine the minimum and maximum possible

Doppler shift in the carrier if the terminal is at a geographical latitude of 45 degrees.

7. Calculate the orbital altitude of the lowest possible geo-synchronous circular polar LEO orbit. (Answer: 256.1 km.)

8. A satellite system is made of 66 satellites in circular LEO at 780 km. If the satellite MTTF is 10 years, the launch success rate 90%, and each satellite contains enough station-keeping fuel for 5 years, what is the satellite availability? Calculate the number of satellites required for 5 year operation. What is the availability and required number of satellites if there is one in-orbit spare?

9. For the satellite system from the previous example calculate the minimum thickness of metal shielding, for the case of CMOS electronics and GaAs electronics on board.

10. For the satellite system from the same example calculate the maximum eclipse time, that is the eclipse time a satellite power subsystem should be designed for.

11. When is an optical intersatellite link a better choice than a microwave intersatellite link?

REFERENCES

1. Morgan W.L., Gordon G.D.: *"Communications Satellite Handbook"*, J.Wiley & Sons, New York, 1989.

2. Markovic Z., Hope W.: "Small LEO Communication Satellites - An Evaluation", *Proc. IREECON 1991*, pp.178-181, Sydney, September 1991.

3. Chakraborty D.: "Survivable Communication Concept Via Multiple Low earth-Orbiting Satellites", *IEEE Transactions on AES*, Vol.25, No.6, pp.879-889, November 1989.

4. Cantafio L.J., ed.: *"Space-Based Radar Handbook"*, Artech House, Norwood MA, 1989.

5. Bousquet M., Maral G.:"Orbital Aspects and Useful Relations from earth Satellite Geometry in the Frame of Future Mobile Satellite Systems", *Proc. 1990 AIAA Conference*, pp.783-789, 1990.

6. Pontano B., De Santis P., White L., Faris F.: "On-board Processing Architectures for International Communications Satellite Applications", *Proc. 1990 AIAA Conference*, pp.655-668, 1990.

7. Goldhirsh J., Vogel W.J.: *"Propagation Handbook for Land-Mobile-Satellite Systems (Preliminary)"*, John Hopkins University, Maryland, April 1991.

8. Binder R., Huffman S., Gurantz I., Vena P.: "Crosslink Architectures for a Multiple Satellite System", *Proceedings of the IEEE*, Vol.75, No.1, pp.74-82, January 1987.

9. Benedetto S., Pent M., Zhang Z.: "Comparison of Advanced Coding and Modulation Schemes for Multicarrier Satellite Communication Systems", *International Journal of Satellite Communications*, Vol.7, pp.165-181, 1989.

10. Vilar E., Austin J.: "Analysis and Correction Techniques of Doppler Shift for Nongeosynchronous Communication Satellites", *International Journal of Satellite Communications*, Vol.9, pp.123-136, 1991.

11. "Study of Land Mobile Communications from Non-Geostationary Orbits", *European Space Agency Contract Report TP-8557*, Volume 1, 2, and 3, 1988.

12. Goodman D.J., Nanda S., ed.: *"Third Generation Wireless Information Networks"*, Kluwer Academic Publishers, Boston, 1992.

13. Maral G. et al.: "Low earth Orbit Satellite Systems for Communications", *International Journal of Satellite Communications*, Vol. 9, pp.209-225, 1991.

14. Griffin M.D., French J.R.: "Space Vehicle Design", *AIAA Education Series*, Washington DC, 1991.

15. Puccio A., Saggese E.: "Identification of requirements for intersatellite links", *International Journal of Satellite Communications*, Vol.6, pp.107-117, 1988.

16. Stassinopoulos E., Barth J.:"Non-Equatorial Terrestrial Low Altitude Charged Particle Radiation Environment", *NASA Report* X-600-87-15, Version 2, Goddard Space Flight Center, October 1987.

17. Jensen J. et al: *"Design Guide to Orbital Flight"*, McGraw-Hill, 1962.

18. *"The 14th International Communication Satellite Systems Conference"*, AIAA 1992, 1992.

Index